IFCoLog Journal of Logics and their Applications

Volume 4, Number 2

March 2017

ISBN 978-1-84890-234-3
ISSN (E) 2055-3714
ISSN (P) 2055-3706

College Publications
Scientific Director: Dov Gabbay
Managing Director: Jane Spurr

http://www.collegepublications.co.uk

Printed by Lightning Source, Milton Keynes, UK

Editorial Board

SCOPE AND SUBMISSIONS

This journal considers submission in all areas of pure and applied logic, including:

pure logical systems
proof theory
constructive logic
categorical logic
modal and temporal logic
model theory
recursion theory
type theory
nominal theory
nonclassical logics
nonmonotonic logic
numerical and uncertainty reasoning
logic and AI
foundations of logic programming
belief revision
systems of knowledge and belief
logics and semantics of programming
specification and verification
agent theory
databases

dynamic logic
quantum logic
algebraic logic
logic and cognition
probabilistic logic
logic and networks
neuro-logical systems
complexity
argumentation theory
logic and computation
logic and language
logic engineering
knowledge-based systems
automated reasoning
knowledge representation
logic in hardware and VLSI
natural language
concurrent computation
planning

This journal will also consider papers on the application of logic in other subject areas: philosophy, cognitive science, physics etc. provided they have some formal content.

Submissions should be sent to Jane Spurr (jane.spurr@kcl.ac.uk) as a pdf file, preferably compiled in LaTeX using the IFCoLog class file.

CONTENTS

ARTICLES

Hilbert's epsilon and tau in Logic, Informatics and Linguistics

Stergios Chatzikyriakidis, Fabio Pasquali and Christian Retoré

Foreword

This special issue of the *Journal of Logics and their Applications* follows a workshop organized in Montpellier in June 12-15 2015.

Both the workshop as well as this issue aim at promoting work on Hilbert's Epsilon in a number of relevant fields: Mathematics, Logic, Philosophy, History of Mathematics, Linguistics, Type Theory, Computer Science, Category Theory among others. Even though the epsilon and tau operators were introduced almost a century ago, many questions on the computational and the mathematical side, as well as their application to philosophy of language and linguistic semantics, remain open and relevant today,

The first two papers, ours and the one by Michele Abrusci, can be seen as complementary introductions. Our paper attempts to show a continuity from Ancient and Medieval philosophy, all the way to type theory and computational approaches to natural language semantics. On the other hand, Abrusci's paper is a historical presentation of the emergence of epsilon and tau on the proof theoretic foundations of mathematics in the works of Hilbert and his followers.

We would like to thank Aïda Diouf and Elisabeth Gréverie from the University of Montpellier for their extremely valuable help with organisational issues. We are grateful to our sponsors, University of Montpellier, the Kurt Gödel Society and its president Matthias Baaz, the Hilbert-Bernays project and its coordinator Michael Gabbay, the ANR project Polymnie and its coordinator Sylvain Pogodalla. Finally, we thank Dov Gabbay, who accepted to publish this issue as a special issue of the *Journal of Logics and their Applications*, and last but not least, Jane Spurr from College Publications, for her invaluable help.

We also thank the program committee, for reviewing the submissions to the workshop and to this present issue:

Daisuke Bekki (Ochanomizu University)

Francis Corblin (University of Paris-Sorbonne & Institut Jean Nicod CNRS)

Michael Gabbay (University of Cambridge)

Makoto Kanazawa (National Institue of Informatics of Tokyo)

Ruth Kempson (King's College, London)

Ulrich Kohlenbach (Darmstadt University of Technology)

Alda Mari (Institut Jean Nicod CNRS & ENS & EHESS)

Richard Moot (CNRS, LaBRI and Université de Bordeaux)

Georg Moser (University of Innsbruck)

Michel Parigot (University of Paris Diderot 7 & CNRS-PPS)

Mark Steedman (University of Edinburgh)

Bruno Woltzenlogel Paleo (Vienna University of Technology)

Richard Zach (University of Calgary)

The workshop and the resulting issue would not have been the same without the good will of the participants who triggered many interesting discussions that benefited the papers in this issue: Vito Michele Abrusci, Bhupinder Singh Anand, Dharini Bhupinder Anand, Federico Aschieri, Stergios Chatzikyriakidis, Viviane Durand-Guerrier, Alain Lecomte, Hans Leiß, Bruno Mery, Wilfried Meyer-Viol, Koji Mineshima, Corey Mulvihill, Thi Minh Huyen Nguyen, Sumiyo Nishiguchi, Michel Parigot, Fabio Pasquali, Namrata Patel, Thomas Powell, Jean-Philippe Prost, Lionel Ramadier, Giselle Reis, Christian Retoré, Georg Schiemer, Hartley Slater, Sergei Soloviev, Federico Ulliana, Shuai Wang, Claus-Peter Wirth, Manel Zarrouk.

While in the process of reviewing and revising the articles for the present volume, a very unfortunate event has occurred: one of our workshop participants, and one of the most prominent scholars on the epsilon calculus, Hartley Slater, passed away (July 2016). A few months ago, he attended our workshop on epsilon and tau, where he gave two talks and had many lively discussions with the participants. This issue includes Hartley's last article and an obituary written by Claus-Peter Wirth in memory of this fine logician and philosopher.

Barry Hartley Slater (October 28th 1936, July 8th 2016)

Adieu, Hartley!

Stergios Chatzikyriakidis (University of Gothenburg)

Fabio Pasquali (University of Padua)

Christian Retoré (University of Montpellier & LIRMM-CNRS)

 Received February 2017

Barry Hartley Slater (1936–2016): A Logical Obituary

Claus-Peter Wirth

Dept. of Math., ETH Zurich, Rämistr. 101, 8092 Zürich, Switzerland
`wirth@logic.at`

1   Obituaries and Curricula Vitae

Under the title "Vale Hartley Slater", on July 14, 2016, `admin` posted on `http://www.croquetwest.org.au/?p=4664`:

> "Hartley Slater passed away on Friday 8th July after a short illness, he was 79. Hartley took up croquet in 1978 and was AC state coach 1994 to 1997. Hartley also won a number of major AC competitions. He was the Croquet Archivist for a number of years and an A/Professor of Philosophy at UWA. His ebullient character will be missed."

This shows true devotion, although it is a bit cryptic. "AC" stands for Association Croquet, a full international version of a very funny garden game, and "UWA" for "The University of Western Australia" in Perth.

Here is another short obituary and curriculum vitae by Slater's colleague Stewart Candlish:

> "Hartley Slater (B. H. Slater) died from cancer on July 8th 2016 at the age of 79. He was born in Keighley, Yorkshire, England. He was educated in England at the universities of Cambridge, Oxford, and Kent (Canterbury). His first academic job was at Kent. He joined the staff at The University of Western Australia in 1976. He took up croquet in 1978 and became a skilled player and coach. He had a deep and informed interest in the arts, especially music and painting. Though he did not perform in public, he was an accomplished pianist. He published extensively in aesthetics as well as logic."

[24] is a long scientific curriculum vitae, probably written by Hartley Slater himself.

2   Hartley Slater's Awards and Degrees

According to Anabelle Jones of the student registry of the University of Cambridge (StudentRegistry@admin.cam.ac.uk), Barry (sic!) Slater attended the university and was awarded the degrees of Bachelor of Arts (1958), Master of Arts (1965), and Doctor of Philosophy by Special Regulations in Divinity (2008). According to [24], Barry Hartley Slater was awarded another Master of Arts (1974) and a PhD (1976) by the University of Kent at Canterbury.

3   Hartley Slater's Academic History

According to [24], he studied at St John's College Cambridge as a mathematics scholar (1955–60) and at Balliol College Oxford as a postgraduate (1960–61), then he taught mathematics in schools (1961–71) until he joined the University of Kent at Canterbury as a postgraduate student (1971–75) and a lecturer in philosophy (1974–76); finally, from 1976 until his death, he was at the Philosophy Dept. of the Univ. of Western Australia as lecturer, senior lecturer, associated professor, and honorary senior research fellow.

4   Meeting Hartley Slater

I met Hartley Slater during Epsilon 2015, a workshop on "Hilbert's Epsilon and Tau in Logic, Informatics and Linguistics", June 10–12, 2015.

Being an hour early for the workshop on the first day, I was standing outside a pretty remote, locked lecture hall at the Campus Triolet of the Université de Montpellier. Nobody else was around. The weather was sunny and the air was very clean after heavy rain of the days before. I had skimmed through some of Slater's publications (in particular [21]), which I found most puzzling and fascinating. I was looking forward to meeting him, and Hartley was the second person who joined me waiting for the door to open.

He was very open and communicative, warm and friendly, with an incredible sharpness and presence of mind.

His critical and to-the-point questions were some of the crucial ingredients that gave this workshop its precious and intensive atmosphere.

During the following months he was very helpful in answering questions and in sending me out-of-print books and papers.

5 Hartley Slater's Mission in Logic

I was deeply impressed by Hartley's talks and discussions at the workshop, and read his publications afterwards with great care. On a deeper study, they turned out to be very well written and systematic, almost flawless, and sometimes even entertaining.

Hartley wrote to me that he sees himself as an opposition against the mainstream of modern logic. From his viewpoint on modern logic, most of its problems, intricacies, and paradoxes[1] result from a wrong modeling of the general human idea of logic as found in the natural languages. For this wrong modeling he somehow blamed mathematics (cf. e.g. the titles of [19; 23]). To solve the problems of modern logic, he suggested to correct its wrong concept formations toward the logic found in natural languages, and abhorred "mathematical" escapes such as intuitionist, substructural, or paraconsistent logics.

For instance, in [18], he explains most carefully that the logic of natural language does not entitle us to turn a reflexive relation into a predicate. So, in a different formulation,
$$R' := \{ \, (x,y) \in R \mid x = y \, \}$$
is justified as a definition of a (binary) relation R' from a relation R, and
$$p(x) := \mathsf{true} \text{ if } (x,x) \in R, \qquad p(x) := \mathsf{false} \text{ otherwise}$$
is justified as a definition of a characteristic function p, but
$$P := \{ \, x \mid (x,x) \in R \, \}$$
is *not* justified as a definition of a *singulary predicate* P. This renders Russell's Paradox as an example of the general problem of consistency of predicate definition.[2] Indeed, the problem is Russell's
$$\{ \, x \mid x \notin x \, \},$$
whereas Slater's
$$\{ \, (x,x) \mid x \notin x \, \}$$
does not produce a contradiction.

---

[1] Historically correct, Slater uses "antinomy" (Latin for "against the law") as a perfect synonym of his preferred "paradox" (Greek for "against the teachings") (cf. e.g. [19, p. 125]), contrary to the modern-logic tradition starting with Zermelo, where an antinomy is a disaster for a theory, whereas a paradox is somewhat weaker and just a criterion for the quality of a theory, cf. e.g. [11, p. 104].

[2] For example, to guarantee object-level consistency of recursive specification with positive/negative-conditional equations [28], the inductive theorem prover QUODLIBET [1] admits only the definition of characteristic functions such as p, equality is the only (predefined) predicate, and the inductive data type of Boolean values has to satisfy $\mathsf{true} \neq \mathsf{false}$, but *not* $\forall x.(x = \mathsf{true} \lor x = \mathsf{false})$, for a Boolean variable x.

Following this observation, in [20] he discusses Frege's Begriffsschrift as one of the sources of modern logic[3] and finally explains why Gödel's fixed-point theorem, which — though not in [7] — can be seen as a main ingredient of the proof of Gödel's first incompleteness theorem and of the main argument in [25], does not hold in the logic of natural languages.

In the tradition of the 19th century, Hartley also used to advocate probability theory as a criterion for logic. See, for instance, [22], where Hartley reveals a clearer understanding of the Aristotelian square of oppositions than found elsewhere.

6 Hartley Slater's Heritage in Logic

Most of Slater's ideas deserve a broader audience and response than they have gained, for instance in the practical application of logic in linguistics and informatics. Especially his highly adequate modeling of intensional logic as found in natural languages and his most skilled mastering of Hilbert's ε ought to become part of the general teaching of logic.

As Slater's ideas are off the beaten track of the mainstream of modern logic, in particular the learned reader will use any occasional ambiguity in Slater's papers to escape the message. Thus, one of the main reasons for a regard lower than deserved is that Slater — to the best of our knowledge — never had a co-author and never wrote a joint paper, "against the realisms of the age" (the title of the book [15]). While there is no way to change this now, the actual problem may still be cured: A freshmen logic textbook that removes the over-simplifications in the abstraction of modern logic from the logic of natural language, by a synergetic combination of Slater's ideas with Frege and Peirce's, would be a crucial step forward in the teaching of logic today.[4]

---

[3]One is tempted to check whether the Frege quotation in [20] is as obviously inappropriate in German as it is in the English translation in [4]. To find out that this is indeed so, see [3, p. 34f.]. The German "Daraus ist zu entnehmen, daß das Verhältniß des Gedankens zum Wahren doch mit dem des Subjects zum Prädicate nicht verglichen werden darf." is actually worse than the English translation "It follows that the relation of the thought to the True may not be compared with that of subject to predicate." because the German is more a "must not" than a "may not" here.

[4]Of course, this textbook would have to leave Gödel's first incompleteness theorem &c. and the paradoxes to more advanced studies, but it would give a much more reasonable, practical, and natural impression of logic to the students, with many further advantages. For instance, informatics teachers would be happy to have a maximal set to construct co-inductive data types from, because Slater's restrictions on predicates discussed above can be seen as a weak version of those found in Quine's New Foundations or Mathematical Logic [12].

If somebody wants to learn about the unique viewpoint of this outstanding scientist, we recommend his short, but strong paper [20] discussed above as an appetizer. A deeper study should start with [13], where his main ideas are fresh and most easy to grasp and evaluate, and then follow his main logic books [14; 16; 19; 23].

7   Conclusion

No doubt, the croquet players truly miss him. I am not competent to judge on his work in aesthetics, architecture, literature, etc., but we know that all logicians miss this wonderful (though sometimes caustic) person, scrutinous scholar, ardent discussion partner, and most creative scientist.

Acknowledgments

We would like to thank Sergei Soloviev and Marianne Wirth for discussions and helpful suggestions.

References

[1] Jürgen Avenhaus, Ulrich Kühler, Tobias Schmidt-Samoa, and Claus-Peter Wirth. How to prove inductive theorems? QUODLIBET! 2003. In [2, pp. 328–333], http://wirth.bplaced.net/p/quodlibet.

[2] Franz Baader, editor. *19th Int. Conf. on Automated Deduction (CADE), Miami Beach (FL), 2003*, number 2741 in Lecture Notes in Artificial Intelligence. Springer, 2003.

[3] Gottlob Frege. Über Sinn und Bedeutung. *Zeitschrift für Philosophie und philosophische Kritik (Fichte-Ulricische Zeitschrift)*, 100:25–50, 1892. English translation: [5, pp. 151–171].

[4] Gottlob Frege. *Translations from the philosophical writings of Gottlob Frege*. Basil Blackwell, Oxford, 1952. Ed. by Peter Geach and Max Black. 1st edn..

[5] Gottlob Frege. *The Frege Reader*. Blackwell Publishing, 1997. Ed. by Michael Beaney.

[6] Dov Gabbay and John Woods, editors. *Handbook of the History of Logic*. North-Holland (Elsevier), 2004ff..

[7] Kurt Gödel. Über formal unentscheidbare Sätze der Principia Mathematica und verwandter Systeme I. *Monatshefte für Mathematik und Physik*, 38:173–198, 1931. With English translation also in [9, Vol. I, pp. 145–195]. English translation also in [10, pp. 596–616] and in [8].

[8] Kurt Gödel. *On formally undecidable propositions of Principia Mathematica and related systems*. Basic Books, New York, 1962. English translation of [7] by Bernard Meltzer. With an introduction by R. B. Braithwaite. 2nd edn. by Dover Publications, 1992.

[9] Kurt Gödel. *Collected Works*. Oxford Univ. Press, 1986ff. Ed. by Sol Feferman, John W. Dawson Jr., Warren Goldfarb, Jean van Heijenoort, Stephen C. Kleene, Charles Parsons, Wilfried Sieg, &al..

[10] Jean van Heijenoort. *From Frege to Gödel: A Source Book in Mathematical Logic, 1879–1931*. Harvard Univ. Press, 1971. 2nd rev. edn. (1st edn. 1967).

[11] Volker Peckhaus. *Hilbertprogramm und Kritische Philosophie – Das Göttinger Modell interdisziplinärer Zusammenarbeit zwischen Mathematik und Philosophie*. Number 7 in Studien zur Wissenschafts-, Sozial- und Bildungsgeschichte der Mathematik. Vandenhoeck & Ruprecht, Göttingen, 1990.

[12] Willard Van O. Quine. *Mathematical Logic*. Harvard Univ. Press, 1981. 4th rev. edn. (1st edn. 1940).

[13] B. Hartley Slater. *Prolegomena to Formal Logic*. Avebury Series in Philosophy. Gower Publ. Ltd (Taylor & Francis), Aldershot (England), 1988.

[14] B. Hartley Slater. *Intensional Logic — an essay in analytical metaphysics*. Avebury Series in Philosophy. Ashgate Publ. Ltd (Taylor & Francis), Aldershot (England), 1994.

[15] B. Hartley Slater. *Against the Realisms of the Age*. Avebury Series in Philosophy. Ashgate Publ. Ltd (Taylor & Francis), Aldershot (England), 1998.

[16] B. Hartley Slater. *Logic Reformed*. Peter Lang AG, Bern, Switzerland, 2002.

[17] B. Hartley Slater. A poor concept script. *Australasian J. of Logic*, 2:44–55, 2004.

Version extd. with a discussion of Gödel's fixed-point theorem is [20].

[18] B. Hartley Slater. Frege's hidden assumption. *Crítica, Revista Hispanoamericana de Filosofía*, 38:27–37, 2006. `http://critica.filosoficas.unam.mx/pg/en/numeros_detalle_articulo.php?id_articulo=154&id_volumen=37`. Also in [19, Chapter 7].

[19] B. Hartley Slater. *The De-Mathematisation of Logic*. Polimetrica, Monza, Italy, 2007. Open access publication, `wirth.bplaced.net/op/fullpaper/hilbertbernays/Slater_2007_De-Mathematisation.pdf`.

[20] B. Hartley Slater. A poor concept script. 2007. In [19, Chapter 8]. Short version is [17].

[21] B. Hartley Slater. Hilbert's epsilon calculus and its successors. 2009. In [6, Vol. 5: Logic from Russell to Church, pp. 365–448].

[22] B. Hartley Slater. Back to Aristotle! *Logic and Logical Philosophy*, 20:275–283, 2011. Also in [23, Chapter 2 (pp. 7–13)].

[23] B. Hartley Slater. *Logic is Not Mathematical*. College Publications, London, 2011.

[24] B. Hartley Slater. Barry Hartley Slater's C.V. Web only: `https://uwa.academia.edu/HartleySlater/CurriculumVitae` (June 22, 2014), 2014.

[25] Alfred Tarski. Der Wahrheitsbegriff in den formalisierten Sprachen. *Studia Philosophica*, 1:261–405, 1936. English translation is [26].

[26] Alfred Tarski. The concept of truth in formalized languages. 1983. In [27, pp. 152–278].

[27] Alfred Tarski. *Logic, Semantics, Metamathematics:*. Hackett Publishing, Indianapolis (IN), 1983. 2nd rev. edn. (1st edn. 1956, Oxford Univ. Press). Ed. by John Corcoran. Translations by J. H. Woodger.

[28] Claus-Peter Wirth. Shallow confluence of conditional term rewriting systems. *J. Symbolic Computation*, 44:69–98, 2009. `http://dx.doi.org/10.1016/j.jsc.2008.05.005`.

Received 23 October 2015

From logical and linguistic generics to Hilbert's tau and epsilon quantifiers

Stergios Chatzikyriakidis[*]
CLASP, University of Gothenburg; Open University of Cyprus
stergios.chatzikyriakidis@gu.se

Fabio Pasquali
IRIF-CNRS, Université Paris Diderot
pasquali@dima.unige.it

Christian Retoré
LIRMM-CNRS, Université de Montpellier
christian.retore@lirmm.fr

Abstract

With our starting point being (universal) generics appearing in both natural language and mathematical proofs, and were further conceptualised in philosophy of language, we introduce the tau subnector that maps a formula F to an individual term $\tau_x F$ such that $F(\tau_x F)$ whenever $\forall x F$. We then introduce the dual subnector $\epsilon_x F$ which expresses the existential quantification since $F(\epsilon_x F) \equiv \exists x F$, and describe its use for the semantics of indefinite and definite noun phrases. Some logical and linguistic properties of this intriguing way to express quantification are discussed — but the reader is referred to the article by Abrusci in this volume for the impact of epsilon on Hilbert's work the logical foundations of mathematics.

Keywords: proof theory; quantifiers; generic objects; philosophy of language; formal linguistics

[*]The author gratefully acknowledges support from the Centre of Linguistic Theory and Studies in Probability (CLASP) in Gothenburg as well as the ANR project Polymnie in France.

1 Presentation

This introduction to the volume presents the epsilon and tau quantifiers that Hilbert came up with in the beginning of the XXth century. [25] Although an introduction to these issues exists on the web, e.g. [5, 46][1] we believe that the current introduction, presenting a slightly different viewpoint, might be of interest to a number of researchers, especially because most of the literature on the topic is in German, in particular voulme II of *Grundalgen de Mathematik* by Hilbert and Bernays [26][2] as well as Ackermann's seminal contribution [2]. We say that tau and epsilon are quantifiers because one can express quantification with them, but given that epsilon and tau map a formula to an individual term, they should rather be called *subnectors* according to Curry's terminology [11]. Given a formulae F, possibly with free variables $\epsilon_x F$ and $\tau_x F$ are terms in which x is bound, and these terms respectively denote the existential and the universal generic objects w.r.t. a formula F. Basically $F(\epsilon_x F)$ means $\exists x.F$ and $F(\tau_x F)$ means $\forall x.F$. Since they belong to a classical, and not an intuitionistic setting, one can be defined from the other: $\tau_x F = \epsilon_x \neg F$. The addition of epsilon and tau to the connectives may seem harmless, but it actually completely changes the logic, i.e. formulae of the epsilon calculus have no counterpart in first order or even higher order predicate calculus. In particular, because of *over binding* (also known as *in situ* binding), these quantifiers are closer to the syntactic behavior of quantifiers in natural language sentences. This is the reason why the epsilon subnectors have been used in the philosophy of language and formal linguistics to model quantifiers of natural language [17, 18, 50, 49, 48, 51, 52, 47, 44]. Symmetrically, the subnector tau could have been used to model universal quantification but it rarely was, exceptions being [41, 34]

2 From the Ancient and Medieval view of universal quantification to *tau*

A long debated question in logic and metaphysics in the Ancient and the Medieval world (starting with Plato, Aristotle and all the way to Porphyry and the scholastics) is the relation between a universal or generic "dog" and the set of individuals "dogs". This is known as the *problem of universals.*[33, 28, 3, 13]

What is a concept like "horse"?

---

[1]The second one was written by Hartley Slater, who authored an article of this volume. Hartley Slater unfortunately passed away before this volume is out, and an obituary by Claus-Peter Wirth is also included in this volume.

[2]There exist a Russian translation (1978), a French translation (2010), but still no English translation.

- a substance, that exists independently of the individuals falling under this concept (realism)

- a name without reality, i.e. a word that stands for the class of all individuals falling under the concept (nominalism)

- a concept, that is a mental construction having an empirical relation to the set of individuals (conceptualism)

In order to illustrate the debate between Abélard and Roscelin (see e.g. [13]) regarding the relation between the concept and the entities that fall under this concept, one could say: *If an illness causes the extinction of all tall dogs, would your concept of dog be altered?* Some have even defended the extreme position that *each dog* is as constitutive of the concept of *dog* as a *wall* or a *roof* is constitutive of the concept of a *house*.

The relation between concept and universal quantification is that, in case the generic object in a concept C enjoys the property P, so do all C. This is of course also related to Aristotle's proof *rule* of *abstraction* (alternatively called *generalisation*): if an integer without any specific property enjoys a property P, then all integers enjoy P. Ancient and Medieval logic was *not* dealing with *models*. These only appeared with Frege at the end of the XIXth century. Philosophers ranging from Aristotle to scholastic logicians were mainly concerned with proofs and rules [29]. Aristotle who was aiming at extracting the logical principles underlying mathematical reasoning, introduced the generalization rule for universal quantification (that he called abstraction).

Using this rule, to establish $F(x)$ for all elements x of a class C one proceeds as follows:

Generalisation (a.k.a Abstraction) Let x be any element of C. A reasoning shows that $P(x)$ holds. The generalization rule asserts that the property holds for any element in C — indeed x does not possess anything special apart from being in C .

This is quite different from another technique used to establish $F(x)$ for all elements x of a known and finite class C. One shows that for each element x_i in the class C the property $F(x_i)$ does hold, and the *conjunction* rule shows that it holds for all $x \in C$: this way to establish a universally quantified statement perfectly matches universal quantification in a given model of first order logic.

Conjunction We can prove $P(c_1)$, then $P(c_2)$, $P(c_3)$, $P(c_4)$ and so on. Once we do so for all elements in C, we can form a conjunction out of all these formulae. This latter proof of a universally quantified statement is in fact, in modern terms, a reduction of universal quantification to conjunction: $\forall x \in \mathcal{C}F(x) \equiv \&_{x \in \mathcal{C}}F(x) \equiv P(c_1)\&P(c_2)\&P(c_3)\&P(c_4)\&\cdots$ Notice, however, that an infinite formula is not a usual first order formula.

This dual nature of universal quantification can be observed in the various versions of universal quantification in natural language. For example, some words refer to each individual in the collection (each), while some others to a (fictive) prototypical or average individual (any):

(1)   a. Each dog has four legs.

    b. All dogs have four legs.

    c. Every has four legs.

    d. A dog has four legs.

    e. Any dog has four legs.

    f. Dogs have four legs.

The distributive reading, which is obligatory when using *each*, has no exceptions, may express a coincidence of properties that can be conjuncted, it is not required that there is a reason (other than probability) to this coincidence and the domain can be complicated. The only good way to refute such a statement is to provide a counter example, i.e. an individual for which the property does not hold (one component of the conjunction fails).

(2)   a. Each bird with both black and white feathers flies.

    b. Not this wound bird. (perfect)

    c. Not autruches. (not a good refutation, since the relation between those two sets is not obvious)

Generic entities rather correspond to ideal and prototypical entities whose properties are derived by reasoning. Compared to noun phrases introduced by *each*, generic noun phrases may accept exceptions, and their domain cannot be complicated. The refutation of a sentence involving a generic is usually performed by another generic, or by a reasoning, i.e. one can go on at an abstract level.

(3)   a. Birds fly.

    b. Not this wound bird. (not a refutation, since generic readings admit exceptions)

    c. Not autruches. (perfect refutation)

The usual rule called generalisation or abstraction, is abbreviated by $\forall_i$ i.e. $\forall$ introduction because it introduces the $\forall$ quantifier. This rule says that when a property has been established for an x which does not enjoy any particular property (i.e. is not free in any hypothesis), one can conclude that the property holds for all individuals[3]

In this paper, we use *sequents*: $H_1, \ldots, H_n \vdash C$ simply means that under hypotheses $H_1, \ldots, H_n$, conclusion C holds. The $\forall_i$ (or generalisation) rule below simply means that from (1) $P(x)$ *holds under hypotheses* $H_1, \ldots, H_n$ *without any free x in any of the* H_i one may deduce that (2) $\forall x.\ P(x)$ *holds under the hypotheses* $H_1, \ldots, H_n$.

$$\frac{H_1, \ldots, H_n \vdash P(x)}{H_1, \ldots, H_n \vdash \forall x.\ P(x)} \ \forall_i - \text{when there is no free occurrence of } x \text{ in any } H_i$$

The rule above can be formulated with a generic element, $\tau_x P(x)$, a virtual element that has no specific relation to the property P. If you think of P as being "to drink", $\tau_x drink(x)$ is the most sober individual you can think about: $\tau_x drink(x)$ drinks if and only if everyone does:

$$P(\tau_x P(x)) \text{ iff } \forall x.\ P(x)$$

From this one easily defines the rules for quantification using the universal generic element w.r.t. P written $\tau_x P(x)$[4], they simply are the standard rules of quantification:

$$\frac{H_1, \ldots, H_n \vdash P(x)}{H_1, \ldots, H_n \vdash P(\tau_x P(x))} \ \tau_i - \text{when there is no free occurrence of } x \text{ in any } H_i$$

There is another rule for universal quantification called the specialisation or instantiation rule, which is easier because it comes without any restriction, and which says that of something holds for any x then it holds for any constant or term.[5]

---

[3]As we shall discuss later on, such a rule is enough to derive that all A are B, from the fact that B holds for an x satisfying A which has no specific property other than A.

[4]Here one sees that τ is a **subnector** according to Curry's terminology i.e. an operator that builds a term (of type individual) from a formula.

[5]This rule can derive its relativised version that says that if all A are B then any particular A is B.

$$\frac{H_1, \ldots, H_n \vdash \forall x.\ P(x)}{H_1, \ldots, H_n \vdash P(a)} \ \forall_e$$

The above rule can also be formulated with the same universal generic element w.r.t. $P\ \tau_x P(x)$:

$$\frac{H_1, \ldots, H_n \vdash P(\tau_x P(x))}{H_1, \ldots, H_n \vdash P(a)} \ \tau_e$$

If $\tau_x P(x)$ enjoys the property $P(\_)$ then any individual does, and vice-versa. So $\tau_x P(x)$ is an ideal entity which is absolutely independent from the property $P(\_)$. This is the reason why when it enjoys $P(\_)$ everything does.

3 Existential generics: from Russell's iota to Hilbert's epsilon

As opposed to generic $\tau_x P(x)$ that enjoys P if and only if every entity does, there is also "this P", "the P", i.e. the unique individual satisfying P, if there is exactly one such individual, denoted as $\iota_x P(x)$. Russell used ι in [45] for definite descriptions (definite noun phrases like the queen of England). Previously, Frege and Peano already used such an operator, in different contexts, Frege with notation / and Peano with notation ι, while Russell did not use any symbolic notation at all. It is also, like τ, a subnector, given that it turns a formula into some individual term. It is the ancestor of Hilbert's ϵ.

However, as argued by von Heusinger, there is little difference between the logical form of definite descriptions and indefinite noun phrases. This is because uniqueness of the noun phrase with a definite article is not always observed:

(4) Recueilli très jeune par les moines de l'abbaye de Reichenau, sur **l'île du lac de Constance**, en Allemagne, qui le prennent en charge totalement; Hermann étudie et devient l'un des savants les plus érudits du XI-ème siècle.

(5) Taken in while very young by monks of the abbey of Reichenau on **the island of the Constance lake**, that fully took care of him, Hermann studied and became one of the most erudite monks of the XIth century.

This example comes from a site dealing with first names and is really a pleasant coincidence:[6] indeed in the original paper by Egli and von Heusinger (1995, hence

---

[6](http://www.prenoms.com/prenom/signification-prenom-HERMANT.html.)

before the example) the case of the islands of the Constance lake is taken as a fictive example in order to exemplify that someone seeing just one of the three islands of this lake could utter **the** *island of the Constance lake* while there are **three** of them.

Given that ι is not expected to have good logical properties — for instance its negation is *zero or more than two entities do not enjoy the property $P(\_)$* — Hilbert introduced the existential ϵ subnector, building the existential generic $\epsilon_x F$:

$$F(\epsilon_x F) \equiv \exists x. \ F$$

A term (of type individual) $\epsilon_x F$ associated with F, the existential generic element w.r.t. F (which usually contains occurrences of x): as soon as an entity enjoys F the term $\epsilon_x F$ enjoys $F(\_)$. The operator ϵ binds in $\epsilon_x F$ the free occurrences of x in F.

The introduction rule is as one expects (a denotes any term):

$$\frac{H_1, \ldots, H_n \vdash F(a)}{H_1, \ldots, H_n \vdash P(\epsilon_x F(x))} \ \epsilon_i$$

The elimination rule for existential quantification in natural deduction is not given now, but rather in the next section that is more formal. For the moment, let us just say that the rule for existential quantification using ϵ-terms simply mimics the introduction and elimination rules for existential quantification.

If $\tau_x \, drink(x)$ is the most sober individual, $\epsilon_x \, drink(x)$ is a soak: he drinks if and only if someone drinks.

4   Syntax of epsilon and tau first order calculus

Terms and formulae are defined by mutual recursion:

- Any constant in $\mathcal{L}$ is a term.

- Any variable in $\mathcal{L}$ is a term.

- $f(t_1, \ldots, t_p)$ is a term provided each t_i is a term and f is a function symbol of $\mathcal{L}$ of arity p

- $\epsilon_x A$ is a term if A is a formula and x a variable — any free occurrence of x in A is bound by ϵ_x

- $\tau_x A$ is a term if A is a formula and x a variable — any free occurrence of x in A is bound by τ_x

- $s = t$ is a formula whenever s and t are terms.

- $R(t_1, \ldots, t_n)$ is a formula provided each t_i is a term and R is a relation symbol of $\mathcal{L}$ of arity n

- $A \& B$, $A \vee B$, $A \Rightarrow B$, $\neg A$ when A and B are formulae.

There is no objection to simultaneously use the usual quantifiers $\forall$ and $\exists$. This superimposition is quite useful to have, since we can show that the epsilon/tau calculus restricted to formulas that are equivalent to usual first order logic, is a conservative extension of first order logic (first and second epsilon theorems, see e.g. Abrusci's paper in the same issue). This means that we can have quantified formulas as well:

- $\exists_x A$ is a formula if A is a formula and x a variable — any free occurrence of x in A is bound by $\exists x$

- $\forall_x A$ is a formula if A is a formula and x a variable — any free occurrence of x in A is bound by $\forall x$

Quantification rules are the ones we already discussed above.

τ rules

introduction The introduction rule of the tau universal quantifier is Aristotle's rule of abstraction (also known as generalisation): from $P(x)$ with x generic (i.e. not present in any hypothesis) infer $P(\tau_x.P(x))$.

$$\frac{H_1, \ldots, H_n \vdash P(x)}{H_1, \ldots, H_n \vdash P(\tau_x P(x))} \tau_i \quad \boxed{\begin{array}{l} \text{Condition:} \\ \textit{no free occur-} \\ \textit{rence of } x \textit{ in any} \\ H_i \end{array}}$$

elimination The elimination rule of universal quantification (also known as instantiation or specialisation) is as usual: from $P(\tau_x.P(x))$ one may infer $P(t)$ for any t.

$$\frac{H_1, \ldots, H_n \vdash P(\tau_x P(x))}{H_1, \ldots, H_n \vdash P(t)} \tau_e$$

ϵ rules

introduction The introduction rule of the epsilon existential quantifier is the usual one: from $P(t)$ where t is any term, infer $P(\epsilon_x P(x)) \equiv \exists x \, P(x)$.

$$\frac{H_1, \ldots, H_n \vdash P(t)}{H_1, \ldots, H_n \vdash P(\epsilon_x P(x))} \epsilon_e$$

elimination The elimination of the epsilon universal quantifier is trickier, as it is in natural deduction for first order logic format: assume that from $P(x)$ and other hypotheses Γ not involving x a conclusion C without x is derivable and that $P(\epsilon_x P(x))$ holds: then C holds under hypotheses Γ.

$$\frac{K_1, \ldots, K_p \vdash P(\epsilon_x P(x)) \qquad H_1, \ldots, H_n, P(x) \vdash C}{K_1, \ldots, K_p, H_1, \ldots, H_n \vdash C} \, \epsilon_e \quad \left| \begin{array}{l} \text{Condition:} \\ \textit{no free occur-} \\ \textit{rence of } x \textit{ in any} \\ H_i \textit{ nor in } C \end{array} \right.$$

the epsiilon/tau calculus is easier in a classical (as opposed to intuitionistic) setting,[7] and in a classical calculus the following are easily derived:

$$P(\epsilon_x P(x)) \overset{\epsilon}{\equiv} \exists x P(x) \overset{classical}{\equiv} \neg \forall x \neg P \overset{\tau}{\equiv} \neg \neg P(\tau_x \neg P(x)) \overset{classical}{\equiv} P(\tau_x \neg P(x))$$

$$P(\tau_x P(x)) \overset{\tau}{\equiv} \forall x P(x) \overset{classical}{\equiv} \neg \exists x \neg P \overset{\epsilon}{\equiv} \neg \neg P(\epsilon_x \neg P(x)) \overset{classical}{\equiv} P(\epsilon_x \neg P(x))$$

Hence:

$$\tau_x P(x) = \epsilon_x \neg P(x) \quad \text{and} \quad \epsilon_x P(x) = \tau_x \neg P(x)$$

Therefore one of the two subnectors/quantifiers ϵ and τ is enough and most people have chosen ϵ, e.g. as in Bourbaki's book on set theory.

The quantifier free epsilon calculus is a strict conservative extension of first order logic.

- Strict: There are formulas that are not equivalent to any formula of first order logic, e.g. $P(\epsilon_x Q(x))$ with P, Q being distinct unary predicate symbols: if P and Q are unrelated predicate letters, this bound formula has no equivalent in first order logic.

- Conservative: With regards to first order formulas, the epsilon calculus derives exactly the same formulas as first order classical logic, i.e. classical predicate calculus.

As shown by Claus-Peter Wirth in this issue, any first order formula can be turned into an equivalent epsilon formula. However, one should have in mind that

---

[7] Given that epsilon induces a weak form of the axiom of choice, it is quite difficult to even define an intuitionistic epsilon calculus, see section 7.3.

such a translation yields formulas that are rather complex and hard to parse. For instance $\forall x \exists y P(x, y)$ can be written has: $\exists y P(\tau_x P(x, y), y)$ which itself can be written as: $P(\tau_x P(x, \epsilon_y P(\tau_x P(x, y), y)), \epsilon_y P(\tau_x P(x, y), y))$!

As discussed in Michele Abrusci's article on epsilon and proof theory in this issue, a major motivation of Hilbert's work since [25], was to establish the logical consistency of arithmetic by elementary means, before Gödel's incompleteness theorem [23]. Following this objective that could not be met because of Gödel's incompleteness theorem, Hilbert proved an ϵ version of quantifier elimination (1st & 2nd ϵ-theorems in [26]) and obtained as a corollary (still in [26]) the first correct proof of Herbrand's theorem (initially published in [24]) — indeed in 1963 Dreben found mistakes in Hebrand's proof [15] and corrected them in 1966 [16]. These issues are more deeply addressed in Abrusci's paper in this volume.

The reader must be aware that the epsilon calculus is *very* different from first order logic. As already said, many formulas of the epsilon calculus are not equivalent to first order formulas. A number of published results on the epsilon calculus are known to be wrong, like cut-elimination and models among other things (see [4, 31] as explained in [7, 36]).[8] As regards the first point, the problem is to take into account the complexity of the formulas inside epsilon terms and as regards the second point, the problem is that if there exist sound and complete models for the whole epsilon calculus, then they must substantially differ from usual models. However, all proposals so far try to provide usual models for a very unusual calculus. Nevertheless, intuitively, epsilon terms might be viewed as a kind of Henkin witnesses, or as choice functions that act simultaneously on all formulas, without any obligation to extend the first order language. Categorical models of typed epsilon calculus are slightly easier to compute as discussed in section 7.3.

The relation between epsilon and choice functions introduced by Skolemisation is complicated. Choice functions are usually introduced when one considers a single formula (or a clause) in prenex form and introduces a function *in the language*, whose arguments are the universal variables, for each of the existential quantifiers. Epsilon introduces all choice functions at once but without reference to a particular formula (except the one defining the epsilon term), nor to universal variables, and the first order language (relations, functions) does not need to be extended. As regards their syntax, there are no specific deduction rules for choice functions: the existential quantifiers are taken into account when interpreting the function — on the syntactic side they do not belong to first order logic.

---

[8]We would like to thank Michele Abrusci, Mathias Baaz and Ulrich Kohlenbach for pointing out these problems to us.

5 Epsilon (and tau) in linguistics

5.1 Some critics of the usual treatment of natural-language quantification

Quantifiers, especially existential ones, are quite common in natural language. There are two sorts of quantifiers w.r.t natural language syntax. The first sort only needs the main predicate (something, someone, everyone, everything,...). The second sort (a, some, every, all,...) first applies to a common noun (a predicate) and then to the main predicate and all of its arguments but one — the verb plus its subject and complement but one.

(6) Something happened to me yesterday.

(7) A man comes on to tell me how white my shirts can be.

(8) Some girls give me money.

(9) Keith played a Beatles song.

The standard treatment of quantifiers initiated by Montague [38] takes place into the framework defined by Church [9] for writing first or higher order formulas of predicate calculi in a way that follows Frege's compositionality principle.

This treatment usually assumes a base type for propositions $\mathbf{t}$ (o in Church writings) and another base type $\mathbf{e}$ (ι in Church writings) for entities, also known as individuals. Usually, there is just one base type for all individuals, as recalled in [39] the logic is single sorted as opposed to many sorted logic of, e.g. [30] used for semantics in [43].

In order to express linguistic semantics one needs logical constants for connectives and quantifiers:

- $\sim$ of type $\mathbf{t} \to \mathbf{t}$ (negation)

- $\supset, \&, +$ of type $\mathbf{t} \to (\mathbf{t} \to \mathbf{t})$
  (implication, conjunction, disjunction)

- two constants $\forall$ and $\exists$ of type $(\mathbf{e} \to \mathbf{t}) \to \mathbf{t}$

Specific constants are needed to represent a first order language:

- R of type $\mathbf{e} \to (\mathbf{e} \to (.... \to \mathbf{e} \to \mathbf{t}))$ (n-ary predicate — n times $\mathbf{e}$)

- f of type $\mathbf{e} \to (\mathbf{e} \to (.... \to \mathbf{e} \to \mathbf{e}))$ (n-ary function symbol — n times $\mathbf{e}$)

As already mentioned, $\forall$ and $\exists$ are constants of type $(\mathbf{e} \to \mathbf{t}) \to \mathbf{t}$ to quantifiers that simply apply to the main predicate. This means that they can interpret natural language quantifiers without a domain. This is the case with *everything, something* but other natural language quantifiers that apply to two predicates like *some, every* need more complex terms:[9]

- existential quantifier (some, a) :

$$\lambda P^{\mathbf{e} \to \mathbf{t}} \lambda Q^{\mathbf{e} \to \mathbf{t}} (\exists \lambda x^{\mathbf{t}}.P(x) \& Q(x)) : (\mathbf{e} \to \mathbf{t}) \to (\mathbf{e} \to \mathbf{t}) \to \mathbf{t}$$

- universal quantifier (every, all) :

$$\lambda P^{\mathbf{e} \to \mathbf{t}} \lambda Q^{\mathbf{e} \to \mathbf{t}} (\forall \lambda x^{\mathbf{t}}.(P\ x) \supset (Q\ x)) : (\mathbf{e} \to \mathbf{t}) \to (\mathbf{e} \to \mathbf{t}) \to \mathbf{t}$$

This modelling suffers from some inadequacies mainly due to a difference in the syntactic structure of the sentence and its logical form.

The standard analysis of the sentence 9 is:

$(a\ (Beatles\ song))(\lambda z.Keith\ played\ z)$[10]

So:

$$\overbrace{(\lambda P^{\mathbf{e} \to \mathbf{t}} \lambda Q^{\mathbf{e} \to \mathbf{t}} (\exists \lambda x^{\mathbf{e}}.(P\ x) \& (Q\ x)))}^{\text{(existential quantifier)}}$$

is applied to

$$\overbrace{(\lambda u^{\mathbf{e}}.wrote(Beatles, u) \& song(u))}^{\text{(being a Beatles song)}}$$

and to

$$\overbrace{(\lambda z.Keith\ played\ z)}^{\text{(being played by Keith)}}$$

As expected, these lambda terms reduce to

$$(\exists \lambda x^{\mathbf{e}}.wrote(Beatles, x) \& song(x) \& (\text{Keith played } x)))$$

i.e. to the expected meaning:

$$(\exists x.\ wrote(Beatles, x) \& song(x) \& ((Keith\ played)\ x)))$$

---

[9]We write $P(x) \& Q(x)$ instead of $\&(P\ x)(Q\ x)$ the pure lambda calculus prefixed notation that is used [39].

[10]Here as well the standard notation of [39] would be $\lambda z.\ played(Keith, z)$.

5.1.1 Difference between the syntactic and semantic structures

The classical treatment of natural language quantifiers infringes the correspondence between syntax and semantics, i.e. the heart of compositionality. Indeed, observing a sentence, its syntactic structure, and its semantic structure:

(10)   a. Keith played a Beatles song.

   b. syntax (Keith (played (a (Beatles song))))

   c. semantics: (a (Beatles song)) $\underline{(\lambda x. \text{Keith played } x)}$

it is clear that the underlined predicate does not correspond to any proper phrase (or subtree) of the sentence. A natural language quantifier is an *in situ* binder (as e.g. *wh*-interrogatives in Chinese): a quantified noun phrase which is deeply nested in the sentence parse tree, may apply to the whole parse tree.

5.1.2 Asymmetry between the domain of quantification and the main predicate

It is easily observed in the following examples that, as opposed to the usual logical formulas representing meaning, one cannot swap the two predicates, for instance in Aristotle's I sentences. However, even when these can be swapped, as it is the case in the last example, the meaning is *not* the same, because the focus is different, and depending on the context (in a university or in a company) only one of the two can be said, the other one begin unnatural.

(11)   a. Some politicians are crooks.

   b. ?? Some crooks are politicians.

(12)   a. Some students are employees.

   b. Some employees are students.

5.1.3 Semantic nature of the quantified noun phrase

According to the usual treatment, a quantified noun phrase is a function that maps a predicate to a proposition. However, intuitively, the type of a quantified noun phrase, and especially the type of an existentially quantified noun phrase should rather be an individual. This is confirmed by the cognitive process: when "an A" is uttered, one actually imagines such an "A", so a quantified noun phrase may have a reference as an individual before uttering the main predicate (if any), as observed in the following nominal sentences:

243

(13)   Cars, cars, cars,... (Blog)

(14)   What a thrill — My thumb instead of an onion. (S. Plath)

5.2   Using epsilon for indefinite noun phrases

Coming back to the linguistic motivation, some researchers have been modelling existential quantification by means of epsilon (rather than universal quantification by means of tau), since the pioneering work of von Heusinger and Egli [17, 18] that has been further pursued in a number of papers [49, 50, 51, 52]. The leitmotiv in these papers, is to model existentially quantified noun phrases like "an A is B" (Aristotle's I sentence) by $B(\epsilon x.\ A(x))$. Nothing in a formula like this says that $\epsilon x.\ A(x)$ has the property A, i.e. that there are some As. Thus, the presupposition $A(\epsilon x.\ A(x))$ could (and should) be added.

Here we should make an important remark on these sentences In general, the formula $B(\epsilon x.\ A(x))$ is not equivalent to any ordinary formula — unless there is a relation between A and B, like $B = A$ or $B = \neg A$. In particular it is *not* equivalent to $\exists x.(B(x)\&A(x))$, but the two formulas are related as follows:

In general $B(\epsilon_\mathbf{x} A(\mathbf{x}))$ does not entail $\exists \mathbf{x}.(B(\mathbf{x})\&A(\mathbf{x}))$ Indeed, it is possible that $\exists x.(B(x)\&A(x))$ is false while $B(\epsilon_x A(x))$ is true. Indeed, let $B(x)$ be $(x = x)$ and let $A(x)$ be $(x \neq x)$ i.e. $\neg B(x)$. Then $B(\epsilon_x A(x)) \equiv B(\epsilon_x \neg B(x)) \equiv B(\tau_x.B(x)) \equiv \forall x.B(x) \equiv \forall x.x = x$ which is clearly true. However, $\exists x.(B(x)\&A(x)) \equiv \exists x.(B(x)\&\neg B(x)) \equiv \exists x.(x = x \ \& \ x \neq x)$ which is clearly false. The argument works with any formula of one variable that is universally true like $B(x) \equiv (x = x)$.

$B(\epsilon_\mathbf{x} A(\mathbf{x}))\&A(\epsilon_\mathbf{x} A(\mathbf{x}))$ entails $\exists \mathbf{x}.\ B\&A(\mathbf{x})$ Indeed, $B(\epsilon_x A(x))\&A(\epsilon_x A(x))$ entails $B(\epsilon_x B\&A(x))\&A(\epsilon_x B\&A(x))$ that is $B\&A(\epsilon_x(B\&A(x))$ which means $\exists x.\ B\&A(x)$.

$\exists \mathbf{x}.A(\mathbf{x})\&\forall \mathbf{y}(A(\mathbf{y}) \Rightarrow B(\mathbf{y}))$ entails $B(\epsilon_\mathbf{x} A(\mathbf{x}))$ Indeed, ϵ-terms are usual terms, a universal quantifier can be instantiated to an epsilon term.

Given the small difference between definite and indefinite descriptions, von Heusinger proposes to model both "a" (introducing indefinite noun phrases) and "the" (introducing definite description) by an epsilon term. The only thing that differentiates them is interpretation: the "a" always refers to **a new individual** in the class, while "the" refers to **the most salient individual** in the context. This context dependent interpretation lead to the indexed epsilon calculus [37]. This is further studied by Hans Leiß in this volume.

244

(15)   A student entered the lecture hall. He sat down. A student (*another student*) left the lecture hall.

(16)   A student arrived lately. The professor looked upset. The student (*the same student*) left.

The epsilon modeling avoids all three aforementioned drawbacks of the standard interpretation, 5.1.3, 5.1.1 and 5.1.2. Indeed, $\epsilon_x F(x)$ is an individual term (as natural language quantifiers) which allows *in situ* binding (as natural language quantifiers):

5.1.3 A quantified noun phrase can be interpreted as an individual even in the absence of the main predicate: indeed $\epsilon_x F(x)$ is an individual term.

5.1.1 The semantics $P(\epsilon_x F(x))$ of an existential sentence of an I sentence like *Some F is P.* follows the syntactic structure: $\epsilon_x F(x)$ is an individual term, which is the standard semantics of a noun phrase, that is inserted into the main predicate P to obtain the sentence semantics: $P(\epsilon_x F(x))$. This is comparable but not equivalent to: *there exists some x satisfying F(x) such that P(x)*.

5.1.2 The asymmetry between subject and predicate is restored
$P(\epsilon_x Q(x)) \not\equiv Q(\epsilon_x P(x))$.

The interpretation of noun phrases with ϵ also solves the so-called E-type pronoun interpretation of Gareth Evans [19] where the semantics of the pronoun is the copy of the semantics of its antecedent:

(17)   A man came in. He sat down.

(18) $[He] = [A\ man] = (\epsilon_x\ Man(x))$.

During the workshop, Slater pointed out that there is a possible problem with the modelling of indefinite nouns phrases by epsilon terms that we just described. The two sentences *A man enters. A man left.* are respectively modelled as $entered(\epsilon_x man(x))$ and $left(\epsilon_x man(x))$ and if you consider them *simultaneously*, you may infer that $(entered\ \&\ left)(\epsilon_x man(x))$. Consequently one has $(entered\ \&\ left)(\epsilon_x(entered\ \&\ left)(x))$ so $\exists x.\ entered(x)\ \&\ left(x)$!

To avoid this problem, Slater proposes to interpret *a man entered* by *entered* $(\epsilon_x man(x)\ \&\ entered(x))$ *a man left* by $left(man(x)\ \&\ left(x))$. This avoids the unpleasant consequence above. But if you allow yourself to interpret the epsilon terms that appear in different propositions in the same way, then using Slater's approach one ends up with another problem. Consider: *The professor presented first order logic. A student left. The professor introduced sequent calculus. A student left.*

Then, the second and fourth sentences yield exactly the same semantic representation, $left(\epsilon_x student(x)$ & $left(x))$, we have just at least one student that left, while we should have at least two. Thus, one must be aware that computing the semantic representation, a term or a formula, is just one step of the interpretation process, the next step being the interpretation of the semantic representation, and one should also be careful with the interpretation of the epsilon terms, *that are not usual first order terms*.

The problem raised by Slater comes from the fact that nothing says that the two propositions should be conjoined by a usual & and so nothing tells us that the epsilon term that appears in both propositions should be interpreted in the same way twice. This is the reason why Mints in [37] introduced and studied the indexed epsilon calculus, where the interpretation is relative to a given context, and contexts are indices for the calculus (see also the paper by Hans Leiß in this volume).

5.3 Tau and universally quantified noun phrases

Coming back to the subnector tau discussed at the beginning of the paper, it is fairly natural and coherent with the treatment of existential quantification to model universally quantified noun phrases introduced by "all", "every" and bare plurals by tau terms, as follows:

(19) a. Every man is mortal.

  b. $mortal(\tau_x man(x))$

This proposal has been explored in [34]. It enables a distinction between *chaque* (*each*) and *tout* (that does not really correspond to *every*) where the first one is a conjunction $\&_{x \in D} P(x)$ over a known domain D and the second one the usual mathematical universal quantifier with the generic object formalised as $\tau_x M(x)$. The analysis is a proof theoretical one, where statements are interpreted as the set of their possible justifications as suggested in [1]

6 Aristotle square of opposition revisited with epsilon and tau

With such an interpretation of quantified noun phrases, existential noun phrases as epsilon terms and universals noun phrases as tau terms, one may reformulate Aristotle's I A E O sentences and revisit Aristotle's square of opposition.

A	Every S is P	$P(\tau_x.S(x))$	Universal Affirmative
I	Some S is P	$P(\epsilon_x S(x))$	Particular Affirmative
E	No S is P	$\neg P(\epsilon_x S(x)) \equiv \neg P(\tau_x.\neg S(x))$	Universal Negative
O	Some S are not P	$\neg P(\tau_x.S(x)) \equiv \neg P(\epsilon_x \neg S(x))$	Particular negative

Aristotle showed that those formulas define a square of opposition, i.e.: $A \equiv \neg O$, $E \equiv \neg I$ A and E are contradictory (they cannot both hold), I and O cannot both fail and are thus subcontraries, A entails I, which is said to be a subaltern of A, and E entails O which is said to be a subaltern of E.

Using standard diagrams, in Hilbert's ϵ-calculus, at least one of the following two squares is a square of opposition provided P *is bivalent with respect to* S this condition being defined as $P(\epsilon_x S(x)) \vdash P(\tau_x S(x))$ or $P(\tau_x S(x)) \vdash P(\epsilon_x S(x))$ (see [41] for more details).

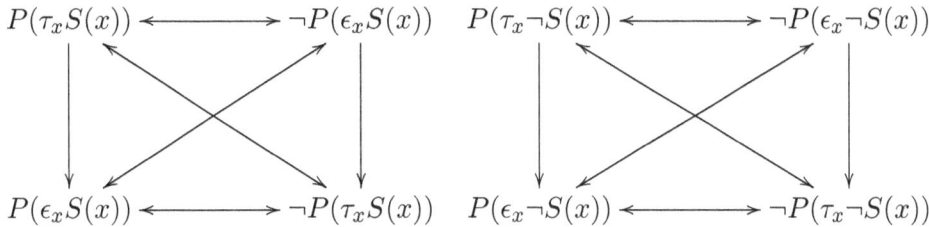

$P(\tau_x S(x)) \longleftrightarrow \neg P(\epsilon_x S(x))$ $P(\tau_x \neg S(x)) \longleftrightarrow \neg P(\epsilon_x \neg S(x))$

$P(\epsilon_x S(x)) \longleftrightarrow \neg P(\tau_x S(x))$ $P(\epsilon_x \neg S(x)) \longleftrightarrow \neg P(\tau_x \neg S(x))$

7 Typed Hilbert operators

7.1 Some critics of the unique universe of Frege

As discussed in [34], there are two ways to describe universal quantification:

universal quantification as a conjunction over the domain (model theoretical view) $\&_{x \in D} P(x)$ (Fench *chaque* English *each*)

universal quantification as a property of the generic member of its class (proof theoretical view) $\forall x P(x)$ or better $P(\tau_x P(x))$ (Fench *tout* without English equivalent, not really *every*)

In the model theoretic view, the domain of quantification is clear, but in the proof theoretic view, the domain is not specified. This means that one cannot write $\&_{x \in D} P(x)$ and a fortiori cannot derive it. The completeness theorem for first order logic [22] makes sure that both notions of quantification agree.

As argued by medieval philosophers, e.g. Abu'l-Barakāt al-Baghdādī, a property is always asserted from an entity, as being a member of some particular class and not as an entity of the class of all entities[12].[11] For quantifying (universally or existentially) a notion of class is needed as well: in standard proof theory (as opposed to type theory) there are no classes, and in standard model theory, only the largest class of all entities is known. Nevertheless quantifying over a class, a sort or a part of the domain is quite common:

(20)   All students passed.

(21)   All students who passed logic passed algebra.

(22)   Numbers ending in 3 have at least one multiple having all 1.

(23)   Some student passed algebra.

(24)   Some student who passed logic passed algebra.

(25)   Some prime numbers are one less than a power of two.

In defining the predicate calculus, Frege [20, 21] used a "trick" to formalize quantification on classes. For instance: Aristotle's A and I sentences $\forall x : S. \ P(x)$ or $\exists x : S. \ P(x)$ [12] can be respectively written as $\forall x. \ \widehat{S}(x) \Rightarrow P(x)$ and $\exists x. \ \widehat{S}(x) \& P(x)$. Given the classical symmetries $\&/\vee, \forall/\exists$, because of the classical definition of $A \Rightarrow B$ as $(\neg A) \vee B$, one obtains that $\neg(\forall x : S. \ P(x))$ is $\exists x : S. \ \neg P(x)$ and $\neg(\exists x : S. \ P(x))$ is $\forall x : S. \ \neg P(x)$

However it is clear that there is no similar trick for generalised quantifiers:

(26)   a.   for 1/3 of the $x : S \ P(x) \not\equiv$ for 1/3 of the $x \ (S(x) \Rightarrow P(x))$

         b.   for few $x \in S \ P(x) \not\equiv$ for few $x \ (S(x) \& P(x))$

In these cases, specific generics could be used also introduced by subnector. This issue has recently started to get explored. [42]

7.2   Quantifiying over a sort and typed epsilon and tau

Sorts and classes with specific quantifiers may be a good direction. It should observed that one can defined a many sorted variant of first order logic, where predicates and functions use sorts fo objects that possibly follow a hierarchy (the powerset of a sort can be a sort, etc.). Such a many sorted logic is precisely defined and studied in, for instance, chapter 5 of [30].

---

[11]This avoids problems like my daughter being both tall (as a girl of her age) and not tall (as a member of the family for taking a group picture). [10]

[12]$x : A$ means that x is in the class A or x is of type A. We do not want to be very precise on this, since we a priori have no type theory, no set theory.

Within the usual Montagovian semantic analysis, the epsilon and tau operators for first order individuals should be of type $\epsilon : (\mathbf{e} \to \mathbf{t}) \to \mathbf{e}$. If we have many sorts or types, there are two possibilities: $\tau^*, \epsilon^* : \Pi\alpha.\ \alpha$ and $\tau, \epsilon : \Pi\alpha.\ (\alpha \to \mathbf{t}) \to \alpha$, where Π is basically quantification over types. [13] The first one maps any type to an element in the type (to be understood as the existential or universal generic element in the type). The second one maps a predicate on α-object o and α-object.

The types for ϵ and τ yield more general subnectors than the ones of ϵ^* and τ^*. This is because types can be mirrored as predicates, but not the converse. It might be useful to apply ϵ to a complicated predicate like *scooter with two front wheels* for interpreting a sentence like *a scooter with two front wheels just passed in the street* while it would be counterintuitive to consider *scooter with two front wheels* as a type.[14] When there is a type T one can introduce a constant $\widehat{T}$ of type $\mathbf{e} \to \mathbf{t}$ where $\mathbf{e}$ is the largest type which includes any other type.

If the predicate S, to whom the subnector is applied, is of type $U \to \mathbf{t}$, then the subnector yields an individual term of type U and nothing asserts that this term enjoys the property P. As natural language quantifiers only appear in a proposition which holds (at least locally), one has to add the presupposition $S(\epsilon\widehat{S})$, and in case S is a type, this can be written as $\epsilon\widehat{S} : S$.

These questions, in particular w.r.t the semantics of determiners, are further discussed in [44]

7.3   Categorical models of typed epsilon calculi

A mentioned above, there is no proper model, in the sense of model theory of epsilon calculus. However, with typed epsilon operators there are categorial models, based on toposes, categories that that generalise the category of sets. In this setting, one can even define categorical models of intuitionistic epsilon calculus as [40] does.

The fact that the entire apparatus of logic can be recast in purely category-theoretic terms is well known and goes back to the pioneering work of F. W. Lawvere. A description of the history of category theory and of its branch known as categorical logic goes well beyond the purposes of this brief section and we refer curious readers to [35] and references therein.

One of the most relevant concepts in category theory is that of elementary topos. The definition of elementary topos as a category with finite limits and power objects was introduced by Lawvere and Tierney as a generalization of the previous notion of

---

[13]No problem of consistency arises with such constants whose type in unprovable (like fix point operator Y). Any of those type entails the other: $\epsilon^* = \epsilon\{\Lambda\alpha.\alpha\}(\lambda x^{\Pi\alpha.\alpha}.\ x\{\mathbf{t}\}) : \Lambda\alpha.\ \alpha$ and $\epsilon = \epsilon^*\{\Pi\alpha.\ (\alpha \to \mathbf{t}) \to \alpha\}$.

[14]However, take a look at [32, 8] for an alternative view.

Grothendieck topos. It was immediately clear that the concept was of fundamental importance and, since its introduction, applications of elementary toposes have been proposed for an extremely large variety of fields [27]. Focusing on logic, we can associate a calculus, often called *internal language of a category*, to every category with finite limits, and therefore to every elementary topos.

The internal language of a category with finite limits is a multi-typed language. By multi-typed language we mean a 5-tuple $(\mathcal{T}, \mathcal{O}, \mathcal{R}, \omega, \rho)$ where $\mathcal{T}, \mathcal{O}$ and $\mathcal{R}$ are collections whose elements are called types, operation symbols and relation symbols respectively, while ω and ρ are arity functions. More specifically ρ sends each element of $\mathcal{R}$ to a finite, possibly empty, list of types $A_1, A_2, ..., A_n$ and ω sends each element of $\mathcal{O}$ to a non-empty finite list of types $A_1, A_2, ..., A_n, A_{n+1}$. A context Γ is a finite list of typed variables, i.e. Γ is of the form $x_1 : A_1, x_2 : A_2, ..., x_n : A_n$. Terms and formulas are written in context and the expressions

$$\Gamma \mid t : B \qquad \Gamma \mid \phi$$

are used to denote that t is a well formed term of type B in the context Γ and ϕ is a well formed formula in the context Γ.

The definition of Hilbert's epsilon calculus in the framework of multi-typed languages does not present many difficulties. What one wants is to require that for every well formed formula $\Gamma, x : A \mid \phi$ there exists a term $\Gamma \mid \epsilon_\phi^A : A$ such that

$$\frac{\Gamma \mid H_1, \ldots, H_n \vdash P(t)}{\Gamma \mid H_1, \ldots, H_n \vdash P(\epsilon_P^A)} \epsilon_i$$

where t is a term of type A in the context Γ, i.e. $\Gamma \mid t : A$, and

$$\frac{\Gamma \mid K_1, \ldots, K_p \vdash P(\epsilon_P^A) \qquad \Gamma, x : A \mid H_1, \ldots, H_n, P(x) \vdash C}{\Gamma, x : A \mid K_1, \ldots, K_p, H_1, \ldots, H_n \vdash C} \epsilon_e$$

If the underling collection of types $\mathcal{T}$ has some special types constructor, it might be convenient, or even necessary, to restrict the existence of terms of the form $\Gamma \mid \epsilon_\phi^A : A$ to those A's that belong to a specific class of types. This caution becomes necessary whenever $\mathcal{T}$ contains the zero type $\mathbf{0}$, which can be thought of as an abstract notion of the empty set. In this case the existence of a term $\Gamma \mid \epsilon_\phi^{\mathbf{0}} : \mathbf{0}$ makes the calculus collapse to an inconsistent calculus. Hence, the typed version of Hilbert's epsilon calculus is a multy-typed calculus where for every well formed formula $\Gamma, x : A \mid \phi$, where A is not the zero type, there exists a term $\Gamma \mid \epsilon_\phi^A$ satisfying the rules ϵ_i and ϵ_e defined above.

The typed Hilbert's tau calculus is defined analogously. In particular, the formation rules of tau terms must obey the constraint that tau terms should not to be of type $\mathbf{0}$. Of course, and within the typed epsilon calculus, the law of excluded middle implies the existence of tau terms and vice versa.

The existence of a categorical model for the classical typed epsilon calculus is not an issue. A standard set-based model, where a non-zero type A is interpreted as a non-empty set $[\![A]\!]$, works. In this model a context Γ is interpreted as the cartesian product of the interpretations of the types in it, a term $\Gamma \mid t : A$ as a function $[\![t]\!] : [\![\Gamma]\!] \to [\![A]\!]$ and a well formed formula in the context Γ as a subset of $[\![\Gamma]\!]$. Entailment is inclusion of subsets. Now suppose that ϕ is a well formed formula in the context $\Gamma, x : A$ and suppose that $[\![\phi]\!]$ is its interpretation. For every g in $[\![\Gamma]\!]$ let $E(g)$ be the set $\{a \in [\![A]\!] \mid (g, a) \in [\![\phi]\!]\}$ and define a set $G = \{g \in [\![\Gamma]\!] \mid E(g) \text{ is not empty}\}$. By the axiom of choice there is a function which sends each g in G to an element $s(g) \in E(g)$. The interpretation of the function $[\![\epsilon_\phi^A :]\!] : [\![\Gamma]\!] \to [\![A]\!]$ is given by the following assignment

$$[\![\epsilon_\phi^A :]\!](g) = \begin{cases} s(g) & \text{if } g \in G \\ x & \text{if } g \notin G \end{cases}$$

where x is any element of $[\![A]\!]$, which certainly exists as $[\![A]\!]$ is not empty.

In the previous interpretation we made use of the axiom of choice as well as the law of excluded middle and since toposes can be seen as universes of sets, it is not surprising that, under some mild hypotheses, the interpretation above can be rephrased in any elementary topos validating the axiom of choice and the law of excluded middle. Note that in toposes, by Diaconescu's argument [14], the axiom of choice implies the excluded middle, so the the second requirement is redundant.

Thus, a more interesting problem is to find an interpretation of the intuitionist typed epsilon calculus. The problem is even more interesting if one considers that in the intuitionist epsilon calculus the axiom of choice is derivable. Therefore, any category in which one can carry out the argument of Diaconescu does not provide the desired interpretation.

To the best of our knowledge, the first study of the typed intuitionist epsilon calculus in categories is due to J. L. Bell [6]. In his work Bell used a special class of elementary toposes as a model of the calculus and he called such toposes Hilbertian. To avoid the excluded middle from holding, Bell considers a calculus in which epsilon terms exists only for those well formed formulas with at most one variable. This is a fragment of the full intuitionist epsilon calculus in which the axiom of choice is no longer derivable, therefore the argument of Diaconescu is no longer valid, and Hilbertian toposes need not collapse to classical ones.

An approach that is in some respects complementary to that of Bell's, is to consider categories which are weaker than toposes. In fact the full power of an elementary topos is required to carry out Diaconescu's argument. Therefore, if one considers other weaker categories, such as Heyting categories, the validity of the axiom of choice does not imply the validity of excluded middle. A study of those categories that provides an interpretation of the intuitionistic epsilon calculus with all epsilon terms can be found in [40].

8 Conclusion

This partial introduction describes quantification with epsilon and tau, in particular existentials that are so commonly found in natural language. This treatment avoids the drawbacks of the standard interpretation: epsilon terms follow the syntactic structure, they refer to individuals and they further avoid the unpleasant symmetry between common noun and verb phrase in existential statements like Aristotle's I sentences.

Epsilon and tau can be typed, thus offering a more refined system for interpreting quantified statements, that can be integrated within syntactic and semantic parsers. Typed epsilon calculi are quite interesting because they admit categorical semantics, while untyped ones do not. As untyped epsilon calculi have no usual models, this is an important property.

Epsilon calculi shed new light to generalised quantifiers: these could be defined syntactically, using a generic as a Hilbert subnector tau and epsilon. Thus, there already exist some first insights for "most" as a typed subnector.

Epsilon calculi suggest intriguing connections between types and properties, type theory and first or higher order logic. Epsilon and tau, which lead to formulas without any equivalent in usual logics, offer a natural treatment of underspecification which should be further explored: which quantifier, existential or universal, comes first in $P(\tau_x.\ A(x), \epsilon x.\ B(x))$?

This issue addresses many facets of the epsilon calculi: history, mathematics (syntax, proof theory, model theory), linguistics and philosophy. We hope the reader will enjoy these articles as much as we enjoyed the meeting.

References

[1] Vito Michele Abrusci, Fabio Pasquali, and Christian Retoré. Quantification in ordinary language and proof theory. *Philosophia Scientae*, 20(1):185–205, 2016.

[2] Wilhelm Ackermann. Begründung des "tertium non datur" mittels der Hilbertschen Theorie der Widerspruchsfreiheit. *Mathematische Annalen*, 93:1–36, 1924.

[3] David Malet Armstrong. *Universals: An Opinionated Introduction.* Westview Press, 1989.

[4] Gunter Asser. Theorie der logischen Auswahlfunktionen. *Zeitschrift für Mathematische Logik und Grundlagen der Mathematik*, 1957.

[5] Jeremy Avigad and Richard Zach. The epsilon calculus. *The Stanford Encyclopedia of Philosophy*, 2008.

[6] John L. Bell. Hilbert's epsilon operator in intuitionistic type theories. *Mathematical Logic Quaterly*, 39:323–337, 1993.

[7] John Thomas Canty. Zbl0327.02013 : review of "on an extension of Hilbert's second ε-theorem" by T. B. Flanagan (journal of symbolic logic, 1975). *Zentralblatt Math*, 1975.

[8] Stergios Chatzikyriakidis and Zhaohui Luo. On the interpretation of common nouns: Types Versus predicates. In Stergios Chatzikyriakidis and Zhaohui Luo, editors, *Modern Perspectives in Type Theoretical Semantics*. Springer, 2017.

[9] Alonzo Church. A formulation of the simple theory of types. *Journal of Symbolic Logic*, 5:56–68, 1940.

[10] Sarah-Jane Conrad. Linguistic meaning and the minimalism-contextualism-debate, 2011. (talk at Cerisy Context Conference May 2011).

[11] Haskell Curry. *Foundations of Mathematical Logic.* McGraw–Hill, 1963. Republished by Dover, 1977.

[12] Alain de Libera. *La philosophie médiévale.* Presses universitaires de France, 1993.

[13] Alain de Libera. *La querelle des universaux de Platon à la fin du Moyen Âge.* Des travaux. Seuil, 1996.

[14] Radu Diaconescu. Axiom of choice and complementation. *Proceedings of the American Mathematical Society*, 51(1):176 – 178, 1975.

[15] Burton Dreben, Peter Andrews, and Stål O. Aanderaa. False lemmas in Herbrand. *Bulletin of the American Mathematical Society*, 69:699–706, 1963.

[16] Burton Dreben and John Denton. A supplement to Herbrand. *Journal of Symbolic Logic*, 31(3):393–398, 1966.

[17] Urs Egli and Klaus von Heusinger. Definite Kennzeichnungen als Epsilon-Ausdrücke. In G. Lüdi and C.-A. Zuber, editors, *Akten des 4. regionalen Linguistentreffens. Romanisches Seminar*, pages 105–115. Universität Basel, 1993.

[18] Urs Egli and Klaus von Heusinger. The epsilon operator and E-type pronouns. In Urs Egli, Peter E. Pause, Christoph Schwarze, Arnim von Stechow, and Götz Wienold, editors, *Lexical Knowledge in the Organization of Language*, pages 121–141. Benjamins, 1995.

[19] Gareth Evans. Pronouns, quantifiers, and relative clauses (i). *Canadian Journal of Philosophy*, 7(3):467–536, 1977.

[20] Gottlob Frege. *Begriffsschrift, eine der arithmetischen nachgebildete Formelsprache des reinen Denkens.* Verlag von Louis Nebert, Halle, 1879.

[21] Gottlob Frege. *Grundgesetze der Arithmetik, Begriffsschriftlich abgeleitet (Band I und*

Band II). Verlag Hermann Pohle, Jena, 1903.

[22] Kurt Gödel. Die Vollständigkeit der Axiome des logischen Funktionenkalküls. *Monatshefte für Mathematik und Physik*, 37:349–360, 1930.

[23] Kurt Gödel. Über formal unentscheidbare Sätze der Principia Mathematicaund verwandter Systeme i. *Monatshefte für Mathematik und Physik*, 38:173–198, 1931.

[24] Jacques Herbrand. *Recherches sur la théorie de la démonstration*. PhD thesis, Faculté des Sciences, Université de Paris, 1930.

[25] David Hilbert. Die logischen Grundlagen der Mathematik. *Mathematische Annalen*, 88:151–165, 1922.

[26] David Hilbert and Paul Bernays. *Grundlagen der Mathematik. Bd. 2*. Springer, 1939. Traduction française de F. Gaillard, E. Guillaume et M. Guillaume, L'Harmattan, 2001.

[27] Peter T. Johnstone. *Sketches of an elephant. A topos theory compendium. Vols. I & II*. Oxford Logic Guides 44; Oxford: Clarendon Press, 2002.

[28] Gyula Klima. The medieval problem of universals. In Edward N. Zalta, editor, *The Stanford Encyclopedia of Philosophy*. Stanford University, fall 2013 edition, 2013.

[29] William Kneale and Martha Kneale. *The development of logic*. Oxford University Press, 3rd edition, 1986.

[30] Georg Kreisel and Jean-Louis Krivine. *Elements of mathematical logic: model theory*. North-Holland, 1967.

[31] Albert C. Leisenring. *Mathematical logic and Hilbert's ϵ symbol*. University Mathematical Series. Mac Donald & Co., 1967.

[32] Zhaohui Luo. Contextual analysis of word meanings in type-theoretical semantics. *Logical Aspects of Computational Linguistics (LACL'2011). LNAI 6736*, 2011.

[33] Mary C. MacLeod and Eric M. Rubenstein. Universals. In *The Internet Encyclopedia of Philosophy*. University of Tenessee at Martin, 2016.

[34] Alda Mari and Christian Retoré. Conditions d'assertion de *chaque* et de *tout* et règles de déduction du quantificateur universel. *Travaux de Linguistique*, 72:89–106, 2016.

[35] Jean-Pierre Marquis. Category theory. In Edward N. Zalta, editor, *The Stanford Encyclopedia of Philosophy*. Winter 2015 edition, 2015.

[36] G. Mints. Zbl0381.03042: review of "cut elimination in a Gentzen-style ϵ-calculus without identity" by Linda Wessels (Z. math Logik Grundl. Math., 1977). *Zentralblatt Math*, 1977.

[37] Grigori Mints and Darko Sarenac. Compelteness of indexed ϵ-calculus. *Archive for Mathematical Logic*, 2003.

[38] Richard Montague. The proper treatment of quantification in ordinary English. In *Formal Semantics: The Essential Readings*, chapter 1. Wiley, 2008.

[39] Richard Moot and Christian Retoré. *The logic of categorial grammars: a deductive account of natural language syntax and semantics*, volume 6850 of *LNCS*. Springer, 2012.

[40] Fabio Pasquali. A categorical interpretation of the intuitionistic, typed, first order logic

with Hilbert's ϵ-terms. *Logica Universalis*, 10(4):407–418, 2016.

[41] Fabio Pasquali and Christian Retoré. Aristotle's square of opposition in the light of Hilbert's epsilon and tau quantifiers. In *Aristotle 2400 world congress*, Thessaloniki, May 2016. https://arxiv.org/abs/1606.08326.

[42] Christian Retoré. Variable types for meaning assembly: a logical syntax for generic noun phrases introduced by "most". *Recherches Linguistiques de Vincennes*, 41:83–102, 2012.

[43] Christian Retoré. The Montagovian Generative Lexicon ΛTy_n: a Type Theoretical Framework for Natural Language Semantics. In Ralph Matthes and Aleksy Schubert, editors, *19th International Conference on Types for Proofs and Programs (TYPES 2013)*, volume 26 of *Leibniz International Proceedings in Informatics (LIPIcs)*, pages 202–229, Dagstuhl, Germany, 2014. Schloss Dagstuhl–Leibniz-Zentrum fuer Informatik.

[44] Christian Retoré. Typed Hilbert epsilon operators and the semantics of determiner phrases (invited lecture). In Glyn Morrill, Reinhard Muskens, Rainer Osswald, and Frank Richter, editors, *Proceedings of Formal Grammar 2014*, number 8612 in LNCS/-FoLLI, pages 15–33. Springer, 2014.

[45] Bertrand Russell. On denoting. *Mind*, 56(14):479–493, 1905.

[46] Barry Hartley Slater. Epsilon calculi. *The Internet Encyclopedia of Philosophy*, 2005.

[47] Barry Hartley Slater. Completing russell's logic. *russell: the Journal of Bertrand Russell Studies*, 27:144–158, 2007.

[48] Wilfried Meyer Viol, Rodger Kibble, Ruth Kempson, and Dov Gabbay. Indefinites as epsilon terms: A labelled deduction account. In Harry Bunt and Reinhard Muskens, editors, *Computing Meaning: Volume 1*, pages 203–218. Springer Netherlands, Dordrecht, 1999.

[49] Klaus von Heusinger. Definite descriptions and choice functions. In S. Akama, editor, *Logic, Language and Computation*, pages 61–91. Kluwer, 1997.

[50] Klaus von Heusinger. *Salienz und Referenz der Epsilonoperator in der Semanlik der Nominalphrase und anaphorischer Pronomen*. De Gruyter, 1997.

[51] Klaus von Heusinger. Choice functions and the anaphoric semantics of definite NPs. *Research on Language and Computation*, 2:309–329, 2004.

[52] Klaus von Heusinger. Alternative semantics for definite nps. In Kerstin Schwabe and Susanne Winkler, editors, *On Information Structure, Meaning and Form – Generalizations across languages*, volume 100 of *Linguistic Today*, pages 485–508. John Benjamins publishing company, 2007.

 Received 4 November 2016

HILBERT'S τ AND ϵ IN PROOF THEORETICAL FOUNDATIONS OF MATHEMATICS: AN INTRODUCTION

V. MICHELE ABRUSCI
Research Group on Logic and Geometry of Cognition,
Dipartimento di Matematica e Fisica, Università Roma Tre,
Roma, Italy

Abstract

In section 1, some basic proof-theoretic features of the operators τ and ϵ are briefly described. Section 2 deals with some additional axioms on these operators that are important for the formalization of arithmetics and Mathematical Analysis. Section 3 is a short introduction to Hilbert's strategy of eliminating ideal elements (quantifiers and the operators τ and ϵ) from the proof of finitary statements, and in particular from a supposed proof of some contradiction (in order to prove the consistency of Arithmetics and Mathematical Analysis). Section 4 is a short presentation of some important proof-theoretical results on the operator ϵ.

1 Basic proof-theoretical features of the operators τ and ϵ

The τ operator has been introduced by Hilbert before the ϵ operator: τ is used by Hilbert in [5] (a lecture given in 1922 and published in 1923), whereas ϵ is used by Hilbert in [6] (a lecture given in 1925 and published in 1926) and in following papers. The ϵ operator is also used in [7].

1.1 Syntactical features

Both operators, i.e. τ and ϵ, apply to a variable v of a given type (usually an *individual* variable x, i.e. a variable for the elements of a set, or a *function* variable f, i.e. a variable for the functions on a set) and a formula A, and produce a term of the same type of the variable v, where all the occurrences of the term's variable v inside the formula A are bound.

This is an invited paper.

The terms produced by τ operator are called τ-terms, whereas the terms produced by the ϵ operator are called ϵ-terms.

The τ-term produced by the τ operator applied to a variable v and a formula A is denoted by $\tau v A$, whereas the ϵ-term produced by the ϵ operator applied to a variable v and a formula A is denoted by $\epsilon v A$; so, in the terms $\tau v A$ and $\epsilon v A$ all the occurrences of the variable v are bound.

1.2   Axioms

The *implicit definition* of the τ and ϵ operators is given by their axioms, i.e. by the τ-axiom and ϵ-axiom.

The implicit definition of the τ operator is given by the τ-axiom, expressed by the following formulations which are equivalent under the substitution rule (the rule allows the replacement of free variables, and in particular the replacement of free predicate variables by formulas):

$$A[\tau v A/v] \rightarrow A[t/v]$$

(where A is a formula, and t is a term of the same type of the variable v) or

$$P(\tau v P) \rightarrow P(v)$$

(where P is a predicate variable).

Analogously, the implicit definition of the operator ϵ is given by the ϵ-axiom, expressed by the following formulations which are equivalent under the substitution rule:

$$A[t/v] \rightarrow A[\epsilon v A/v]$$

(where A is a formula, and t is a term of the same type of the variable v) or

$$P(v) \rightarrow P(\epsilon v P)$$

(where P is a predicate variable).

1.3   Meaning

The *meaning* (in Hilbert's words, the *content*) of the τ and ϵ operators is given by the natural reading of the τ-axiom and ϵ-axiom.

The natural reading of the τ-axiom is: when the object represented by the τ-term $\tau v A$ has the property expressed by the formula $A[v]$, then this property holds for each object of type v. So a proof of $A[\tau v A/v]$ is like a proof of $A[a/v]$ where a is considered a generic object of type v : the proof of $A[\tau v A/v]$ allows one to conclude that every object of type

v has the property A, and the same conclusion is obtained from the proof of $A[a/v]$ where a is a generic object of type v. Therefore, τ-terms are like *generic objects* in the proofs of *universal statements*.

The natural reading of the ϵ-axiom is the following: when an object of the type of the variable v has the property expressed by the formula $A[v]$, then this property holds for the object represented by the ϵ-term $\epsilon v A$. So, a proof from the hypothesis $A[\epsilon v A/v]$ is like as a proof from the hypothesis $A[a/v]$, where a is considered a generic object of type v: a proof from the hypothesis $A[\epsilon v A/v]$ gives a proof from the hypothesis that some object of type v has the property A, and this proof is given also from a proof of $A[a/v]$, where a is considered as a generic object of type v. Therefore ϵ-terms are like *generic objects* in the proofs from *existential statements*.

Moreover, when the property expressed by the formula $A[v]$ is not empty (i.e. when this property holds for some object of type v), the ϵ-term $\epsilon v A$ may be considered as an object selected from the class of all objects with the property expressed by $A[v]$, i.e. as a *representative* of this property. Therefore, the ϵ-axiom also includes a form of the *choice principle* (cf. [2]).

Hilbert considered τ-terms and ϵ-terms as *ideal objects*, i.e. objects that are *useful* in mathematics but need to receive a *justification* by means of finitary mathematics. Therefore, the τ-axioms ϵ-axioms belong to the ideal so to say part of mathematics. The justification of τ-axioms and ϵ-axioms by means of finitary mathematics, i.e. the finitary proof of the consistency of these axioms, is the justification of the mathematical existence of what is expressed by the natural meaning of τ-terms and ϵ-terms.

1.4 Mutual definability

Hilbert very soon noticed that the τ and ϵ operators are *mutually definable*, i.e. one may define the ϵ operator by means of the τ operator as follows:

$$\epsilon v A[v] = \tau v \neg A[v]$$

and one may define the τ operator by means of the ϵ operator as follows:

$$\tau v A[v] = \epsilon v \neg A[v]$$

The mutual definability of these operators is based on *logical duality*, and in particular on the the identities $C = \neg\neg C$ and $C \rightarrow D = \neg D \rightarrow \neg C$. Indeed, each τ- axiom $\neg A[\tau v \neg A[v]/v] \rightarrow \neg A[t/v]$ by duality is the same as $A[t/v] \rightarrow A[\tau v \neg A[v]/v]$ that is a ϵ- axiom by putting $\epsilon v A[v] = \tau v \neg A[v]$; each ϵ-axiom $\neg A[t/v] \rightarrow \neg A[\epsilon v \neg A[v]/v]$ by duality is the same as $A[\epsilon v \neg A[v]/v] \rightarrow \neg A[t/v]$ that is a τ-axiom by putting $\tau v A[v] = \epsilon v \neg A[v]$.

Therefore, we may restrict us to use the τ-operator only (and so we obtain ϵ-operator as a defined one) as in the first Hilbert's papers (e.g. [5]), or to use the ϵ-operator only (and so we obtain τ-operator as a defined one) as in the other Hilbert's papers (e.g. [6], [7]).

On the basis of these inter-definability, Hilbert decided to use only one of the two symbols (ϵ, τ), and he preferred to use ϵ for reasons linked to his method of elimination of ideal elements from a proof of a sentence without ideal elements.

Perhaps, Hilbert's use of only one of the two symbols is not the optimal choice. Indeed, the experience in proof-theory shows that - when we have two dual concepts - it is better to use both: e.g. conjunction and disjunction, universal and existential quantifier. So, a better formulation of the τ and ϵ operators would be in classical logic:

- simultaneously defining the class of formulas (with negation on atomic formulas, and connectives $\wedge$ and $\vee$) and the class of terms, by stating:

$$\text{if } A[v] \text{ is a formula, then } \epsilon v A[v] \text{ and } \tau v A[v] \text{ are terms of the type of } v$$

- defining negation $\neg A$ of each formula A in the metalanguage, and defining $\neg A[v]$ for each formula:

$$\epsilon v \neg A[v] = \tau v A[v], \tau v \neg A[v] = \epsilon v A[v]$$

- to use as logical axioms (each axion gives the other one):

$$\neg A[\tau v A/v] \vee A[t/v] \ (\tau\text{-axiom}) \qquad \neg A[t/v] \vee A[\epsilon v A/v] \ (\epsilon\text{-axiom}).$$

where A is a formula, and t is a term of the same type of v.

1.5 Definability of usual quantifiers

Hilbert (with the help of Paul Bernays) very soon noticed the possibility of defining the usual quantifiers - the universal quantifier $\forall$ and the existential quantifier $\exists$, on individual variables (i.e. first-order quantifiers) and on function variables (i.e. second order quantifiers) - by means of the τ and the ϵ operator:

- the universal quantifier $\forall$ may be defined by means of the τ-operator as follows

$$\forall v A = A[\tau v A/v]$$

- the existential quantifier $\exists$ may be defined by means of the ϵ-operator as follows

$$\exists v A = A[\epsilon v A/v]$$

260

The definability of the usual quantifiers by means of the operators τ and ϵ is based on the fact that:

- the usual axioms and rules for the universal quantifier $\forall$- when $\forall v A$ is translated as $A[\tau v A/v]$ - become derivable from the τ-axiom,

- the usual axioms and rules for the existential quantifier $\exists$ - when $\exists v A$ is translated by $A[\epsilon v A/v]$ - become derivable from the ϵ-axiom.

Since the τ and ϵ operators are mutually definable,

- in case the formulation with τ only is used, the usual quantifiers are defined as follows by means of τ:

$$\forall v A = A[\tau v A/v] \qquad \exists v A = A[\tau v \neg A/v]$$

- in case the formulation with the ϵ operator only is used, the usual quantifiers are defined as follows by means of ϵ:

$$\forall v A = A[\epsilon v \neg A/v] \qquad \exists v A = A[\epsilon v A/v]$$

Therefore, first-order as well as second-order predicate logic may be formulated standardly by means of the usual quantifiers, but also by means of the either the τ operator or the ϵ operator: indeed, every formula of predicate logic based on usual quantifiers may be translated into a formula of predicate logic based on the τ operator on the ϵ operator. Furthermore, the translation of every theorem of predicate logic based on usual quantifiers is a theorem of predicate logic based on the τ operator or the ϵ operator.

Note that these results - obtained at the first steps of proof-theory within the Hilbertian tradition - have been *the first results concerning the translations between formal languages and formal theories*: this kind of results is now very common in proof-theory.

Let us consider some examples of translation of formulas of usual predicate logic, as well as the translation of Aristotle's categorical propositions, into a formalism based on the τ or ϵ operators (cf. [10]):

- the predicate logic representation of the categorical proposition "some Q is P"

$$\exists x (Q(x) \wedge P(x))$$

is translated (by using the ϵ operator) into the formula:

$$Q(s) \wedge P(s) \text{ with } s = \epsilon x (Q(x) \wedge P(x))$$

- the predicate logic representation of the categorical proposition "every Q is P"

$$\forall x(Q(x) \rightarrow P(x))$$

is translated (by using the τ operator) into the formula

$$Q(t) \rightarrow Q(t) \text{ with } t = \tau x(Q(x) \rightarrow P(x))$$

Clearly, the τ and ϵ operators give a new representation of universal statements and existential statements, different from the old representation given in the ancient logic and from the more usual one based on usual quantifiers introduced by Frege. The new representation of universal statements and existential statements is based on the notion of *generic object* and on the notion of *representative* of each property (even for empty properties) expressed in the language.

But in predicate logic based on the operators τ or ϵ there are formulas that are not translations of formulas of predicate logic with usual quantifiers: e.g., the following formulas where P and Q are unary predicate variables:

$$P(\epsilon x Q(x)) \qquad P(\tau x Q(x))$$

The natural reading of the first formula is: the predicate P holds of the representative of the predicate Q i.e. another possible way to express the categorical proposition "some Q is P". The natural reading of the second formula is: the predicate P holds of the generic object for the universal validity of the predicate Q i.e. another way to express the categorical proposition "every Q is P".

Remark that the ϵ-axiom and the τ-axiom may be applied also to function variables or predicate variables, in a way analogous to a second-order quantification. An example of τ-axiom applied to a function variable will be considered in 2.3.

2   Other axioms on the operators τ and ϵ

When other axioms are added on the τ and ϵ operators, the implicit definition of these operators changes. We shall list and comment some additional axioms on the τ and ϵ operators.

2.1   Extensionality axioms

The τ and ϵ operators produce a term for each formula A and each variable v. So, if A and B are different formulas, the τ-term $\tau v A$ is different from the τ-term $\tau v B$ and the ϵ-term $\epsilon v A$ is different from the ϵ-term $\epsilon v B$; when A and B are different but equivalent formulas, the proof of the equivalence between A and B does not give a proof of $\tau v A = \tau v B$ and $\epsilon v A = \epsilon v B$.

In some cases, it is really useful to require that the equivalence between two formulas implies the identity between τ-terms and ϵ-terms associated to these formulas: when two formulas are syntactically different but one has strong reasons to see that these formulas must have the same content, i.e. when these formulas correspond to two different ways to express the same content E.g., the formulas $A \wedge (B \wedge C)$ and $(A \wedge B) \wedge C$ (associative law), the formulas $A \wedge (B \vee C)$ and $(A \wedge B) \vee (A \wedge C)$ (distributivity law), the formulas $\exists x \exists y A[x, y]$ and $\exists y \exists x A[x, y]$ (permutation of existential quantifiers), or $\forall x \forall y A[x, y]$ and $\forall y \forall x A[x, y]$ (permutation of universal quantifiers). In these cases, when A and B, it is natural to think that the generic object $\tau v A$ is equal to the generic object $\tau v B$ and the representative of the property expressed by $A[v]$ i.e. $\epsilon v A$ is equal to the representative of the property expressed by $B[v]$ i.e. $\epsilon v B$.

In other cases, when the equivalence between two formulas depends on specific axioms of the theory (e.g. Peano's Axioms), it is more delicate to require that this equivalence implies the identity between τ-terms and ϵ-terms associated to these formulas.

The general *extensionality axiom* - which was never used by Hilbert and his school - states that always, when the equivalence between these formula A and B holds, $\tau v A$ must be equal to $\tau v B$ and $\epsilon v A$ must be equal to $\epsilon v B$.

$$\forall v(A[v] \leftrightarrow B[v]) \to \tau v A[v] = \tau v B[v] \qquad \forall v(A[v] \leftrightarrow B[v]) \to \epsilon v A[v] = \epsilon v B[v]$$

As you see, the formulation of the general extensionality axiom requires the use of usual quantifiers, in order to express the equivalence between two formula. Of course, a general extensionality axiom for ϵ implies general extensionality axiom for τ, and vice versa.

A more limited form of extensionality axiom is used by Hilbert and his school: the limitation to the case when two formulas $A[v, a]$ and $A[v, b]$ differ for two terms a and b which are syntactical different and the equivalence between these formulas is given by the equality between the terms a and b:

$$a = b \to \tau v A[v, a] = \tau v A[v, b] \qquad a = b \to \epsilon v A[v, a] = \epsilon v A[v, b]$$

These axioms are called τ-identity or ϵ-identity (in German, ϵ-*Gleicheit*) axioms. We can of course prove τ-identity from ϵ-identity and viceversa.

2.2 Arithmetical axioms

Hilbert - and its school - investigated this new way to axiomatize and formalize Number Theory inside a logical formalism containing the operators τ or ϵ, and in particular inside a logical formalism containing the ϵ operator only (without the usual quantifiers since these are definibale by means of the ϵ operator).

The methodological approach is to avoid the *formal* introduction of a *specific* way to prove and use universal and existential statements on natural numbers, i.e. to avoid the *formal* introduction of the *induction* rule or axiom, and to express the induction rule as well as the induction axiom as a consequence of other arithmetical axioms together with some specific axioms for ϵ-terms.

The result has been the discovery that - inside a formalism using ϵ and the ϵ-axiom - the induction axiom and the induction rule may be replaced by very simple arithmetical axioms (on the predecessor function and successor function) together with a new specific axiom on ϵ-terms where the bound variable is an individual variable, i.e. by the following axiom:

$$A[a/x] \to \epsilon x A[x] \neq (a)'$$

where $()'$ is the symbol fro the successor function.

This axiom is called the *second ϵ-axiom*, in German *zweite ϵ-Formel*; whereas the name *first ϵ-axiom* is given to the ϵ-axiom

$$A[a/x] \to A[\epsilon x A[x]/x]$$

where x is an individual variable - i.e. a variable for natural numbers, and a is a term of the same type of the variable x i.e. an arithmetical term.

Remark that:

- the second ϵ-axiom is not the translation of a formula belonging to the formalisms with usual quantifiers,

- the second ϵ-axiom says that $\epsilon x A[x]$, if there is a natural number with the property expressed by A, this is the first natural number enjoying the property expressed by $A[x]$,

- it is possible to formulate a corresponding second τ-axiom simply by translating ϵ-term $\epsilon x A[x]$ by the corresponding τ-term $\tau x \neg A[x]$.

2.3   Axioms for Mathematical Analysis

It is well-known that Hilbert's aim has been to give the foundations of Mathematical Analysis (and as a consequence, or a preliminary step, the foundations of the arithmetics of natural numbers), by stating the consistency of the axioms for Mathematical Analysis.

In supplement IV of [7], one finds the most precise formalization of Mathematical Analysis by means of the ϵ operator. In this formalization, the axioms are: the tautologies of propositional logic, the axiom of equality, the following arithmetical axioms (where $()'$ is the symbol of successor function and $\delta()$ is the symbol for the predecessor function)

$$a' = b' \rightarrow a = b$$

$$a' \neq 0$$

$$a \neq 0 \rightarrow (\delta(a))' = a$$

and the following axioms on the ϵ operator where x is an individual variable (i.e. a variable for natural numbers), f is a function variable (i.e. a variable for functions on natural numbers), a is term of the type of x and ϕ is a term of the type of f:

$$A[a/x] \rightarrow A[\epsilon x A/x] \text{ (first ϵ-axiom)}$$

$$A[a/x] \rightarrow \epsilon x A \neq (a)' \text{ (second ϵ-axiom)}$$

$$A[\phi/f] \rightarrow A[\epsilon f A/f] \text{ (third ϵ-axiom)}$$

So, the first ϵ-axiom is the ϵ-axiom for the individual variables, the third ϵ-axiom is the ϵ-axiom for function variables, and the second ϵ-axiom is a limited form of an extensionality axiom for the operator ϵ (cf 2.1.).

A formalization of Mathematical Analysis by means of the τ operator may be obtained by replacing the ϵ-terms by the corresponding τ-terms and the ϵ-axioms by the corresponding τ-axioms.

In [5] Hilbert gives a very nice example of a definition - by means of the τ-axiom - of a very strong and useful *function of functions*, i.e. a kind of functions whose existence has been criticized by Brouwer and Weyl.

Let us consider this very simple formula

$$f(x) = 0$$

were f is a function variable (a variable for functions on natural numbers) and x is an individual variable (a variable for natural numbers). The τ-axiom for this formula is:

$$f(\tau x(f(x) = 0)) = 0 \rightarrow f(x) = 0$$

Now, $\tau x(f(x) = 0)$ denotes a functional Φ such that for every function ϕ from N to N gives a result $\Phi(\phi) = \tau x(\phi(x) = 0)$, and

- $\Phi(\phi) = 0$, if for every n $\phi(n) = 0$

- $\Phi(\phi) \neq 0$, if for some n $\phi(n) \neq 0$.

3 The use of the operator ϵ: Hilbert's strategy for the elimination of *ideal elements* inside the proofs of finitary statements

Inside a proof, *ideal elements* - in Hilbert's philosophical approach - are closed formulas containing quantifiers or the τ and ϵ operators. Closed formulas not containing ideal elements are finitary statements. Among false finitary statements we find the closed formula $0 \neq 0$: a proof ending with $0 \neq 0$ is a proof of a contradiction.

Hilbert, in *Die Grundlagen der Mathematik* (1927, published in 1928), explained his strategy of eliminating all *ideal elements* inside a proof of a finitary statement, (e.g. from a "proof" of a contradiction). This strategy is at the core of research in Hilbert's school. This is well described in [7] (pp. 92-129) as well.

A more recent study of this strategy is contained in [8].

3.1 Preliminary steps

Preliminary steps of this strategy comprise very interesting procedures - at least historically these are the first proof-theoretical procedures - that are now standard in proof-theory.

Some of these procedures are fully syntactical, whereas some others also refer to the content of the formalism on the basis of the Hilbert philosophical approach: to use only finitary contents, and in particular finitary arithmetics, as the framework where proof-theory has to be developed and where one has to build the common foundations of logic and mathematics.

The preliminary steps are the following procedures designed to be applied to any given proof of a finitary statement, in a suitable formal system:

1. represent *the proof in a tree form*;

2. *remove quantifiers* by using by using ϵ-terms and τ-terms,

3. *remove free variables* (this is possible since the conclusion of the proof is a closed formula, and so does not contain free variables), by replacing each free variable by a term belonging to finitary arithmetics;

4. *interpret* each formula not containing ϵ-terms as a finitary arithmetical proposition.

The first procedure is a very important one, since it introduces the representation of proofs in a tree form in proof-theory, a representation that is now very familiar and has been very useful for the development of proof-theory.

The second procedure is the replacement - proposed by Hilbert - of usual quantifiers $\forall$ and $\exists$ by means of the ϵ operator.

The third step is another important proof-theoretical tool: in the formalisms containing not the usual quantifiers but the τ or the ϵ operator, a proof of a closed formula (i.e. a formula without free variables) may be transformed into a proof without free variables.

The last step is the finitary interpretation - as "true" or "false" - of each formula without free variables and without ϵ-terms (or τ-terms).

3.2 Replacement of ϵ-terms

3.2.1 The general aim

After the preliminary steps, each proof of a finitary statement (and in particular, every proof of a contradiction) is transformed into a proof that contains no free variable and may contain two kinds of terms:

- finitary or real terms, i.e. closed terms not containing the operator ϵ (and not containing the τ operator);

- ideal terms, i.e. closed terms containing the ϵ operator.

The aim of the strategy proposed by Hilbert is to replace, in such a proof, each ϵ-term by a finitary or real term (ϵ-terms of the form $\epsilon x A$ are replaced by specific natural numbers, whereas ϵ-terms of the form $\epsilon f A$ by specific finitary functions on natural numbers), in such a way that all formulas of the given proof become true statements of finitary arithmetics.

3.2.2 The first case

The first case considered by Hilbert in this strategy is the simplest one: in the given proof there is *only one ϵ-term* (with one or more occurrences), this ϵ-term has the form $\epsilon x A$ where x is *an individual variable*, and there is *only one occurrence of the first ϵ-axiom*.

In this first case, let us suppose that the unique first ϵ-axiom occurring in the proof is

$$A[t/x] \to A[\epsilon x A/x]$$

The strategy for replacement of such a ϵ-term - in the first case - is the following one (and perhaps this procedure is one of the motivations of the Hilbert's preference for the ϵ operator) :

- one begins by replacing all the occurrences of $\epsilon x A$ inside the given proof by the term 0, and so the unique occurrence of the ϵ-axiom

$$A[t/x] \to A[\epsilon x A/x]$$

becomes the finitary closed formula

$$A[t/x] \rightarrow A[0/x]$$

- if this formula is true, then we stop: every formula in the proof becomes a true finitary formula;

- otherwise, $A[t/x]$ is true and $A[0/x]$ is false, so that we change the replacement as follows: each occurrence of $\epsilon x A$ is replaced by t, and so

$$A[t/x] \rightarrow A[\epsilon x A/x]$$

becomes

$$A[t/x] \rightarrow A[t/x]$$

that is a true finitary formula.

3.2.3 Second case

The second case, in the strategy for the elimination of ideal elements inside a proof of a finitary statement, is a little more complicated in comparison with the first one: in the given proof, there is *only one ϵ-term* (in a finite number of occurrences), this ϵ-term has the form $\epsilon x A$ where x is *an individual variable*, and there are in the proof the following occurrences of ϵ-axioms

$$A[t_1/x] \rightarrow A[\epsilon x A/x] \, , \, ..., \, A[t_n/x] \rightarrow A[\epsilon x A/x]$$

In this case:

- one begins by replacing $\epsilon x A$ by 0, and so each $A[t_i/x] \rightarrow A[\epsilon x A/x]$ becomes $A[t_i/x] \rightarrow A[0/x]$ which is a finitary formula;

- if all these formulas are true, then we stop: every formula in the given proof becomes a true finitary formula;

- otherwise, there is j such that $A[t_j/x]$ is true and $A[0/x]$ is false, so that we change the replacement: $\epsilon x A$ is replaced by the term t_j, and so each $A[t_i/x] \rightarrow A[\epsilon x A/x]$ becomes $A[t_i/x] \rightarrow A[t_j/x]$ i.e. a true finitary formula (since the consequent is true).

3.2.4   A more general case

A more general case considered in the strategy for the elimination of the ideal elements inside a proof of a finitary statement, is the following one: in the given proof, there are more ϵ-terms and more ϵ-axioms, but all the ϵ-terms have the form $\epsilon x A$ where x is a individual variable and there is no third ϵ-axiom.

In this case, the idea is to start by replacing the *simplest ϵ-terms* as done in the previous case, i.e. to define a strategy of elimination *by induction on the complexity of ϵ-terms*. Remark that the strategy designed by Hilbert for ϵ elimination is the first case of a method ("by induction on the complexity of ...") that is now very usual in proof-theory.

Of course, the first task is to define the complexity of ϵ-terms.

It is evident that the complexity of ϵ-terms t is given by the number of the other ϵ-terms occurring inside t, and that the simplest ϵ-terms are ϵ-terms where no other ϵ-term occurs.

But a ϵ-term may occur inside a ϵ-term in two very different ways:

1. as a term *inserted* (in German, *eingelagert*): the ϵ-term $\epsilon y B$ is *inserted* inside the ϵ-term

$$\epsilon x A[x, \epsilon y B]$$

   when the variable x does not occur in B;

2. as a term *subordinated* (in German, *untergeordnet*): the ϵ-term $\epsilon y B[x, y]$ containing the variable x is *subordinated* inside the term

$$\epsilon x A[x, \epsilon y B[x, y]]$$

The number of the ϵ-terms inserted inside an ϵ-term t is the *degree* of t, and the number of ϵ-terms subordinated inside a *epsilon*-term t is the *rang* of t. The procedure of the elimination of ϵ-terms is given by induction on the complexity of the ϵ-terms, defined by taking into account both the degree and the rank of the ϵ-terms.

Remark that the framework becomes extremely complicated, since the way ϵ-terms are constructed, leads one to consider *improper* subordinated terms. For example:

- there is no valid ground to consider as a the order of the existential quantifiers in the formula $\exists x \exists y P(x, y)$ as a relevant one. Thus, we must consider this formula as *equal* to the formula $\exists y \exists x P(x, y)$

- $\exists x \exists y P(x, y)$ becomes the formula with subordinated ϵ-terms where the dependency does not mean anything:

$$P(\epsilon x P(x, \epsilon z P(x, z)), \quad \epsilon y P(\epsilon x P(x, \epsilon z P(x, z)), y))$$

- $\exists y \exists x P(x, y)$ becomes the formula with other subordinated ϵ-terms where the dependency does not mean anything:

$$P(\epsilon x P(x, \epsilon y P(\epsilon z P(z, y), y)), \quad \epsilon y P(\epsilon z P(z, y), y))$$

3.3   Replacement of ϵ-terms: concluding remarks

The development of Hilbert's strategy for the elimination of ϵ-terms inside the proofs of finitary statements has been the first topic of proof-theory and has been investigated by excellent logicians (in particular by Ackermann (1924) and also by Bernays).

The aim of these researchers has been to get a finitary proof that, in the case of the formal systems of Mathematical Analysis or in the case of the formal systems for Arithmetics, the procedure of the elimination of ϵ-terms inside a proof of a finitary statement ends in a finite number of steps. But, we know that this aim cannot be reached by second Gödel's Incompleteness Theorem, if finitary arithmetics is contained inside the formal system for Arithmetics.

Some partial results obtained in a finitary way by using Hilbert's strategy are considered and discussed in the book *Grundlagen der Mathematik, II.*

Here I wish to point out the importance of Hilbert's strategy for the development of proof-theory.

Firstly, Hilbert's strategy aims to find - inside a given proof - for each ϵ-term $\epsilon v A[v]$ a term ϵ-free t in order to replace inside the proof the occurrences of $\epsilon v A[v]$ by occurrences of the term t. We may consider this strategy as a strategy for the substitution of terms.

The *Substitution method*, introduced by Herbrand in his very important theorem, may be considered as something linked to Hilbert's strategy: indeed, Herbrand's theorem states that - given a proof of an existential theorem $\exists x A$ - it is possible to find a finite number of terms $t_1, \cdots, t_n$ and to transform the proof of $\exists x A$ into a proof of the disjunction $A[t_1/x], \cdots, A[t_n/x]$. On the relationships between Herbrand's substitution method and ϵ calculus, cf [9].

Secondly, Hilbert's strategy is motivated by this methodological point of view: the quantifiers (and the operators ϵ- and τ) are what is problematic in logic and mathematics. Logic and mathematics without quantifiers (and without the ϵ and τ operators) is not problematic, so in order to justify logic and mathematics we have to prove that we can eliminate quantifiers (and ϵ-terms and τ-terms) from the proofs of statements not containing quantifiers (and ϵ-terms).

The development of proof theory (firstly, the important contributions of Gentzen on the role of cut-rule, and more recently the investigations on the role of structural rules) has showed that this methodological point of view may be refined: also the use of the cut-rule and structural rules in the proofs is very problematic, and in particular (at least for first-order logic) the use of quantifiers is not problematic in cut-free proofs not containing structural rules.

4   The operator ϵ: proof-theoretical theorems

Main theorems of proof-theory concerning the operator ϵ - in the framework of first-order logic - are well exposed in the second volume of [7]: the first and the second ϵ-theorems, and the extensions and applications of these theorems.

Of course, these theorems concern also the τ operator, by the fact that the ϵ and τ operators are mutually definable.

4.1   First ϵ-theorem

The *First ϵ-theorem* is the theorem stating that Hilbert's strategy of elimination of ideal elements inside the proofs of finitary statements *is successful* when we restrict ourselves to first-order logic and to proofs from finitary statements.

Finitary statements of first order logic are formulas without quantifiers and ϵ-terms, i.e. formulas without bound variables.

More precisely, the first ϵ-theorem concerns

- every first-order formal system F,

- its restriction F^-, obtained from F by leaving out the quantifiers (so, the proofs in F^- are proofs without the use of the axioms and rules concerning first-order quantifiers),

- its extension $F\epsilon$, obtained from F by adding the operator ϵ restricted to individual variables, i.e. by adding :

    - to the language, ϵ-terms of the form $\epsilon x A$ where A is a formula;

    - to the axioms, first ϵ-axiom

$$A[a/x] \to A[\epsilon x A/x]$$

    where A is a formula, x is an individual variable and a is an individual term.

The first ϵ-theorem says that for any first-order formal system F: if $A, A_1, \cdots, A_n$ are formulas of F^- (i.e. formulas without bound variables), and the formula A is provable from the formulas $A_1, \cdots, A_n$ in the formal system $F\epsilon$, then A is provable from $A_1, \cdots, A_n$ in the formal system F^- (i.e. without the use of bound variables and without the use of axioms and rules on the operator ϵ and quantifiers $\forall$ and $\exists$).

The proof of the theorem is done using Hilbert's strategy exposed in the previous section, inside the finitary mathematics.

The second volume of [7] contains not only the proof of this theorem, but also the proof of an extension of the first ϵ-theorem and the application of first ϵ- theorem to obtain consistency proofs of first-order formal systems.

Of course, the first order formal system for Arithmetics does not satisfy the conditions of the first ϵ-theorem, since first-order axioms of Arithmetics contain bound variables.

4.2   Second ϵ-theorem

Second ϵ-theorem is a theorem stating that first-order logic with the ϵ-operator is a *conservative extension* of first-order logic: this means that the formulas of standard first-order logic that are provable by means of the ϵ operator may be proved also in the usual first-order logic.

In order to appreciate the second ϵ-theorem, let us consider the fact that by using the operator ϵ we may produce formulas that are not the translation of usual first-order logic: the second ϵ theorem says that this great power of the operator ϵ does not produce the provability of unprovable formulas of first-order logic.

More precisely, this theorem concerns every usual formal system F and its extension $F\epsilon$ defined as above for the first ϵ theorem, and says that: if $A, A_1, \cdots, A_n$ are formulas of F and A is provable from $A_1, \cdots, A_n$ in the formal system $F\epsilon$, then A is provable from $A_1, \cdots, A_n$ in the formal system F (i.e. without the use of the operator ϵ).

The proof of the second ϵ-theorem is performed by means of finitary mathematics, and is obtained by using the extension of the first ϵ-theorem.

The second volume of [7] contains not only the proof of this theorem but also the extension of the second ϵ-theorem to the predicate logic with identity: from this extension, a proof of Herbrand's theorem is obtained.

Conclusion

Hilbert's operators, τ and ϵ, and Hilbert's strategy for the elimination of these operators inside proofs of formulas without τ and ϵ, have been the first investigations in proof theory.

The investigations contained in [8], in [9] and in [10] show that the interest in the proof-theory using these operators is still present and active.

We hope that further developments on Hilbert's operators τ and ϵ in a strong connection with current proof-theoretical investigations will be made possible. These developments would be highly desirable given that these operators open an alternative approach to the treatment and to the understanding of logical quantifiers.

References

[1] Ackermann, W. : Begründung des "Tertium non datur" mittels der Hilbertschen Theorie der Widerspruchsfreiheit, Math. Ann., 93, (1924).

[2] Bell, J.L. : The Axiom of Choice, College Publications, London (2009).

[3] Gentzen, G.: Untersuchungen über das logische Schliessen, Math. Zeits., 39 (1935), pp. 405-431

[4] Herbrand, J.: Recherches sur la théorie de la démonstration, Travaux de la Soc, des Sciences et Lettres, Varsovie, (1930)

[5] Hilbert, D.: Die logischen Grundlagen der Mathematik, Math. Ann, 88 (1923), pp. 151-165

[6] Hilbert, D.: Über das Unendliche, Math. Ann. 95 (1926), pp. 161-190

[7] Hilbert, D. and Bernays, P.: Grundlagen der Mathematik, Springer, vol. 1 (1934), vol. 2 (1939).

[8] Mints, G.: Epsilon substitution for first- and second-order predicate logic. Qnn. Pure Appl. Logic, 164-6 (2013), pp. 733-739.

[9] Moser, G. and Zach, R. : The Epsilon Calculus and Herbrand Complexity. Studia Logica, 82(2006), pp. 133-155.

[10] Pasquali, F. and Retoré, C.: Aristotle's square of oppositions in the light of Hilbert's epsilon and tau quantifiers, Aristotle 2400 wordl congress (2016) (https://arxiv.org/abs/1606.08326).

 Received 26 October 2015

$$(\exists y)(y = \epsilon x F x)$$

HARTLEY SLATER

Abstract

Not many people realise just how different the epsilon calculus is from standard predicate logic, even while the one is a conservative extension of the other. In terms of the main philosophical debate that has taken place in modern times it moves thought away from Empiricism and towards Rationalism. I shall run through a number of specific differences in this paper, although there are many more. I shall be mentioning discourse referents, directly referential terms, fictions, the *a prioricity* of logic, access to necessary beings, the debate between D.K. Lewis and Haecceicists, the necessary existence of Timothy Williamson, and Arthur Prior's concerns with the temporal version of the Barcan Formula.

1 Discourse Referents

Prior does not seem to have known of epsilon terms as such, but he was well aware of their linguistic correlates, which he called 'Russellian names'. These are what we would now call 'directly referential terms' or 'definite descriptions used referringly', in the manner of Donnellan [12, 152]:

> If we [are] using an expression as a Russellian name, we may find this concept exemplified in unexpected ways. For example, it may be that phrases of the form 'The ϕ-er' can be used as Russellian names as well as having the quite different use that Russell assigns them. We may, for example, so use the phrase 'The man over there', in a sentence like 'The man over there is clever', that its sole purpose is to identify the individual of whom we wish to say that he is clever, and the sentence may be being used simply to say that that particular individual is clever, and not at all to say, for example, that the individual is a man, or that he is 'over there'. The sentence used would then be true if and only if the individual meant was clever, and it would still be true if it turned

This is an invited paper.

out that the individual was not a man but a woman or a Robot, or that he had moved into quite a different position without our noticing it....Where the phrase is used as a Russellian description, the case of course is different ...And if phrases of the form 'the ϕ-er' can be used as Russellian names, no doubt ordinary proper names can be so used also. We might use 'Johnny Jones' in 'Johnny Jones has measles' simply to identify a certain individual and say of him that he had measles, so that this would be true if this individual had measles, even if for example, his name were not really 'Johnny Jones'.

Prior's hesitancy about the confirmation of the existence of such names and descriptive phrases was in tune with his time, indeed in tune with the whole of twentieth century predicate logic. Russell himself had only made vague suggestions about his logically proper names', and no mainline textbook, or academic study, has subsequently taken the matter much further. In fact the proof that they exist rests on a piece of predicative language which standard predicate logic cannot handle: discourse referents. The case is sufficiently illustrated when someone says 'There is a mouse in the room', we may reply to this with 'Where is it?', and first of all use an expression in language, namely the pronoun 'it', which standard predicate logic cannot symbolise. One needs a referential term, derivable from the initial existential remark, as a discourse referent, to keep track of the subject of discussion. In this case 'that mouse in the room' is the descriptive replacement for the 'it', i.e. this referential phrase is the discourse referent derivable from the introductory statement 'There is a mouse in the room' The phrase contains a demonstrative 'that', indicating the indefiniteness of its referent, but the most crucial point is that the reference of the 'it' is obtained quite independently of whether the first speaker speaks truly or falsely. The reference is to *that mouse in the room*, whether or not there is a mouse in the room. In the first case the object under discussion is properly described by the phrase. In the second case the referential phrase refers to a fiction, although it still may be applied ironically to whatever might have occasioned the initial remark. Maybe the so-called 'mouse in the room' was merely a shadow on the carpet, which can be taken, formally, to be a counterpart in this world of the fiction involved. As we shall see, it is quite clear that epsilon terms formalise such referential phrases ([2], [14]).

Hans Kamp, however, has tried to tackle the discourse referents problem in another way; and there are other treatments of the problem in the literature, for instance [4], [5]. The advantage of the epsilon calculus over these other approaches lies simply in its capacity to explicitly formulate demonstrative referential phrases like 'that mouse in the room' In Kamp's 'Discourse Representation Theory', by contrast, (see [6]) the required demonstrative phrases cannot be formulated. For instance, in the case

A farmer owns a donkey $[(\exists z)(\exists t)(Fz.Dt.Ozt)]$,

Kamp's 'Discourse Representation Structure' is

$$[x, y : \mathrm{farmer}(x), \mathrm{donkey}(y), \mathrm{owns}(x, y)].$$

But that merely says that there are two discourse referents, x and y, with the supposed properties, and does not formalize them explicitly. There is no representation of the fact that x is *that farmer who owns a donkey* $[x = \epsilon z(\exists t)(Fz.Dt.Ozt)]$, and y is *that donkey that x owns* $[y = \epsilon t(Fx.Dt.Oxt)]$. Moreover, it is quite possible that $\neg Fx$ or $\neg Dy$ or $\neg Oxy$, i.e. that the object lacks the stated properties, since there is no guarantee that the initial statement is true.

The eternal objects referred to by such phrases are, as a result, not necessarily empirical objects, but simply subjects of discourses whether those discourses are factual or fictional. One of the theoretical misconceptions that has arisen through the lack of familiarity with Hilbert's Epsilon Calculus is that the individual terms in predicate logic cannot refer to fictions (even though predicate logic clearly can apply to fictional remarks, and many examples widely used to illustrate it are of fictions). But, as the case of the mouse illustrates, a topic of discussion may well be a fiction, because individual terms are features of discourses that may be factual or fictional indifferently. The empirical world of science is certainly the only world we move around in, but in so moving and operating we make assumptions that by their nature may be wrong, on occasion. This engagement with fictions is in addition to our engagement with those in Literature and other arts. So it is clearly an error to believe that only in the arts and not in science do we have any connection with otherworldly beings. Standard predicate logic, through limiting itself to the formalisation of single sentences cannot enter into the central area that clarifies the matter — the area of multiple sentence discourse where cross-referencing pronouns occur linking single sentences together on a common topic. But the resultant possibility that the discourse is fictional gives us a simple proof that any individual term within it must be an epsilon term. For even if 'that mouse in the room' does refer to a mouse, which mouse it refers to is still to be decided. And, in any case, this is all a contingent matter, and in other circumstances it could have a different referent, such as a shadow.

Hilbert's Epsilon Calculus is a conservative extension of the predicate calculus, and contains individual terms of the form $\epsilon x F x$ for every predicate 'F' in the language ([7], [10], [17]). The standard epsilon calculus is based on the axiom

$(\exists x)Fx \supset F\epsilon xFx$, from which one can naturally obtain the equivalence between the two sides. That means that the epsilon term's reference is not further defined, and in just the right way to formalize the various types of discourse referent above. For if $(\exists x)Fx$, then ϵxFx is some individual amongst the Fs, being *the* F, if there is just one F, and *that* F if there is more than one F (so the epsilon term in the latter case is a demonstrative referential phrase rather than a definite description). But if $\neg(\exists x)Fx$, then ϵxFx is a fiction, which means it is simply a pragmatically chosen individual in the whole world at large. Naturally it cannot then satisfy the description in the epsilon term, which is revealed to be all that its being a fiction really involves. Here the epsilon term refers to some so-called 'F', i.e. to something merely said to be F. Thus, as before, if there is no mouse in the room then maybe 'that mouse in the room' refers to just a shadow on the carpet; and if there is no man with martini in his glass then maybe 'that man with martini in his glass' refers to a man with water in his glass. In the latter case, i.e., in Donnellan's historic case with the phrase 'the man with martini in his glass' used referringly, the speakers are just selecting a referent for the epsilon expression $\epsilon x(Mx.Gx)$ when $\neg(\exists x)(Mx.Gx)$, in line with the general semantics for epsilon terms. So what epsilon terms formalise more generally are demonstratives, i.e. terms with referents given not by description but by context, or 'acquaintance' as Russell's might put it.

One can deduce from this that logic is, after all, an *a priori* discipline. For commonly this is doubted, since individual terms are normally illustrated by proper names, but by just those proper names which have (or are supposed to have) a single bearer in the physical world. But whether there is a unique someone named 'Johnny Jones' is an empirical matter, making logic wait upon the verification of facts about the world. Certainly it is an empirical matter whether $(\exists y)(y = \iota x(x$ is called 'Johnny Jones'), i.e. whether there is in fact a unique person called 'Johnny Jones'; but it is not an empirical matter whether $(\exists y)(y = \epsilon x(x$ is called 'Johnny Jones'). In line with the general semantics of epsilon terms above, if there are people properly called 'Johnny Jones' then the epsilon term refers to a chosen one of them, but if there are no people called 'Johnny Jones' then the epsilon term's referent is arbitrary, allowing the name to become a nickname for whatever one chooses. So it invariably has a referent, and it is the element of mental intention, or choice, in determining that referent that the mainline tradition overlooks. Indeed the opinion has even been expressed that it is a blemish or worse in natural language that there might be many people, or no-one at all with a given name, so that in a 'perfect language', such as one that modern logic has aimed at producing, there should be no such latitude. But everything is in order as it is, in natural language. An individual subject is always an intended object either actually with the given name, or merely supposedly with that name. Thus one can say

There was a man called 'Johnny Jones'. He has measles, i.e.

$$(\exists x) J x . M \epsilon x J x,$$

and the $\epsilon x J x$ then refers to that 'Johnny Jones' brought up into the discourse by the introductory phrase, and, as before, does so whether or not the discourse is factual or fictional. Given that $\epsilon x J x$ could be anything in the latter case, it could be true or false that $M \epsilon x J x$. But this arbitrariness is of no consequence, since it is false that $(\exists x) J x$, so the conjunction as a whole has to be false. Russell's logic, when completed in this way with epsilon terms, is clearly a logic of intensional objects (c.f. [17]).

2 Necessary Existence

The central point to make in correcting the mainline tradition on the above matters is that while the individuals that 'x' ranges over have, as Russell knew, a necessary existence (c.f. [12, 149]), they must be separated from any entities that merely have 'existence' in this world, or some other. For what, in connection with individuals, has 'existence' just in this world, or just in some other (making them 'physical objects', and 'fictions', respectively) are not the individuals themselves, but their identifying properties. The difference between the two kinds of entity (one might dub one 'Platonic' the other 'Aristotelian') is illustrated most clearly in the epsilon calculus theorem (c.f. [17, 417-8]) that shows that 'A sole king of France exists and is bald' i.e.

$$(\exists x)(K x . (y)(K y \supset y = x) . B x),$$

is equivalent to

$$(\exists x)(K x . (y)(K y \supset y = x)) . B \epsilon x (K x . (y)(K y \supset y = x)),$$

i.e. 'A sole king of France exists. He [the king of France] is bald' The first conjunct in the second expression is then about certain identifying properties being instantiated which we can call an 'Aristotelian' matter. That is what must hold for a sole king of France to exist (contingently). The second conjunct in the second expression, however, is about a certain eternally existing individual, which is therefore a 'Platonic' object. And that object is a sole king of France, i.e. is the individual with the identifying properties, if there is such a thing, i.e. if there is a thing with the sole king of France character; but still exists *as a fiction* even if there is no such thing, i.e. if there is no thing with this character. The individual is a noumenal 'thing in itself', in Kant's terms, and must be separated very distinctly from any phenomenal appearance.

The point is central in connection with the shift required for a proper understanding of individuals. For the same point is involved in understanding how eternally real objects are accessed, which has seemed a perennial difficulty with Platonic entities. Paradigmatically the situation is represented again in the epsilon variant to Russell's analysis of 'The king of France is bald' For the first conjunct in the second expression above, as before, is itself equivalent to a conjunction:

$$K\epsilon x(Kx.(y)(Ky \supset y = x)).(y)(Ky \supset y = \epsilon x(Kx. (z) (Kz \supset z = x))).$$

So access to the individual $\epsilon x(Kx.(z) (Kz \supset z = x))$, i.e. the king of France, is provided entirely by means of the linguistic act of supposing there is a sole king of France, and through its then being invariably possible to cross-refer to the same individual from within further assertions. Eternal objects, as we saw, are simply subjects of discourse.

The consequences for Prior's own account of individuals we will see later. But it is important to see that the same consequences arise for another well-known theory of individuals: David Lewis'. Indeed, the point that has to be made against Prior is thoroughgoing against many twentieth century stories about individuals: it is *that they confused Platonic and Aristotelian entities* as above. For David Lewis' theory of world-bound individuals and their counterparts in other worlds (see, for instance, [8], [9]) is opposed by Haecceicists who allow trans-world individuals with next to no recognisable character. The above points provide a defence of the latter through showing that the former involves a wrong view of identity.

The error can possibly be traced to a feature of Russell's logic. For in Russell's logic iota terms are used rather than epsilon terms, and specifically $(\exists x)(x = \iota y Ky)$ is equivalent to

$$(\exists x)(Kx.(y)(Ky \supset y = x)).$$

So it looks like the identity of an individual is inseparable from the presence of certain properties. In the absence of some of these properties it therefore becomes plausible that a 'counterpart' might be recognisable if sufficient of the others are present. Thus we seem to be able to accommodate 'Quine might not have been a logician', 'Quine might not have been such a traveller', etc. But what about more extreme thoughts like 'Quine will become a centipede' as might be found in Buddhist accounts of re-incarnation? Indeed it is said there was a man who thought his wife was a hat! There is nothing too ladylike about a hat, and nothing recognisably Quine-like about a centipede. Indeed which one will he be, they all look much the same!?

The point holds in reverse with respect to items that are fictional in our world. Someone in the world of Myth might think 'I wish Pegasus had been a zebra!', or 'Pegasus would have had an extra pair of flippers, if he had been a seal'. These could be handled through the presence of enough recognisable qualities in a counterpart. But with stranger imaginings the possibility of something recognisable becomes weaker and weaker. That is one reason why we have no reason to deny that Pegasus exists in this world though not in his mythological form. For he need not have wings, or be a horse, here. But the philosophical tradition in this area has been preoccupied with a linguistic version of this issue. For one can hardly say 'Pegasus does not exist', since 'Pegasus' is a referential term, and so its use presumes it has a referent. Not so with 'A winged horse does not exist', of course, because 'a winged horse' is descriptive and not referential. The specific case of this matter that influenced Russell was 'The king of France does not exist' which has 'the king of France' as a referential phrase. Russell, to avoid the seeming contradiction, chose to replace the whole with 'A single king of France does not exist' Here was the start of the error that David Lewis made: the identification of an individual with a bundle of properties. But as we have seen, there are ways of correcting this error, and getting a formal account of the property-free 'thisness' much beloved by Haecceicists.

The point, of course, applies not just to trans-world individuals but also to individuals as they are extended in time, and the material transformations they undergo, even with humans and their eventual dissolution back to their inorganic parts. And these further considerations have a very direct, contemporary relevance. For Timothy Williamson has recently been concerned with an argument that seemingly proves that he exists. It goes as follows (see [19]): (1) Necessarily, if I do not exist then the proposition that I do not exist is true. (2) Necessarily, if the proposition that I do not exist is true then the proposition that I do not exist exists. (3) Necessarily, if the proposition that I do not exist exists then I exist. (4) Necessarily, if I do not exist then I exist. So (5) I necessarily exist. Williamson says, amongst other things, that parallel arguments would have to be equally sound, such as those that replace 'necessarily' with 'at all times', and 'I' with 'this body'.

But the further conclusions then obtained, namely that he exists at all times, and that his body necessarily exists, he finds 'counterintuitive'. So the matter has become the subject of considerable debate. Clearly the above shows that the conclusion of Williamson's argument is undoubtedly true, and in a way that shows the further conclusions that might be drawn in parallel arguments are also true. One does not need to debate, like Williamson, the individual worth of his premises, since we have arrived at his conclusion(s) another, more direct way. The central distinction that needs to be made is between logical existence and other forms of 'existence', such as 'being alive', 'being present', and 'being actual'. But the required distinction is not readily made using just the Predicate Calculus. Instead what is wanted, as we have seen, is its conservative extension, Hilbert's Epsilon Calculus.

3 Presupposition Theories

As we saw at the start Prior knew of, or at least sensed, the account of individuals that has been developed above. In fact he considered many views of individuals, but he also formulated a specific temporal logic Q that was pre-suppositional with respect to individuals, thinking that since these sometimes exist, and sometimes do not, a statement about them when they did not exist was unavailable, i.e.. was 'gappy'. Considering two times as an illustration that gave him a six valued logic covering possible combinations of 'true', 'false', and 'unstatable' ([11, 14]). But what he was thinking about were not individuals properly so called, but instead individuals in a certain form, for instance a person when alive, as opposed to when dead. That would be taking a collection of properties (such as heart beating, breathing, and brain activity) to define an individual whereas the individual is instead the entity which has these properties at a given time, and which persists even when they are lacking. I have given some details of the tense logic that results from acknowledging this fact in [15]. Its overall distinction is that it is not 'gappy' but is instead a straightforward two-valued logic where an arbitrary assignment of 'true' or 'false' replaces Prior's 'unstatable'.

What happens to a person when they die? An Empiricist might prefer to refrain from saying anything; but human history is full of unverifiable, though firmly believed stories on the matter, which differ widely from one culture to the next. The crucial point is that such fictions can be as logical and illogical, as consistent and inconsistent as factual writings.

Of course, Strawson, directly against Russell, had preceded Prior with a pre-suppositional account again in the temporal area. In *On Referring* ([18]) Strawson wanted 'The king of France is bald' to lack a truth-value when there was nothing with the characteristics of a king of France. But the supposition that there is not one and just one king of France merely liberates propositions about the individual concerned from having a determinate truth-value. For, directly against Strawson, we can still make the claim that the king of France is bald. If we suppose (contrary to fact) that there is one and only one king of France then we can go on making up a story about this character. The continuing sentences still may have a truth-value, but one based entirely on choice — commonly on the choice of a literary author, who makes up the story about the character. So the pre-suppositionists error here is to forget fiction, and think that truth is not attributable to fictional statements, when instead, what distinguishes 'factual truth' is merely that it is (in principle) verifiable and so determinate.

In his formal work Frege was not a pre-suppositionist about individual terms, constructing a way of handling 'empty names' with some resemblance to what we have seen above. But in his philosophical essay 'The Thought' Frege agonised about whether 'this lime tree is just my idea' might be true, not wanting to give any statement about the lime tree a truth value if that tree was just an idea, i.e. was fictional. Indeed he makes the point about fictions quite explicitly. He says ([3, 300]):

> Is that lime tree my idea? By using 'that lime tree' in this question I have really already anticipated the answer, for with this expression I want to refer to what I see and to what other people can also look at and touch. There are now two possibilities. If my intention is realised when I refer to something with the expression 'that lime tree' then the thought expressed in the sentence 'that lime tree is my idea' must obviously be negated. But if my intention is not realised, if I only think I see without really seeing, if on that account the designation 'that lime tree' is empty, then I have gone astray into the sphere of fiction without knowing it or wanting to. In that case neither the content of the sentence 'that lime tree is my idea' nor the content of the sentence 'that lime tree is not my idea' is true, for in both cases I have a statement which lacks an object.

But there is no difficulty talking about Frege's lime tree, whether or not it is fictional. Indeed we have just done it!! We can still ask for properties of it, even if it is a fiction, indeed someone determined enough not to be thought to be in error about his imaginings, when in fact mistaken, could construct an enormous story about his dream object, to prop up his view that it is not a fictional object. But alas, despite the confabulation, and no matter how persuasive the story, the story might still be a fiction, like Prior's story about his system Q.

Prior, of course, being the thorough scholar that he was, was aware that he was treading on perilous ground. Indeed he knew in some detail what the contrary, proper treatment might be. Talking about the associated Barcan formula 'If it will be the case that something (ϕ's then there is something which will (ϕ he says ([11, 2930]) :

> ... if [this formula] is laid down as a logical law, i.e. as yielding with all concrete substitutions for its variables a statement which is true whenever it is made, it can only be justified by the assumption that whatever exists at any time exists at all times, i.e. the assumption that all real individuals are sempiternal. ... It may be that this assumption is capable of metaphysical justification. With regard to our counterexample — that perhaps there will be someone flying to the moon although it will not be anyone now existing — it may be argued that persons are not

individuals in the sense in which the x's and y's of quantification theory stand for names of individuals, and that all genuine individuals do exist at all times. On this view, the point about the flight to the moon is that although the collection of genuine individuals which will perform the flight has not yet come to constitute a person, these genuine individuals electrons or whatever they might be do exist now, and always have done and always will. And this would save the Barcan formula, for the difference appealed to now turns out to be one between forms more like 'It will be the case that something is a person and flies' and 'Something is a person of which it will be the case that he flies'.

He goes on 'But I doubt whether this story about sempiternal electrons is good physics, and am sure it is not good logic. That is to say, even if it be true that whatever exists at any time exists at all times, there is surely no inconsistency in denying it, and a logic of time distinctions ought to be able to proceed without assuming it' Nevertheless, almost immediately, he counters this, seeing that his Barcan formula does not need any empirical backing. It could be supported with an account of terms that refer to necessary existents. He says ([11, 30]):

It must be admitted, however, that when this assumption is driven out the front door it is liable to return through the back. In discussing our counterexample to the Barcan formula we might, for example, say that the flight to the moon may be accomplished by 'someone who does not exist yet but will exist later on' And this way of talking seems to imply that *there is* an x such that x does not exist yet but will exist later on. But what sort of x could this be? An object, apparently, which does not yet exist but nevertheless can already be talked about, or at all events can be the value for the variables bound by our quantifiers. And if this object, although it does not yet exist, can already be talked about, or can be a value for our bound variables, presumably it is in this position at all times — it is at all times an object, even if it is not at all times an existent object. And, of course, if the bound variables in the Barcan formula are supposed to range over all objects in this wide sense of 'object' — all the items in this supposed permanent pool of things that are, have been, or will be — it can again be justified.

It is by getting a proper account of these eternal objects that Prior here had some sense of that we have improved on, indeed corrected his presuppositional temporal logic Q.

References

[1] Bealer, G. 1982, *Quality and Concept*, Clarendon, Oxford.

[2] Evans, G. 1977, 'Pronouns, Quantifiers, and Relative Clauses 1', *Canadian Journal of Philosophy* 7, 467-453.

[3] Frege, G. 1956, 'The Thought: a Logical Enquiry' *Mind* 65, 289-311.

[4] Groenendijk, J. and Stokhof, M. 1991, 'Dynamic Predicate Logic', *Linguistics and Philosophy* 14.1, 39-100.

[5] Hintikka, J. and Kulas J. 1985, *Anaphora and Definite Descriptions*, Kluwer, Dordrecht.

[6] Kamp, H. 1995. 'Discourse Representation Theory' in J. Verschueren, J.-O. Östman & J. Blommaert (eds.), *Handbook of Pragmatics*, Benjamins, Amsterdam, 253-257.

[7] Leisenring, A.C. 1969, *Mathematical Logic and Hilbert's Epsilon Symbol*, Macdonald, London.

[8] Lewis, D. 1973, *Counterfactuals*, Blackwell, Oxford.

[9] Lewis, D. 1986, *On the Plurality of Worlds*, Blackwell, Oxford.

[10] Meyer Viol, W.P.M. 1995, *Instantial Logic*, ILLC, Amsterdam

[11] Prior, A.N. 1957, *Time and Modality*. O.U.P., Oxford.

[12] Prior, A.N. 1971, *Objects of Thought*. Edited by P.T. Geach and A. J. P. Kenny, O.U.P., Oxford.

[13] Schiffer, S. 2003, *The Things we Mean*, Clarendon, Oxford.

[14] Slater, B.H. 1986, *E*-type Pronouns and ?-terms' *Canadian Journal of Philosophy* 16, 27-38.

[15] Slater, B.H. 1987, 'Hilbertian Tense Logic' *Philosophia* 17.4, 477-488.

[16] Slater, B.H. 2007, 'Completing Russell's Logic' *Russell* 27, 78-92, also in N. Griffin, D. Jacquette and K. Blackwell (eds) *After 'On Denoting'*, The Bertrand Russell Research Centre, Hamilton.

[17] Slater, B.H. 2009, 'Hilbert's Epsilon Calculus and its Successors' in D. Gabbay and J. Woods (eds) *Handbook of the History of Logic*, Vo15, Elsevier Science, Burlington MA, 385-448.

[18] Strawson, P. F. 1951, 'On Referring', *Mind* 59, 320-344.

[19] Williamson, T. 2002, 'Necessary Existents', in A. O'Hear (ed.), *Logic, Thought and Language*, C. U. P., Cambridge, 233-251.

[20] Williamson, T. 2013, *Modal Logic as Metaphysics*, O.U.P. Oxford.

   Received 8 September 2015

Buying Logical Principles with Ontological Coin:
The Metaphysical Lessons of Adding Epsilon to Intuitionistic Logic

David DeVidi
University of Waterloo *
david.devidi@uwaterloo.ca

Corey Mulvihill
University of Ottawa
corey.mulvihill@uottawa.ca

Abstract

We discuss the philosophical implications of formal results showing the consequences of adding the epsilon operator to intuitionistic predicate logic. These results are related to Diaconescu's theorem, a result originating in topos theory that, translated to constructive set theory, says that the axiom of choice (an "existence principle") implies the law of excluded middle (which purports to be a logical principle). As a logical choice principle, epsilon allows us to translate that result to a logical setting, where one can get an analogue of Diaconescu's result, but also can disentangle the roles of certain other assumptions that are hidden in mathematical presentations. It is our view that these results have not received the attention they deserve: logicians are unlikely to read a discussion because the results considered are "already well known," while the results are simultaneously unknown to philosophers who do not specialize in what most philosophers will regard as esoteric logics. This is a problem, since these results have important implications for and promise significant illumination of contemporary debates in metaphysics. The point of this paper is to make the nature of the results clear in a way accessible to philosophers who do not specialize in logic, and in a way that makes clear their implications for contemporary philosophical discussions. To make the latter point, we will focus on Dummettian

---

*The financial support of the SSHRC of Canada during the work on this paper is gratefully acknowledged.

discussions of realism and anti-realism.

Keywords: epsilon, axiom of choice, metaphysics, intuitionistic logic, Dummett, realism, antirealism

The goal of this paper is to argue for the philosophical, and in particular meta-physical, importance of certain results involving Hilbert's ε-operator. The results involve the addition of ε to intuitionistic (rather than classical) logic. More pre-cisely, we will suggest that these results offer illumination of debates about realism and anti-realism, and so help us get a clearer picture of what it means to say that an individual or property is real, or objective, or is in some related way attributed a special status of this sort.

We suspect that the sort of discussion we will offer here tends to fall between the stools in contemporary philosophy. The formal results are not new, and so mathematicians and formal logicians might look past the philosophical discussion surrounding the results and conclude that "all this is already known." On the other hand, the formal results in question are in an area that many contemporary philoso-phers are likely to regard as the esoteric reaches of deviant logic, potentially inclining them not to read the paper either. We will suggest that this is a case where the interesting philosophy starts when the formal proof is completed—the philosophical importance is not something that can be read straight off the proof, but takes some showing. We hope that it repays the effort it might take to overcome whatever impatience (with philosophical niceties or with technical details, depending on one's background) one might bring to the task. Since this is a presentation of a basic idea that, if worthwhile at all, is worthy of a deeper study, we will attempt in the present paper not to overly try the patience of either sort of reader, keeping both philosophical and mathematical details to the minimum that will allow us to try to make our point clear.

The formal results in question show that the addition of ε to intuitionistic predi-cate logic is non-conservative.[1] As is well known, intuitionistic logic is a sub-system of classical logic in the sense that all validities of intuitionistic logic are also classi-cally valid, but not conversely. As is also well known to people already familiar with the ε-operator, its addition to classical logic is conservative—this is the upshot of the so-called ε-theorems. To introduce some terminology that will facilitate discussion later on, we will call principles that are classically but not intuitionistically valid *superintuitionistic*.[2] We call the system that results by adding the ε-operator to in-tuitionistic predicate logic the *intuitionistic ε-calculus*. The basic form of the results

---

[1]The results we review below are from [1, 2, 3]; Bell's work was motivated by [4] and [5].

[2]This is fairly standard terminology, though we note that we intend to focus on consistent systems and to ignore consistent extensions of intuitionistic predicate logic that are not consistent

of interest is that certain superintuitionistic principles are valid in the intuitionistic ε-calculus, and that when we add in other, seemingly innocuous assumptions to that calculus we increase the number of interesting superintuitionistic principles that are valid, until we eventually make them all valid. That is, the addition of ε to intuitionistic logic is *always non-conservative*, and with the addition of some modest-seeming additional assumptions ε is sufficient to make over intuitionistic into classical logic.

The philosophical importance of these results is not, we submit, immediately apparent. The most direct and digestible way to make them clear, we think, is to consider the implications of the results for Michael Dummett's highly influential account of the relationship between realism and anti-realism. We will try to show that they offer us a way to improve Dummett's account, shoring up what might be regarded as a "soft spot" in his story. But we hope (and will suggest) that the lessons we draw are not entirely dependent on the details of Dummett's program for their interest, and that they offer more general lessons for how to think about notions such as reality and objectivity.

So far, we've described the project at a pretty high level of abstraction. In slightly more detail, we shall proceed as follows. First, we will present a sketch of Dummett's framework for understanding debates between realists and their opponents in different areas of philosophical dispute—realists and nominalists about universals, realists and behaviourists about mental states, realists and constructivists in mathematics, and so on. In our selective sketch we will draw attention to some key features of Dummett's framework for our purposes. One is his suggestion that all these debates are best re-cast as debates in philosophical logic, in the sense that the correctness of the principle of bivalence, and so the law of excluded middle, and so of classical logic, for a particular domain of discourse is a criterion for realism being correct for that domain. Another is that intuitionistic logic has a special status—it is, in fact, logic properly so-called, and so is metaphysically neutral. The classically but not intuitionistically valid principles, *i.e.* the superintutionistic principles, have a status akin to mathematical induction or laws of physics in that they can be employed perfectly legitimately in certain domains, but not others.

Secondly, we shall argue that the case Dummett makes for the link between realism and classical logic, while not unpersuasive, is a soft spot in his general account because it relies on a metaphor to link the metaphysical notion of a *mind-independent reality* to the acceptance of superintuitionistic principles. We will argue that the technical results to which we want to draw attention help fill in the details

---

with classical logic. We find the terminology, for instance, in [6, p.103]: "Extensions [of intuitionistic logic] are called superintuitionistic logics. Superintuitionistic logics which are contained in the classical logic are said to be intermediate. An intermediate propositional logic is the same as a consistent superintuitionistic logic; it is not true for predicate logics."

in Dummett's argument by providing a more direct link between metaphysical assumptions and the acceptance of superintuitionistic principles, including excluded middle.

The case in the second part has a few sub-parts. First, we note that the ε-operator, being a sort of choice principle, involves an "existence assumption." Choice principles and other metaphysically loaded principles, such as axioms of infinity, are commonly (if not universally) taken to be, for this very reason, non-logical. Next, we present a sketch of how the addition of these non-logical principles to intuitionistic logic make various superintuitionistic principles provable. Finally, we turn to the job of drawing philosophical lessons. We will contend that the results establish a clearer connection between metaphysical assumptions and the superintuitionistic principles than do vague suggestions about reality "fixing the truth values" of well-formed claims. We also argue that there is philosophical mileage in the fact that the superintuitionistic principles don't necessarily come as a package deal. As we shall see, adding ε together with different choices of additional assumptions yields different intermediate logics between the metaphysically neutral basis of intuitionistic logic and the classical logic that, according to Dummett, corresponds to full-blown realism. We will argue that these way-stations between the intuitionistic basis that antirealists should have no complaint about and full classical logic, and the assumptions that suffice to reach these different stations, link up in interesting ways to our intuitions surrounding notions like objectivity and reality. It is this final point, we think, that makes these results of interest whether or not one approves of the details of Dummett's story, since it offers the prospect of a more fine-grained categorization of metaphysical options than is available if one supposes that realism is an all or nothing matter, and it offers a way to discuss the question of reality and unreality, at least sometimes, at the level of individual, fairly homey, properties, rather than in terms of nebulous notions such as "discourses."

1 Realism and Its Opponents

Michael Dummett was one of the most influential philosophers of the second half of the 20th Century, so many philosophers are likely to have at least a vague idea of his views. As well-known as any of these would be his commitment to the idea that the acceptance of the principle of bivalence, and so of the law of excluded middle, and therefore of classical logic, was a "criterion of realism."[3] Given what we said in the introduction, it will not surprise anyone that this is an important part of

---

[3]While commentators on Dummett perhaps used the phrase "criterion of realism" more often than Dummett did himself, he does use it occasionally, for instance in [7, p.379,467].

Dummett's view for our purposes. We will provide only a quick sketch—we hope not a caricature—of the case Dummett makes for this view, highlighting strands that will be of use in the ensuing discussion. The argument was presented many times with differences of emphasis and detail throughout Dummett's long career, and the full-dress presentation involves discussion of theories of meaning, the learnability of language, and more, that are fascinating, much disputed, and far too intricate for what we are hoping to do in this paper. We ask forgiveness from Dummett scholars who regard our simplified presentation as too simple. We rely largely on Dummett's own restrospective descriptions of what was central to his account of realism and antirealism from late in his career, especially his inaugural address upon taking up the Wykeham Chair at Oxford, [8].

Realism has two parts:

- First, the idea that the things we say in a particular area of discussion, a *discourse*, are properly regarded realistically involves a commitment to the idea that we are making claims about a reality that is in an important way "independent of us," and in particular is independent of our ability to know about it. Of course, few would deny that in some sense reality is independent of us. What is distinctive about realism is the suggestion that if realism is correct for a discourse, in spite of this independence our language somehow links up with the reality in a special way: the truth values of our statements in the discourse are fixed by that reality, independently of whether we can come to know those truth values. Indeed, in discourses about which realism is correct, the prospect that there are claims whose truth values we *cannot* come to know, even in principle, cannot be ruled out. While there is some fussing to be done about difficult details (for instance, what to say about vague statements that don't appear to be either definitely true or definitely false), the presence of a mind- and language-independent reality to fix the truth values of our statements is the link between metaphysics and bivalence.

- Secondly, some of our claims are true.[4] We may regard the language we use to discuss unicorns as purporting to refer to a mind- and language-independent reality, but think that there are no unicorns. Presumably that should suffice for us to count as anti-realists about unicorns.

Since realism, according to this story, has a two part definition, speaking at a very general level there are two ways of rejecting realism. As the example of unicorns

---

[4]More precisely, this should be formulated as the requirement that some of the atomic claims we make in the discourse in question must be true, since we don't want the condition to be met simply because of vacuous quantificational claims or negative claims turning out true.

suggests, if none of the (atomic) statements of a domain are true, then realism is not true for that domain. But there are, of course, more interesting versions of this sort of antirealism, such as the various "error theories" in ethics (most famously [9]) or mathematics (*e.g.*, Field's account of arithmetic in [10]). Various versions of fictionalism in these domains (starting with Mackie's and Field's own positive accounts) are plausibly viewed as antirealist for similar reasons—the claims in the domain are not *literally* true (or true *when taken at face value*), but there is some other story about what makes the claims involved seem so important to us in spite of their (literal) falsity. We set aside the question of whether a proper formulation of realism requires that we find a way to unpack the work that *literalness* (or some similar notion) plays in explaining why fictionalism is not a sort of realism.

Anti-realisms of another sort for some reason reject the other component of the definition of realism. Emotivists in ethics say that our statements are not in the business of saying true and false things at all, and kindred *expressivist* accounts in various areas of philosophy (about conditionals, laws of nature, *etc.*) similarly deny that the apparent statements of some domain are properly regarded as statements (if by statement we mean a claim that could be either true or false) at all. Others, though, do not want to go so far as saying that the apparent statements of a domain are not actually in the true/false game, but deny in some way that their truth values are suitably "independent" of us. Such, for instance, are the views of constructivists in mathematics who hold that a mathematical statement is true precisely if it is provable, and false if it is refutable. Such a view certainly fails to make the truth values of statements in this domain independent of our ability to know them (presuming that by "provable" we have in mind, somehow, provability by agents like us and not, for instance, mathematicians with infinite capacities of some sort), but it allows that the statements in this domain are in the game of making true or false claims.

Finally, many antirealist views have traditionally taken on the guise of reductionism—for instance, the phenomenalist claim that claims about the physical world were somehow indirect ways of talking about perceptions, or logical behaviourist claims that talk of mental states were really indirect descriptions of behaviour.

2 The status of superintutionistic principles

We are now in a position to see Dummett's reasons for emphasizing the role of superintuitionistic principles in debates between realists and antirealists. If, as has been suggested, realism involves a commitment to bivalence, then it is (barring some esoteric further maneuvering) a short step to the acceptance of classical logic, since bivalence more-or-less implies classical logic. (See [11] for some of the nuances ignored in this statement.)

From the other direction, Dummett has little good to say about expressivist or reductionist versions of antirealism. Indeed, he sometimes suggests that, due to the prevalence of such versions of antirealism in the history of philosophy, his review of historical debates between realists and antirealists might have seemed to not even be worth pursuing, for the merits of the respective views "threatened that all the contests would end in victory for the realist before the comparative study began" [7, p.470]. Constructivist antirealism about mathematics, he judged, was the only antirealism not to fall prey to arguments likely to be fatal to expressivist and reductionist views. Presuming that there were substantial philosophical insights behind traditional antirealist positions in other realms to which their expressivist or reductionisist presentations failed to do justice, Dummett suggested that the best road forward was to recast the antirealist views in all the traditional disputes along the lines suggested by constructivism in mathematics. That is, they should take as their starting point the idea that the truth of a claim in the disputed domain consists in whatever counts as conclusively establishing that claim, just as constructivism takes truth to be provability, while falsity amounts to the possibility of conclusively ruling the claim out. Dummett's suggestion is that evidentially constrained notions of truth of this sort will share the same basic logic as one finds in the mathematical antirealism that serves as its model—namely, intuitionistic logic.

It is important to recognize that on this account the question of realism and antirealism is one that varies by domain, or at least could do so. It is an open possibility that one ought to be a realist about tables and chairs but not about mental states, for instance. Since intuitionistic logic is a subsystem of classical logic, we can say that the intuitionistically valid logical principles are *metaphysically neutral* in the sense that they are not in dispute between the participants in any of the realism/antirealism debates, once those debates are recast according to Dummett's recommendations. The superintuitionistic principles, on the other hand, arguably do not deserve the label "logic" at all, on this account. The case for saying so turns one of the grounds Frege used to defend the claim that the truths of arithmetic are logical to the opposite purpose. Frege's case appeals to the idea that properly logical inferences are the ones that apply in *every realm of human thought*, something he claimed was true of arithmetic priniciples. If Dummett is right, superintuitionistic principles are correct in domains where realism is the correct view, but not in general. Nowadays we take the same criterion to show that, for instance, mathematical induction is not a principle of logic because it holds when talking about countable, discrete objects but not when talking about real numbers. Similar reasoning seems to show that superintuitionistic principles should be regarded as non-logical, and, as we have noted above, can be seen as similar to mathematical induction, or the laws of physics which apply to inferences about physical objects but not in every domain.

There is another line of argument to be found in Dummett's writings for the claim that logic, properly so called, is intuitionistic logic. Probably the most explicit discussion of this matter occurs in *The Logical Basis of Metaphysics* [12]. The discussion there considers candidates for the status of "logical operator" in natural deduction terms, taking the meaning of any proposed operator to be specified in terms of its introduction and elimination rules.[5] Building from the intutitive idea that a properly valid deductive inference must not allow us to infer in a conclusion anything that was not already "contained in the premises," Dummett argues that it is a necessary condition of being a properly logical operator that there be "harmony" between the introduction and elimination rules for the operator. That is, if P is a sentence with a particular operator as its main logical operator, all and only the things required to make an inference of P using the introduction rule(s) are things we can extract from P using the elimination rule(s).[6] In *Logical Basis* we get Dummett's most explicit argument that the problem with some of the classical logical operators is a lack of harmony: that is, the introduction and elimination rules for, for instance, classical negation are not in harmony (and so are not properly logical). The principles of intuitionistic logic are by contrast, he suggests after considerable discussion, in harmony.

To summarize: we find in Dummett two different lines of argument for the view that, in spite of what we teach students in their first formal logic courses, such classically valid principles as $\forall x(Px \lor \neg Px)$, $P \lor \neg P$, $(P \to Q) \lor (Q \to P)$, and $\neg(P \land Q) \to (\neg P \lor \neg Q)$ are not really *logical* principles at all. Instead, Dummett argues, the principles of intuitionistic logic are logic, properly so-called.

It follows that in circumstances in which the superintuitionistic principles are correct, their correctness must have some further basis besides logical correctness, for they are not logically correct. As we have seen, Dummett's suggestion is that what justifies them, when they are justified, is the existence of a suitably mind-independent realm to which the statements of the discourse in question hook on in the right way so that the principle of bivalence is true for that domain. Given their dependence on a metaphysical truth, it is perhaps not too much of a stretch to call them "metaphysical laws."

But perhaps this is too hasty. One might buy the suggestion that intuition-

---

[5]The reasons for this approach have primarily to do with a particular sort of theory of meaning that Dummett also argues for in the book. We think it is harmless to leave those details aside.

[6]We set aside interesting details again here: for instance, whether an introduction/elimination pair is in harmony depends on what logical operators are presumed to be in place already.

istic logic is logic properly so-called without accepting the suggestion that extra metaphysical commitment warrants additional principles of reasoning. Dummett suggests that the presence of a "mind-independent reality," onto which our language hooks appropriately, suffices to establish the correctness of the superintuitionisic principles—because in those circumstances reality is supposed to determine one of two truth values for each well-formulated statement, and bivalence implies classical logic. This picture has not universally been found to be compelling. For instance, one might well be puzzled, on roughly Dummettian grounds, by how we could come legitimately to think that we've successfully latched onto reality in this way. Occasionally, commentators on Dummett's work raise this concern in the form of a question: how, if the standard antirealist arguments he canvasses ever work, can they fail to *always* work—can Dummett be arguing for anything but *global* antirealism? [13] We next will suggest alternative grounds that help bolster Dummett's suggestion that metaphysical, and so non-logical, commitments can ground the acceptance of these metaphysical laws.

3   Existence and Logic

The basic structure of the help we propose to offer Dummett is this: certain formal results make clear that adding in clearly non-logical, plausibly metaphysical (because ontological) principles to "logic properly-so-called" make various superintuitionistic principles correct. We therefore have a more direct, rigorous link between metaphysical assumptions and the superintuitionistic principles than is provided by a suggestive detour through talk of mind-independent realities. As a next step, we consider the question of whether we're right to say that the principles being added are metaphysical and non-logical.

We begin by noting that the principles in question all assert the existence of something, and it is quite a prevalent view that existence claims are *ipso facto* not logical. As long ago as his 1919 *Introduction to Mathematical Philosophy* [14], Russell complained about the "impurity" of sentences like $\exists x.x = x$, which are valid in usual formulations of predicate logic but which, he complains, are true only if at least one object exists (and so are not truly logical). Carnap, in *The Logical Syntax of Language* [15, §38a], sketches a method for constructing a logical system which, he says, does not make such assumptions. We find here early steps towards the development of what became known as *free logic*.[7]

---

[7]Neither Russell nor Carnap made provision for names (properly so-called) which do not refer to existent objects, preferring to explain away "Pegasus" and "The King of France" as not genuine names. It wasn't until the 1950s and 1960s that a sustained effort was made to provide suitable

Now let us consider what we take to be the standard attitude towards free logics. While some of the advocates of free logic were quite militant about the lessons to be drawn from their work—and correspondingly militant in their opposition to continued use of the usual classical predicate calculus—few logicians work in free logic nowadays except in special cases where the role of the existence assumptions involved in standard predicate logic are especially salient (*e.g.*, type theories in which some types might be empty, or in some quantified modal logics). It is easy enough to keep in mind that the validity of $\exists.x = x$ in standard predicate logic is merely an artifact of a simplifying assumption. We work in a system that considers only models with non-empty domains in order to simplify our system of rules; nobody is thereby tempted to such reasoning as "aha, something necessarily exists, so let's call it 'God' ..." so the simplification is harmless. We are all clear that not every validity in the first order predicate calculus as usually presented is *really* a *logical truth*.

Consider next a perhaps more familiar example, a potted version of the standard story of the demise of logicism in the philosophy of mathematics.[8] The story comes in two parts. First, Frege's logicism. It had the considerable virtue of deriving the existence of the natural numbers from self-evident principles. Alas, it also had the even more considerable demerit of being inconsistent, and thus serves as an early pothole in the rough ride the 20th Century provided for the notion of self-evidence. The second chapter is Russell and Whitehead's logicism, where they make a valiant attempt to formulate all of mathematics within a type-theoretic logic in *Principia Mathematica*. But their formidable technical achievement does not count as a vindication of logicism because along the way they must appeal to certain axioms which are manifestly non-logical—the usual culprits pointed to being the axioms of infinity, of reducibility, and of choice. And the reason these principles (especially the first and third) are regarded as manifestly non-logical is that they imply the existence of particular entities.

It is worth pointing out that it is not the unanimous opinion in the history of logic and philosophy that having existential implications is enough to rule a principle out as a principle of logic. Indeed, any vindication of logicism as it seems to have been conceived in the late 19th and early 20th Centuries seems to have

---

formulations of classical predicate logic innocent of existential assumptions and which allowed for singular terms which do not refer to existing objects, and nowadays "free logic" usually refers to systems meeting both those conditions. An impressive cast of logicians contributed to the effort to develop free logics, including Henry Leonard, Hugues Leblanc, Theodore Halperin, Jaako Hintikka, Dana Scott, Bas van Fraassen, Robert Meyer, Karel Lambert and many others.

[8]We are well aware that the actual history of logicism is longer, more complicated, and more interesting than it would be useful to detail here.

presupposed that we would be able to prove the existence of infinitely many objects (*e.g.*, the natural numbers) on purely logical grounds, and Frege famously argued for the existence of these "logical objects." Frege has at least two distinguishable methods for getting to this claim. First, he *derives* the axioms of arithmetic from "self-evident" principles. Alas, among these was the notorious Basic Law V, which rendered his system inconsistent. On the other hand, he argues that the truths of arithmetic are logical because they are among the principles that "extend to everything that is thinkable; and a proposition that exhibits this kind of generality is justifiably assigned to logic" (Frege, "On Formal Theories of Arithmetic," as quoted in [16, p.44]). The first path clearly does not have many advocates today. Nor, though, does the latter: as discussed above, at least some of the basic principles of arithmetic, including mathematical induction, are standardly viewed as applying in some domains and not others, and so these principles are categorized as non-logical by appeal to essentially the same criterion Frege uses to classify them as part of logic. So while there may be grounds for believing in "logical objects," not many today are likely to think they find those grounds in Frege—and we know of no other compelling alternative arguments.

It's tempting to state the lesson as follows: there had to be something wrong with Frege's account, since it extracted such rich ontological information from putatively logical principles; and we ought not to be surprised that to get mathematics out of logic Russell would have had to smuggle the non-logical existence assumptions in somewhere.

As a final remark for this section, we note that the results we will consider begin with principles that are all versions of (or relatives of) the axiom of choice. The axiom of choice, of course, has its own long and contentious history in the philosophy of mathematics. Its legitimacy has at times been hotly disputed, usually because of its awkward or implausible consequences—for instance, it's classical equivalence to the well-ordering principle implies that there is a well-ordering of the real numbers, whose well-ordering is somewhat difficult to imagine, and it is the key to proving the theorems that lie behind things with names like "The Banach-Tarski Paradox" and "Skolem's Paradox."[9] What matters for us, though, is that the Axiom of Choice

---

[9]In standard form, AC is the claim that for any family of non-empty, disjoint sets there is a function that chooses an element from each. Most famously, this turns out to be classically equivalent to the well-ordering theorem (the claim that every set can be well ordered), and Zorn's Lemma. But it is also equivalent to the upward and downward Löwenhiem-Skolem Theorem, and to the claim that every onto function has a "section" (*i.e.*, "epis split"): that is, if $h : A \to B$ is an onto function, there is a function $s : B \to A$ such that $h \circ s$ is the identity function on B. There are weaker choice principles that are also much studied, and which also come in classically equivalent families: König's Lemma is equivalent to the completeness of first order logic, which is equivalent to the Prime Ideal Theorem, *etc.* Many of the principles which are equivalent in classical set theory

is an existence principle; depending on formulation, it might assert the existence of functions or of sets (*e.g.* a set that includes a single member from each of a family of sets). But it is an "existence principle" in the sense of asserting that for each of one sort of thing that exists, there is a thing of another sort that exists, too. At the risk of too much repetition, it was precisely this existential import that marked the version of Russell's axiom of choice off as properly mathematical, and so not logical, when people point to the need to invoke it as a sign of the failure of *Principia Mathematica* to achieve its logicist goals.[10]

4 Choice Principles and Classical Logic

We turn, at last, to the technical results. The results in question are all relatives of a result that has been known for a while: that in intuitionistic set theory and related mathematical systems, the axiom of choice implies the law of excluded middle (and hence all of classical logic).[11] That the axiom of choice in a constructive setting implies the law of excluded middle, and so all of classical logic, is often called Diaconescu's Theorem. Diaconescu's original proof was in the context of Topos Theory, and so required somewhat formidable mathematical machinery to formulate and explain. It has since become clear that the heavy machinery is not necessary to get this result. Starting in the mid-1990s with the work of John Bell [1, 2], more illuminating versions of and variations on this result began to appear in the philosophical literature.

Let's look first at the "stripped down" version of the proof of Diaconescu's The-

---

are not equivalent in intuitionistic systems.

[10]It is not uncommon to hear it claimed that the Axiom of Choice is, in fact, a principle of constructive logic, since its truth follows from the meaning of the constructive existential quantifier. ("A choice is implied in the very meaning of existence," as Bishop and Bridges say in *Constructive Analysis*.) Indeed, in the early 1990s there were two very different research programs travelling under the name "intuitionistic type theory," in one of which the axiom of choice implied classical logic while in the other the axiom of choice was said to be a principle of constructive logic. It would take us too far afield to review this fascinating history here. See [17, 18] for discussion. While we would contend that the principle that goes by the name AC in the constructive systems where it is said to be logically valid doesn't really deserve the name, the key point for the present is that the case made for calling that principle logical involves showing that it does not have existential import in the relevant sense (*i.e.*, the claim is that the existence of the choice function is implied by the truth of the existential claim in the antecedent because the existential quantifier requires the existence of a "witness" for its truth).

[11]While true, this claim hides some hedging in the "related systems" clause. As noted in an earlier footnote, some systems of constructive mathematics can't be counted as related systems since in them a version of AC is valid—unless, of course, one argues instead that the valid principle itself isn't really the Axiom of Choice. We steer clear of this debate for present purposes.

orem presented in [1]. One virtue of this version of the theorem for present purposes is that it removes complications about the specific version of intuitionistic set theory or the axiom of choice in question—those versions where something called "choice" holds in an intuitionstic set theory without implying excluded middle somehow do not satisfy the assumptions of the theorem. A second virtue it will ease our transition from the set theoretic to the logical context.

Theorem 4.1 (Diaconescu's Theorem). *The core of the argument*

Proof. Assume the following:

> (1) There are two terms c and d such that $\vdash c \neq d$
>
> (2) For any A, we can find an s and t, such that:
>
>> (a) $\vdash A \rightarrow s = t$
>> (b) $\vdash (s = c \vee A) \wedge (t = d \vee A)$

We can then reason as follows:

$$\vdash (s = c \wedge t = d) \vee A \qquad\qquad\qquad \text{(distributivity)}$$
$$\vdash (s \neq t) \vee A \qquad\qquad\qquad\qquad\qquad \text{(from 1)}$$
$$\vdash (s \neq t) \rightarrow \neg A \qquad\qquad \text{(2 (a), contraposition)}$$
$$\vdash A \vee \neg A$$

□

Since the validity of excluded middle is enough to make all of classical logic valid, this proof provides us with an easy way to show that a particular intuitionistic theory is powerful enough to prove all the principles of classical logic—we need only show that it allows us to prove conditions (1) and (2) from theorem 4.1.

Consider, for instance, why this should be expected to hold in an intuitionistic set theory.[12] If we assume that we have the Axiom of Choice, and assume that sets and functions behave in what will strike classical mathematicians as a natural way, then we have LEM.

Theorem 4.2 ("Intuitionisitic set theory" plus (ε) implies LEM).

---

[12] We do not present this discussion in terms of any particular intuitionistic set theory, instead simply flagging important assumptions that should be familiar to anyone with a passing acquaintance with classical set theory. Once again, we justify this approach as a way to avoid getting bogged down in details that do not advance the narrative.

Proof. Clearly, in any reasonable set theory we'll have $\vdash 0 \neq 1$, so $c \neq d$ is very easy. To get (2), we choose a y not free in A and define $B(y)$ to be $A \vee y = 0$ and $C(y)$ to be $A \vee y = 1$. This defines two non-empty subsets of $\{0, 1\}$, namely

$$z = \{y \in \{0, 1\} | B(y)\}$$

and

$$w = \{y \in \{0, 1\} | C(y)\}.$$

Since we assume the axiom of choice, let f be a choice function on the power set of $\{0, 1\}$. Then $f(z)$ and $f(w)$ will serve as the terms s and t in (2). For if A is true then by extensionality $z = w$, and so since f is a function we have $A \rightarrow f(z) = f(w)$. Moreover, since $\vdash f(z) = 0 \vee f(z) = 1$ and $f(z) = 1 \rightarrow A$, we have $\vdash f(z) = 0 \vee A$, and similarly $\vdash f(w) = 1 \vee A$, so we have (b) as well. $\square$

Note that we didn't require the full power of the Axiom of Choice to get s and t, only a choice function on $\mathcal{P}(\{0, 1\})$. There are two obvious lessons in this fact. First, we might expect other principles weaker than the Axiom of Choice to give us the required terms. Secondly, the existence of a choice function on the power set of a two-element set cannot be the same triviality in intuitionistic set theory that it is in classical set theory—where the only non-empty subsets are $\{0\}$, $\{1\}$ and $\{0, 1\}$ after all—since there are perfectly good intuitionistic set theories in which the law of excluded middle does not hold, even though all the other elements of the proof just sketched are in place.

For the purposes of drawing metaphysical lessons, though, it will be helpful to move from the mathematical to a more straightforwardly logical setting. The tool that will allow us to do is is Hilbert's ε-operator. Loosely speaking, the ε–operator adds, for each predicate Φ of a language, a new term $\varepsilon x \Phi$ to the language, one in which x does not occur free.[13] This makes ε a (variable-binding) term-forming operator of a familiar sort, similar to a definite description operator, for instance. What distinguishes one such operator from another are the logical rules governing them. The logical rules for ε are give by *the epsilon axiom*, which is the following scheme:

$$\exists x \Phi(x) \rightarrow \Phi(\varepsilon x \Phi(x)), \quad \text{for all } \Phi(x). \tag{ε}$$

A moment's reflection will make clear why ε is sometimes called a "logical choice function," and so one might expect that it would be a useful tool for translating

---

[13]More precisely, a clause to this effect needs to be added to the recursive definition of the well-formed expressions of the language, because we want to allow for the presence of ε–terms in the formulas from which new such terms are formed.

facts about the axiom of choice from a mathematical to a logical setting. As we shall see, it allows us to do more than that.

It will be useful for what follows if we specify (and in some cases re-introduce) some terminology. We will use "IPC" to refer to the intuitionistic predicate calculus (with identity), and will often refer to the addition of the epsilon axiom to a theory using locutions such as "with (ε)" or "+ (ε)." The logical theory that results from adding (ε) to IPC we call the *intuitionistic epsilon calculus*, and we sometimes designate it as "IPC(ε)." We will often have occasion to refer to *Ackermann's extensionality principle* as "(Ack)."

$$\forall x(\Phi(x) \leftrightarrow \Psi(x)) \rightarrow \varepsilon x \Phi = \varepsilon x \Psi, \tag{Ack}$$

IPC(ε) + (Ack) we will refer to as the *extensional intuitionistic epsilon calculus*.

Theorem 4.3 (The extensional intuitionistic epsilon calculus + the existence of two provably distinct individuals implies LEM). *Let T be a theory in the extensional intuitionistic epsilon calculus in which we can prove $c \neq d$. Then $T \vdash LEM$.*

Proof. For the present, we write $\vdash$ for $T \vdash$. Recall that, according to theorem 4.1, to prove LEM it suffices that the following hold:

(1) $\vdash c \neq d$

(2) For any A, we can find an s and t, such that:

　(a) $\vdash A \rightarrow s = t$
　(b) $\vdash (s = c \vee A) \wedge (t = d \vee A)$

We have assumed that (1) holds.

To establish that we also have (2), first, for any A, choose a variable y not free in A and define:

$$B(y) \equiv (A \vee (y = c)) \text{ and } C(y) \equiv (A \vee (y = d))$$

Let $\varepsilon y B(y) = s$ and $\varepsilon y C(y) = t$. Since obviously $\vdash \exists x B(x)$ and $\vdash \exists x C(x)$, using the epsilon axiom we readily derive:

$$\vdash (s = c \vee A) \wedge (t = d \vee A),$$

that is we get (2b).

To see that we also have (2a), note that $\vdash A \rightarrow \forall y(B(y) \leftrightarrow C(y))$, so by (Ack) we have that $\vdash A \rightarrow s = t$.

The result follows by theorem 4.1. □

The parallel between this proof and the proof of the preceding proof that the axiom of choice implies excluded middle in intuitionistic set theory is obvious. There are virtues, though, in the present proof. It makes clear, for instance, the role that the assumption of extensionality plays in the proof of excluded middle. In a set theoretic context, especially if one's familiarity with set theory is based on the classical versions, appeals to extensionality hardly need to be noticed. We shall return to this point in our philosophical discussion below.[14]

Thus the addition of (Ack) and (ε) to first-order intuitionistic logic is non-conservative in a most striking way (in any situation in which there are provably distinct objects). However, even the addition of (ε) without (Ack) is non-conservative in the sense that in intuitionistic logic with (ε) we can prove ε–free formulas we cannot prove in ε–free intuitionistic logic. For instance, it is easy to see that the (ε) principle implies the validity of the scheme

$$\exists x(\exists y \Phi(y) \rightarrow \Phi(x)), \tag{†}$$

which is not provable in the usual formulations of intuitionistic logic. This is a striking contrast to the classical case, where Hilbert's 'Second ε–Theorem' tells us that adding (ε) to classical first-order logic is a conservative extension. So, summarizing roughly, ε and extensionality is dramatically non-conservative, but ε alone is (less dramatically) non-conservative.

Of course, one striking difference is that in the presence of (Ack) and the modest assumption that two provably distinct entities exist we make valid both additional quantifier laws and additional propositional principles, while our only example of an additional principle made valid by epsilon alone is a quantificational law. For the philosophical discussion to follow it is interesting that ε without extensionality also implies new propositional laws. To get the result we again need some assumptions. We continue to assume the existence of two provably distinct objects, and while we no longer assume (Ack) we replace it with the assumption that one of the terms is "decidable," i.e. that $\forall y(y = c \vee y \neq c)$. With these assumptions we can no longer prove excluded middle, but we *can* prove important superintuitionistic principles, including the intuitionistically invalid De Morgan's law, $\neg(A \wedge B) \rightarrow (\neg A \vee \neg B)$.

---

[14] But it is perhaps worth noting immediately that in constructive mathematical settings in which something called "Choice" is provable, its failure to imply excluded middle can often be traced to some failure of extensionality. See, for instance, [19, 20].

Indeed, we can prove the stronger principle sometimes called "Linearity" or "Dummett's scheme,"

$$(P \to Q) \vee (Q \to P),$$

(LIN)

from which DeMorgan's law follows.

Theorem 4.4. *IPC(ε) plus two provably distinct objects, one decidable, implies LIN.*

Proof. Assume:

(1) $\vdash c \neq d$
(2) $\vdash (\forall x)(x = c \vee x \neq c)$

Now, chooosing an x free in neither P nor Q, we define:

$$A(x) \equiv (P \wedge x = c) \vee (Q \wedge x \neq c)$$

(*)

We have:

$$A(c) \leftrightarrow P, \text{ and } x \neq c \vdash A(x) \to Q$$

Since: $(\exists x)A(x) \leftrightarrow P \vee Q$, we have:

$$P \vee Q \leftrightarrow A(\varepsilon x A(x))$$

and by (2) we have:

$$[(P \vee Q) \to (A(\varepsilon x A(x)) \wedge \varepsilon x A(x) = c)] \vee [(P \vee Q) \to (A(\varepsilon x A(x)) \wedge \varepsilon x A(x) \neq c)]$$

by the definition of $A(x)$ and (*) we have:

$$((P \vee Q) \to P) \vee ((P \vee Q) \to Q)$$

and so:

$$((P \to P) \wedge (Q \to P)) \vee ((Q \to Q) \wedge (P \to Q))$$

Simplifying we have:

$$(Q \to P) \vee (P \to Q)$$

$\square$

Of course, as it stands these proofs only show that epsilon plus these other conditions are *sufficient* to get these results, not that they are necessary. To show necessity, we want a semantics for intuitionistic epsilon calculus. Unfortunately, this can be a tricky business. We will therefore satisfy ourselves with a few simple remarks, and point readers to, for instance, [3] for details.

In, for instance, [21], a very simple form of semantics is employed for the classical ε-calculus. We add a choice function f to each interpretation, and $\varepsilon x A(x)$ is interpreted by whatever f chooses from the "truth set" for $A(x)$, *i.e.* the set of elements of the domain that make $A(x)$ true when x is assigned to them under the interpretation in question; if the truth set is empty, then the epsilon term gets assigned to an arbitrary but fixed element of the domain. This is problematic in the intuitionistic case for several reasons.

First, it is not hard to see that such an approach will make (Ack) come out valid. In the classical case this arguably doesn't matter very much, since ε, with or without (Ack), is conservative over classical logic. We have seen, though, that ε + (Ack) is as far from conservative over intuitionistic logic as anybody is going to want to go.

A second problem is that intuitionistic logic cannot have a bivalent semantics. Whatever semantics we use for intuitionistic logic, one way or another we are going to have to confront the prospect that many formulas (for a given interpretation) are not "completely true" nor are they "completely false." Since all such formulas will have the same "truth set" as, for instance, $P(x) \wedge \neg P(x)$, namely $\emptyset$, they will all have the same object as the referent of their ε term.

Relatedly, to get the ε principle to come out valid, we need to ensure that for each φ, the truth value of $\varphi(\varepsilon x.\varphi)$ is always equal to the truth value of $\exists x \varphi$. It is easy to see how the Leisenring semantics can ensure this in the classical case, since there is a sufficient supply of saturated models in classical predicate logic. We can therefore restrict attention to interpretations under which if $\exists x \varphi$ is true, then there is some element of the domain d that makes $\varphi(x)$ true when x is interpreted as d, *i.e.*, by restricting attention to interpretations where φ's truth set is non-empty. In the intuitionistic case we obviously can't restrict attention to truth sets, given what was said in the preceding paragraph, but we do need to ensure that some element d gives $\varphi(x)$ the same truth value as $\exists x \varphi$, *i.e.*, we need to ensure that the "as true as possible" set is non-empty.

Solving these problems in detail is messy. The approaches we will focus on are built on a standard algebraic semantics for intuitionistic logic. The basic idea is this: interpretations of predicates in classical logic take them to be "propositional functions" in the sense of taking tuples of members of the domain of interpretation into the set $\{0, 1\}$. But the standard truth tables for the classical $\wedge$, $\vee$ and $\neg$ operations correspond exactly to the algebraic operations of meet, join and complement if we set $0 < 1$ and consider this a two element Boolean algebra. Algebraic semantics, generally speaking, starts with the question "what's special about the two element Boolean algebra?" If we allow our interpretations to be other Boolean algebras, we get *Boolean valued semantics*, but this turns out not to change which principles count as valid. But if we allow *other types of algebras* besides Boolean ones, we get more interesting, non-classical logics. In particular, if we allow the algebra of truth values to be Hetying algebras (of which Boolean algebras are a special case), we have a semantics for intuitionistic logic.[15]

Bell solves the third problem by restricting attention to interpretations under which the algebra of truth values is an inversely well-ordered set—that is, every subset of the set of truth values has a maximal element. The result is a sound but not complete semantics for intuitionistic ε calculus. It allows him to prove several interesting independence results, including that while ε + (Ack) implies the law of excluded middle, ε alone does not. His relatively simple semantics also makes (Ack) turn out valid. To get a non-extensional semantics, [3] makes the value of $\varepsilon x \varphi$ depend not only on the truth values φ takes when the various members of the domain are used to interpret x, but also on the syntax of φ.

We do not need to pursue the details here. For our purposes it is enough to note that this sort of semantics allow us to demonstrate independence results that establish that the proofs above don't just give us sufficient conditions for, *e.g.*, deriving excluded middle, but that the various suppositions in the proofs each play an essential role (*e.g.* Theorem 4.3 doesn't go through without (Ack)). Thus, the algebra of truth values for ε + (Ack) + two-provably-distinct-objects must be a Boolean algebra, while ε plus two provably distinct objects and one decidable object assures that the truth values form and L-algebra, *i.e.*, a Heyting algebra in which $(a \rightarrow b) \vee (b \rightarrow a) = 1$ for all a, b.

The most obvious examples of non-Boolean L-algebras are chains; if a chain has more than two members, it is Heyting but not Boolean. However there are other more interesting examples, Horn, for example, constructs an L-algebra that is neither Boolean nor a simple linear ordering by considering a lattice composed of a selection of infinitely long sequences of 0s, $\frac{1}{2}$s and 1s compared component-wise [24, p.404]. We can construct a much simpler lattice of ordered pairs to illustrate the basic idea

---

[15]A Heyting algebra is sometimes defined as a Brouwerian Lattice with a bottom element. A Brouwerian Lattice, or implicative lattice, is a lattice with relative pseudo-complementation. However the terminology is not uniform, in some of the literature a Brouwerian Algebra is taken to mean the same thing as a Co-Heyting Algebra, the dual of Heyting Algebra. Heyting Algebras, Co-Heyting Algebras, and Brouwerian Algebras are all also referred to collectively as Pseudo-Boolean Algebras. For a comprehensive explication of Heyting algebras see [22, pp.58*ff*.] or [23, pp.33*ff*. and pp.128*ff*.]

Horn expands on. Consider a algebra of ordered pairs compared component wise (*i.e.*, $\langle a, b \rangle \leq \langle c, d \rangle$ iff $a \leq c$ and $b \leq d$). The pairs in question include all elements of $(\{1, 2, 3\} \times \{1, 2, 3\}) \cup \{\langle 0, 0 \rangle\}$. Clearly, the bottom element $\bot = \langle 0, 0 \rangle$, while the top $\top = \langle 3, 3 \rangle$.

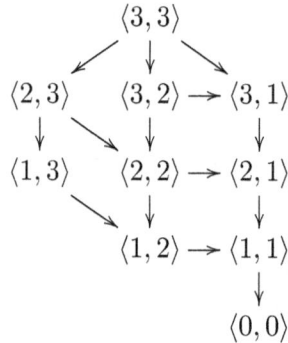

$$
\begin{array}{ccccc}
& & \langle 3, 3 \rangle & & \\
& \swarrow & \downarrow & \searrow & \\
\langle 2, 3 \rangle & & \langle 3, 2 \rangle & \rightarrow & \langle 3, 1 \rangle \\
\downarrow & \searrow & \downarrow & & \downarrow \\
& \langle 1, 3 \rangle & \langle 2, 2 \rangle & \rightarrow & \langle 2, 1 \rangle \\
& \searrow & \downarrow & & \downarrow \\
& & \langle 1, 2 \rangle & \rightarrow & \langle 1, 1 \rangle \\
& & \downarrow & & \\
& & \langle 0, 0 \rangle & &
\end{array}
$$

For every pair $x \neq \bot$, $\neg x$ is $\bot$ (while $\neg \bot = \top$), and yet $x \vee \neg x = x$ (*e.g.* $\langle 2, 3 \rangle \vee \neg \langle 2, 3 \rangle = \langle 2, 3 \rangle$) and for any two pairs x, and y we get $(x \rightarrow y) \vee (y \rightarrow x) = \top$.[16] We shall discuss such L-algebras and their philosophical interest briefly below.

5   More philosophy

We return now to a more explicitly philosophical discussion, trying (briefly) to make good our suggestion that there are metaphysical lessons in these formal results—both lessons for how to fill in some gaps in Dummett's story linking metaphysics to logical principles and more general lessons for those not persuaded of the details of Dummett's account.

Let us draw together some strands of the discussion. First, we will recall some key features of (our potted version of) Dummett's account. Intuitionistic logic is logic properly-so-called, and so is metaphysically neutral in the sense that everyone should accept it, regardless of their metaphysical commitments. Superintuitionistic principles, if justified, must be justified on extra-logical grounds. Dummett suggests

---

[16] For example consider two non-comparable elements $\langle 2, 3 \rangle$ and $\langle 3, 2 \rangle$ of the algebra presented above:

$$
(\langle 2, 3 \rangle \rightarrow \langle 3, 2 \rangle) \vee (\langle 3, 2 \rangle \rightarrow \langle 2, 3 \rangle) = \bigvee \{x | x \wedge \langle 2, 3 \rangle \leq \langle 3, 2 \rangle\} \vee \bigvee \{y | y \wedge \langle 3, 2 \rangle \leq \langle 2, 3 \rangle\}
$$
$$
= \langle 3, 2 \rangle \vee \langle 2, 3 \rangle
$$
$$
= \langle 3, 3 \rangle
$$

that these grounds, if they ever are available, will be metaphysical ones, *i.e.* reason for believing that the truth-values of the statements in some domain are fixed by reality in some way that is suitably "independent of us." In such cases, the mind- and language-independent reality will justify a commitment to bivalence, and so to classical logic.

What the formal results above show is that there are ways to make the connection between metaphysical commitment and logical principles less nebulous. Choice principles, and in particular the epsilon principle, are metaphysical assumptions, because they encode claims about conditions under which we can assert the existence of "objects" of some sort, and we can see from the results that these lead fairly directly to the validity of superintuitionistic principles. This strikes us as less metaphorical than the detour through "mind-independent reality fixing truth values," and so already as more philosophically illuminating. But the proofs that show the role of an extensionality assumption in getting all of classical logic, while weaker assumptions lead us to superintuitionistic but non-classical systems, allow us to make connections between metaphysical assumptions and logical principles that give interesting ways of seeing that the question of realism is not an all-or-nothing thing.

As a preliminary step, consider what the ε axiom says. Recall that when teaching classical logic, it's not uncommon to have to try to explain what $\exists x(\exists y A(y) \to A(x))$ is saying, by way of trying to convince students that it's not crazy that it's valid. A common way to do so is to use examples like "suppose $A(x)$ means 'x will pass the test'; then the formula is saying that there is someone who will pass the test if anyone does." And, indeed, this is what the ε-axiom says: that for any property there is an object which is *the likeliest thing* to have the property, or perhaps the $\varepsilon x A(x)$ is the *paradigm example* of the As. As noted, this is already a constructively invalid principle, for interesting reasons that we cannot pursue here. (See [18] for discussion.) For the present, it is more important to ask: do our intuitions about when we find it reasonable to think that *there is always a likeliest* and when we don't track our thoughts about the reality or objectivity of the subject under discussion? We think it does. We will not argue for the claim, but only offer what we hope are some suggestive comments. It is no accident that we use an example like "will pass the test if anyone does" because in most classes there is a student or a small group of students who are more diligent in preparing for tests, and diligence is a good predictor of success on tests. On the other hand, there are reasons we don't instead use examples like "suppose $A(x)$ means 'x will win the lottery,'" as there is no reason to think in advance of the draw that there is (already) someone who is will win if anyone does, for if we think the lottery is fair we don't think there is any fact grounding such a claim. We think similar intuitions can be generated for other standard examples where antirealist intuitions are reputed to be especially common—is there really a "likeliest to be funny" joke?

What, then, do we make of theorem 4.3? It shows that in domains where we not only assume that every property is such that some object is likeliest to have it, but that that "likeliest to have" is determined *extensionally* (and there are at least two provably distinct objects), then we have classical logic. For instance, if the students who pass are precisely the students who study, then the likeliest-to-pass and the likeliest-to-study will be the same student. If Dummett is right about the link between realism and classical logic, this result shows us that discourses in which we have grounds to believe that all the properties come with extensionally-determined "likeliest" objects are ones about which we have reason to accept realism. In this connection, it is worth noting that it is actually not news to think that there is a link between extensionality and *objectivity*, a notion clearly important to our thinking about realism. Famously, in the middle of the past century the need for the grammar of our attributions of intentional states to be non-extensional was regarded by some as reason to question the appropriateness of intentional states for inclusion among the features of the world apt for scientific description. But there are examples that are both homier and more current. When teaching decision theory, it is important to draw students' attention to reasons for doubting whether "preference functions" are tracking something real. Would you prefer chocolate or broccoli? Would you prefer something that will give you a heart attack or broccoli? Since the answers to such questions depend on the description and not just what is described, we have reason for scepticism about whether the answers about what someone prefers are "objective" or not.

What do we get if we (continue to assume that there are two distinct objects and) remove extensionality while assuming that one of the objects is "decidable"? We no longer have classical logic, and so according to the Dummettian account we must accept some sort of antirealism for the domain in question. But we think this intermediate way-station is one which gives us grounds for saying "well, maybe not realism, but not really antirealism either."

Consider the models as described in the previous chapter, for the "shape" of the "algebras of truth-values" can provide us with some idea how the properties in a domain must behave. As noted, the obvious L-algebras are linear. While the two-valued L-algebra case is precisely the one Dummett pointed to as encoding realist assumptions, there is some reason to regard any situation in which the truth-values of claims are arranged linearly as one where something "objective" is in question. For it is natural in such cases to think in terms of "degrees of truth,"[17] so for instance

---

[17]Though one needs to be cautious not to transfer over ideas from other discussions where that

for any pair of objects there will be a fact of the matter about which of them has any property P to the greater extent. But more complex L-algebras suggest other interesting possibilities to do with "multi-dimensional" properties, where each dimension is "objective," but taken together they generate a property with an in-between status, as reflected in our example of a non-linear L-algebra.

Intelligence, for instance, might be this sort of thing. While not uncontroversial, it is common to hear people speak of intelligence as having various dimensions.[18] Maybe both culinary smarts and strategic ability are real things, and each is part of what we mean by "intelligence." And perhaps Yotam Ottolenghi has more culinary smarts, but less strategic ability, than Magnus Carlsen. In that case, perhaps there is just no answer to the question of which of the two is smarter—to be smarter means being at least as smart on every dimension and smarter on some. Both, though, might be smarter than the present authors, having both more strategic sense and more culinary ability than we do. (Of course, there might be other dimensions that are part of intelligence on which we can pin our hopes for blunting this judgment: we couldn't be *more intelligent* than those two, but were we to rank ahead on another dimension we could at least be judged *non-comparable* with them with respect to intelligence, rather than less intelligent.)

If we accept that domains in which superintuitionistic principles are valid are ones in which some, so-to-speak, realistically-inclined metaphysical presuppositions are legitimate, and if the discussion above shows that discussions of intelligence are such a domain, we should regard intelligence in ways different from how we judge discourses where realism is truly implausible—for instance, perhaps, humour or beauty. And yet we should not regard it in the same way in which we regard domains about which we are fully realists, either. This strikes us as very much how intelligence *is* regarded by those who defend multi-dimensional views. Critics of the view, on the other hand, often argue that there is a single factor ("general intelligence") that underlies strong performance in any dimension, and so that the apparent multi-dimensionality is an illusion—in effect, arguing that reducing it to a single dimension is to show that intelligence is a "real thing."[19]

We think this discussion does a few useful things for discussions of realism and antirealism in the Dummettian tradition. First, as noted, it puts some additional

---

phrase is used; for instance, for all formulas P with a non-$\perp$ truth value in a linear L-algebra, the truth value of $\neg P$ is $\perp$, rather than 1 minus the truth-value of P, as in probability semantics, for instance.

[18]We deliberately choose somewhat flippant "dimensions" rather than opting for some seriously offered, for instance, by advocates of Multiple Intelligence theories. We do not intend to be wading into this debate, merely using obvious aspects of what is presumably a familiar example to most readers.

[19]There is more to be said here. [25] contains a fuller discussion.

flesh on the bones of Dummett's suggestion that metaphysical commitments can give rise to commitment to logical principles. It also gives us good reason to say that the question of realism is not an all-or-nothing thing—intelligence may not be as objective as mass, but it's not as ephemeral as humour, either, even if the Multiple Intelligence folks are right. Finally, this sort of discussion can help bring the discussion of realism and antirealism down a couple of levels of abstraction. Rather than discussing things like what logic applies to a "domain of discourse" (whatever that is), this approach offers a way to discuss relatively familiar concepts and to see what is at issue in ways that reflect debates as we actually see them occurring between theorists we can actually see debating the status of those familiar concepts.

6   Conclusion

Of course, there are other ways we can get from intuitionistic to classical logic than via choice principles, or indeed to get part way from one to the other. Indeed, there are other ways to make the transition using variants on ε (such as the one encoding "dependent choice" included in [2]), or using different term-forming operators. What such formal results offer, we think, is a variety of pathways for investigating relationships between metaphysical commitments and principles of reasoning. By focusing on just a few such results we've tried to sketch one way one might try to spell out a strategy for showing how realist commitments imply logical principles, but it is just one among many. One thing that we hope attention to such matters would do is put some flesh on the bones of the Dummettian suggestion that realists about a particular domain must commit themselves to at least one of these pathways.

References

[1] J. Bell, "Hilbert's ε-operator and classical logic," *Journal of Philosophical Logic*, vol. 22, pp. 1–18, 1993.

[2] J. Bell, "Hilbert's ε-operator in intuitionistic type theories," *Mathematical Logic Quarterly*, vol. 39, pp. 323–337, 1993.

[3] D. DeVidi, "Intuitionistic ε- and τ-calculi," *Mathematical Logic Quarterly*, vol. 41, pp. 523–546, 1995.

[4] R. Diaconescu, "Axiom of choice and complementation," *Proceedings of the American Mathematical Society*, vol. 51, no. 1, pp. 176–178, 1975.

[5] N. D. Goodman and J. Myhill, "Choice implies excluded middle," *Zeitschrift fur Mathematische Logik und Grundlagen der Mathematik*, vol. 24, p. 461, 1975.

[6] D. Gabbay and L. Maksimova, "Interpolation and definability," in [26], Springer, 2011.

[7] M. Dummett, *The Seas of Language*. Clarendon Press, Oxford, 1993.

[8] M. Dummett, "Realism and anti-realism," in *Seas of Language*, pp. 462–472, Clarendon Press, Oxford, 1993.

[9] J. Mackie, *Ethics: Inventing Right and Wrong*. Penguin UK, 1977.

[10] H. Field, *Science Without Numbers: The Defence of Nominalism*. Princeton University Press, 1980.

[11] D. DeVidi and G. Solomon, "On confusions about bivalence and excluded middle," *Dialogue: Canadian Philosophical Review*, vol. 38, pp. 785–899, 1999.

[12] M. Dummett, *The Logical Basis of Metaphysics*. Cambridge, Mass.: Harvard University Press, 1991.

[13] G. Rosen, "Review article: The shoals of language," *Mind*, vol. 104, pp. 599–609, 1995.

[14] B. Russell, *Introduction to Mathematical Philosophy*. London: George Allen and Unwin, 1919.

[15] R. Carnap, *The Logical Syntax of Language*. London: Routledge and Kegan Paul, 1937. transl. by A. Smeaton.

[16] M. Dummett, *Frege: Philosophy of Mathematics*. Cambridge, Mass.: Harvard University Press, 1991.

[17] D. DeVidi, "Choice principles and constructive logics," *Philosophia Mathematica*, vol. 12, pp. 222–243, 2004.

[18] D. DeVidi, "Assertion proof and the axiom of choice," in [27], Springer-Verlag, 2006.

[19] M. Maietti, "About effective quotients in constructive type theory.," in *Types for Proofs and Programs, International Workshop "Types 98"* (T. Altenkirch, W. Naraschewski, and B. Reus, eds.), no. 1657 in Lecture Notes in Computer Science, pp. 164–178, Springer-Verlag, 1999.

[20] M. Maietti and S. Valentini, "Can you add power-set to Martin-Löf intuitionistic type theory?," *Mathematical Logic Quarterly*, vol. 45, pp. 521–532, 1999.

[21] A. Leisenring, *Mathematical Logic and Hilbert's ε-Symbol*. New York: Gordon & Breach Science Publishers, 1969.

[22] H. Rasiowa and R. Sikorski, *The Mathematics of Metamathematics*. Monografie Matematyczne, Państwowe Wydawn. Naukowe, 1963.

[23] B. Davey and H. Priestly, *Introduction to Lattices and Order*. Cambridge University Press, second ed., 2002.

[24] A. Horn, "Logic with truth values in a linearly ordered Heyting algebra," *Journal of Symbolic Logic*, vol. 34, pp. pp. 395–408, 1969.

[25] C. Mulvihill, *Existence Assumptions and Logical Prinicples: Choice Operators in Intuitionistic Logic*. PhD thesis, University of Waterloo, 2015.

[26] D. Gabbay and F. Guenthner, eds., *Handbook of Philosophical Logic*, vol. 15. Kluwer Academic Publishers, 2011.

[27] D. DeVidi and T. Kenyon, *A Logical Approach to Philosophy Essays in Honour of Graham Solomon*. Springer-Verlag, 2006.

Received 23 October 2015

A Proof Theory for First-Order Logic with Definiteness

Nissim Francez

Computer Science dept., the Technion-IIT, Haifa, Israel

`francez@cs.technion.ac.il`

Bartosz Więckowski

Institut für Philosophie, Goethe-Universität, Frankfurt am Main

`wieckowski@em.uni-frankfurt.de`

Abstract

The paper presents a proof-theory (in the form of a natural-deduction (ND) proof-system) for *definiteness*, expressed as the ι-subnector, extending 1-st order logic. The ND-system is proposed as *meaning conferring* and is shown to qualify as such by being *harmonious* and *stable*. Some relationship of this proof-theory with the *presupposition* of definiteness is pointed out.

Keywords: Definiteness, iota operator, proof-theoretic semantics, harmony

1 Introduction

The purpose of this paper is to provide a proof-theory (in the form of a natural-deduction (ND) proof-system) for *definiteness*, expressed as the ι-subnector[1] extending 1-st order logic (FOL). This ND-system is shown to have certain advantages over previous such systems in the literature. In Section 4 we compare our proposed ND-system for definiteness with some other proposals of such ND-systems.

We first explain the definiteness problem itself. We chose the interpretation of definiteness following Russell's contextual definition. We assume the usual object language for *FOL* with identity and individual constants (but without function symbols), with the usual definition of free/bound variables. Some of the material in this paper was presented in [2].

We thank Steven Kuhn for a discussion of nested definiteness, and Koji Mineshima for discussing presupposition in proof-theory. The second author gratefully acknowledges support by DFG-grant WI 3456/2-1.

[1]A *subnector* is an operator that turns an open formula to a variable binding term.

2 Definiteness

We start from Russell's contextual definition of definiteness in [9] (in his famous example[2] (2.1))

The (current) king of France is bald   (2.1)

consisting of three components:

1. **existence**: in the example, there is at least one (current) king of France.

2. **uniqueness**: in the example, there is at most one (current) king of France.

3. **predication**: in the example, the only (current) king of France is bald.

This naturally generalizes to a general statement about the only φ being ψ, usually expressed as

$$\exists x(\varphi(x)\wedge\forall y(\varphi(y)\rightarrow y=x)\wedge\psi(x))$$ (2.2)

For later use, we find it useful to consider instead the following equivalent formulation.

$$\exists x.\varphi(x)\wedge\forall u.\forall v.\varphi(u)\wedge\varphi(v)\rightarrow u=v\wedge\forall w.\varphi(w)\rightarrow\psi(w)$$ (2.3)

(avoiding a quantifier scoping over the conjunction). The actual formulation which we use is yet another equivalent formulation, as follows.

$$\exists x.\varphi(x)\wedge\psi(x)\wedge\forall u.\forall v.\varphi(u)\wedge\varphi(v)\rightarrow u=v$$ (2.4)

Here the existential quantifier can be presented as a binary, *restricted* quantifier. So, we extend FOL by means a term-forming operator (subnector) $\iota x.\varphi(x)$, where each of (2.2), (2.3) and (2.4) is expressed as the formula

$$\psi(\iota x.\varphi(x))$$ (2.5)

where, at a first stage, ψ is assumed to be a unary atomic[3] predicate, and φ is assumed to have exactly one free variable[4], say x, and all free occurrences of x in φ are bound by ιx. As usual, bound variables can be renamed, and $\iota x.\varphi(x)$ is the same as $\iota y.\varphi(y)$.

---

[2] All natural language examples are displayed in san-serif font and are always mentioned, not used.

[3] In principle, one may have ψ as any unary predicate, not necessarily atomic. The atomicity assumption does make the presentation technically simpler.

[4] One can extend the theory also to φ containing *no* free occurrences of x, like vacuous quantification in FOL; we shall not bother doing so here.

Thus, the regimentation of (2.1) in this extended language is expressed as

$$B(\iota x.K(x)) \tag{2.6}$$

with the obvious interpretation of the predicate symbols used. Denote this extension of FOL by FOL_ι.

Under the model-theoretic interpretation of this subnector, when (2.3) holds, such a subnector term is *proper*, a referring expression (denoting an element in the domain of the model). However, there is a problem when such a term is *improper*, i.e., when (2.3) does not hold, in which case the term cannot refer without some further stipulations. As we shall see below, under our proof-theoretic meaning definition, the issue of being improper never arises. Whenever $\psi(\iota x.\varphi(x))$ is asserted, its introduction rule (I-rule) guarantees that (2.3), being the premises of this I-rule, holds. The semantic focus shifts from referentiality to grounds for assertion. We consider this shift as a most important message of our presentation. This shift is elaborated upon further in Section 3.1, when relating it to *presupposition*.

Before presenting the proof-system, we first extend, in stages, the generality of the basic formulation in (2.5).

2.1 Parameterized Definiteness

The first relaxation of the above assumptions about the expression of definiteness arises from a need to regiment examples like the following.

$$\text{The king of France is bald and the king of Spain is bald} \tag{2.7}$$

For that purpose, it is clear that king cannot be considered a unary predicate symbol, but a binary one, known as a *relational noun*, expressing a relation between, in this case, a person and a country. This would render the regimentation of (2.1) as

$$B(\iota x.K(x, France)) \tag{2.8}$$

where $France$ is an individual constant standing for France. Thus, (2.7) is expressed as

$$B(\iota x.K(x, France)) \wedge B(\iota x.K(x, Spain)) \tag{2.9}$$

Now, suppose that a situation of a plague of "royal baldness" occurs, where one needs to express the fact that every country is such that its king is bald. For that, the second argument of K needs to be a variable, quantifiable from outside the ι-boundary. So, we relax the assumption that φ is unary, and extend the language to contain parameterized ι-terms such as

$$\iota x.\varphi(x, \overline{a}) \tag{2.10}$$

where $\bar{a}$, the parameters, is a list of free variables in φ, not bound by ιx, or constants. The analogue of the Russellian contextual definition of ι-terms becomes

$$\exists x.\varphi(x,\bar{a}) \wedge \forall u.\forall v.\varphi(u,\bar{a}) \wedge \varphi(v,\bar{a}) \rightarrow u = v \wedge \forall w.\varphi(w.\bar{a}) \rightarrow \psi(w) \qquad (2.11)$$

We thus can express the royal baldness situation by

$$\forall y.B(\iota x.K(x,y)) \qquad (2.12)$$

2.2 Parallel Definiteness

Next, we transcend Russell's original definition of definiteness with what we call *parallel definiteness*. Consider the sentence

The king of France loves Marie $\qquad (2.13)$

$$L(\iota x.K(x, France), Marie) \qquad (2.14)$$

where the context is a binary relation love. This means we have to relax the assumption that ψ is unary, and let it be n-ary (for an arbitrary n, where $n = m + 1$ and $m \geq 0$). Thus, the general form of a formula with a ι-term becomes

$$\psi(\iota x.\varphi(x,\bar{a}), \bar{b}) \qquad (2.15)$$

where $\bar{b}$ is a list of m additional parameters. Its Russellian contextual definition now becomes

$$\exists x.\varphi(x,\bar{a}) \wedge \forall u.\forall v.\varphi(u,\bar{a}) \wedge \varphi(v,\bar{a}) \rightarrow u = v \wedge \forall w.\varphi(w,\bar{a}) \rightarrow \psi(w,\bar{b}) \qquad (2.16)$$

However, the story does not end here. Consider the sentence

The king of France loves the queen of Spain $\qquad (2.17)$

The regimentation of (2.17) should look like

$$L(\iota x.K(x, France), \iota y.Q(y, Spain)) \qquad (2.18)$$

In this case, one of the parameters is itself a ι-term, also requiring the imposition of existence and uniqueness, w.r.t. the same predication ψ (L in the example). This calls for the following modification of the contextual definiteness for parallel definiteness, expressed for brevity for a binary parallelism.

The contextual definition of

$$\psi(\iota x_1.\varphi_1(x_1,\overline{a}_1), \iota x_2.\varphi_2(x_2,\overline{a}_2), \overline{b}) \tag{2.19}$$

is

$$\exists x_1.\varphi_1(x_1,\overline{a}_1) \wedge \exists x_2.\varphi_2(x_2,\overline{a}_2)$$
$$\wedge \forall u_1.\forall v_1.\varphi_1(u_1,\overline{a}_1) \wedge \varphi_1(v_1,\overline{a}_1) \rightarrow u_1 = v_1 \wedge \forall u_2.\forall v_2.\varphi_2(u_2,\overline{a}_2) \wedge \varphi_2(v_2,\overline{a}_2) \rightarrow u_2 = v_2$$
$$\wedge \forall w_1.\varphi_1(w_1,\overline{a}_1) \wedge \forall w_2.\varphi_2(w_2,\overline{a}_2) \rightarrow \psi(w_1,w_2,\overline{b}) \tag{2.20}$$

The extension to more than two paralel ι-terms is obvious and left to the reader.

2.3 Nested Definiteness

The ι-operator can be nested (embedded within another ι-term). This innocently looking construct raises several problems. A thorough study of nested definiteness can be found in [5]. For example, a regimentation of

$$\text{The girl that the boy loves smiles} \tag{2.21}$$

would be

$$S(\iota x.(G(x) \wedge L(\iota y.B(y), x))) \tag{2.22}$$

with the obvious interpretation of the predicate symbols used. By the Russellian analysis, (2.22) can be interpreted as the proposition that there exists exactly one girl that is loved by exactly one boy, and that girl smiles. Note that (2.22) does not preclude there being other girls (loved by no boy or loved by more than one boy). The general form of nested definiteness is

$$\psi(\iota x.\chi(\iota y.\varphi(y,\overline{a}), x), \overline{b}) \tag{2.23}$$

See [5] for a criticism of the Russellian reading of (2.23) (and (2.21)) and a proposal of an alternative reading, claimed to be more plausible. Note that [5] does not consider n-ary ψs.

2.4 Definiteness and Scope

As Russell was aware, there is a scopal problem regarding the ι-terms. Consider the special case of

$$\neg\psi(\iota x.\varphi(x)) \tag{2.24}$$

given by

$$\neg B(\iota x.K(x)) \tag{2.25}$$

There is a scopal ambiguity in (2.25): it can be read as a regimentation of each of the following two sentences.

$$\text{The (current) king of France is not bald} \tag{2.26}$$

$$\text{It is not the case that the (current) king of France is bald} \tag{2.27}$$

Clearly, (2.26) and (2.27) are not equivalent. For example, (2.27) is true (by its Russellian reading) if there is no king of France, while (2.26) is not. To amend the situation, Whitehead and Russell (in [16]) introduced explicit scoping of the ι-term, using the following form.

$$[\iota x.\varphi(x)]\psi(\iota x.\varphi(x)) \tag{2.28}$$

This allows for (2.26) to be regimented as

$$[\iota x.K(x)]\neg B(\iota x.K(x)) \tag{2.29}$$

while (2.27) is regimented as

$$\neg[\iota x.K(x)]B(\iota x.K(x)) \tag{2.30}$$

However, this amended notation precludes the representation of nested definiteness. See [5] for a criticism of the amended notation. We will adhere here to the original notation, reading it as expressing narrow scope of negation.

As for the relative scope of ι-terms w.r.t. other quantifiers, the issue does arise. Here ι-terms behave like quantifiers regarding scope. The general form of which (2.12) is an instance is

$$\forall y.\psi(\iota x.\varphi(x, y)) \tag{2.31}$$

by which the universal quantifier on y has a higher scope than the definiteness binding x. Thus, the uniqueness of x is given per y. Hence, (2.31) differs in meaning from its "cousin"

$$\psi(\iota x.\forall y.\varphi(x, y)) \tag{2.32}$$

in which the scope of the definiteness binding on x is higher than the universal quantification on y. The instance $B(\iota x.\forall y.K(x, y))$ would mean that there is a unique person that is the king of every country, and that person is bald.

318

$$\frac{\Gamma\vdash\exists x.\varphi(x,\overline{a}) \quad \Gamma\vdash\forall u.\forall v.\varphi(u,\overline{a})\wedge\varphi(v,\overline{a})\to u = v \quad \Gamma\vdash\forall w.\varphi(w,\overline{a})\to\psi(w,\overline{b})}{\Gamma\vdash\psi(\iota x.\varphi(x,\overline{a}),\overline{b})} \ (\iota I)$$

$$\frac{\Gamma\vdash\psi(\iota x.\varphi(x,\overline{a}),\overline{b})}{\Gamma\vdash\exists x.\varphi(x,\overline{a})} \ (\iota E_1)$$

$$\frac{\Gamma\vdash\psi(\iota x.\varphi(x,\overline{a}),\overline{b}) \quad \Gamma\vdash\varphi(u,\overline{a}) \quad \Gamma\vdash\varphi(v,\overline{a})}{\Gamma\vdash u = v} \ (\iota E_2), \quad u,v \text{ fresh for } \Gamma$$

$$\frac{\Gamma\vdash\psi(\iota x.\varphi(x,\overline{a}),\overline{b})}{\Gamma\vdash\forall z.\varphi(z,\overline{a})\to\psi(z,\overline{b})} \ (\iota E_3)$$

Figure 1: I/E-Rules for single definiteness

3 An ND-System for Definiteness With ι-Terms

The standard ND-system for FOL is modified in two ways:

- It is extended with I/E-rules for ι-terms, using Gentzen's 'logistic' notation with sequents, as presented in Figure 1 for single definiteness. Refer by **pr$_1$**, **pr$_2$** to the first two premises of the (ιI)-rule, expressing existence and uniqueness. The second premise occurs also in the I-rule suggested by Hilbert and Bernays [3], discussed below, but without explicit parameterizations. Whenever not needed, parameterization will be omitted.

- The rules allowing non-introduced uses of a ι-formula (as assumptions, or premises of an E-rule), are modified so as to enforce such an introduction. This is elaborated in Section 3.1.

Note that the premises of (ιI) enforce that whenever $\iota x.\varphi(x)$ *can* be introduced, existence and uniqueness obtain. This way, the need to appeal to a free logic (see Section 4.3 below) is avoided.

For example, $\psi(\iota x.\varphi(x)\wedge\neg\varphi(x))$ cannot be introduced, since it would need as premise $\exists x.\varphi(x)\wedge\neg\varphi(x)$, not derivable in FOL from a consistent Γ. The I/E-rules for parameterized definiteness are obtained from those in Figure 1 by adding the $\overline{z}$ parametrization to every occurrence of φ in the rules.

A *general elimination* rule $GE\iota$ (with parameters omitted) is presented in (3.33).

319

$$\frac{\Gamma \vdash \psi(\iota x.\varphi(x)) \quad \Gamma, \exists x.\varphi(x), \forall u.\forall v.\varphi(u) \wedge \varphi(v) \to u = v, \forall w.\varphi(w) \to \psi(w) \vdash \xi}{\Gamma \vdash \xi} \ (\iota GE)$$

(3.33)

The three regular (ιE)-rules are easily derivable from (ιGE) (assuming the structural rule of *weakening* is present). GE-rules emerged independently for allowing a better correspondence between normal ND-derivations and CUT-free derivations in sequent-calculi (see, for example, [14, 15]).

As an instance of a derivation involving the ι-rules, consider the following example.

Example 3.1. *We show* $\psi(\iota x.\varphi(x)), \forall x.\psi(x) \to \chi(x) \vdash \chi(\iota x.\varphi(x))$. *To fit the page, the derivation is displayed in pieces. Let* $\mathcal{D}_2$ *be the following sub-derivation.*

$$\frac{\psi(\iota x.\varphi(x)) \quad \dfrac{\dfrac{[\varphi(u) \wedge \varphi(v)]_2}{\varphi(u)} \ (\wedge E_1) \quad \dfrac{[\varphi(u) \wedge \varphi(v)]_2}{\varphi(v)} \ (\wedge E_2)}{\dfrac{u = v}{\varphi(u) \wedge \varphi(v) \to u = v} \ (\to I^2)}}{\forall u \forall v.\varphi(u) \wedge \varphi(v) \to u = v} \ (\forall I \times 2)$$

Also, let $\mathcal{D}_3$ *be the following sub-derivation.*

$$\frac{\dfrac{[\varphi(w)]_1 \quad \dfrac{\dfrac{\psi(\iota x.\varphi(x))}{\forall x.\varphi(x) \to \psi(x)} \ (\iota E_3)}{\varphi(w) \to \psi(w)} \ (\forall E)}{\dfrac{\psi(w)}{\dfrac{\chi(w)}{\varphi(w) \to \chi(w)} \ (\to I^1)}} \quad \dfrac{\forall x.\psi(x) \to \chi(x)}{\psi(w) \to \chi(w)} \ (\forall E)}{\forall w.\varphi(w) \to \chi(w)} \ (\forall I)$$

Then, the main derivation is the following.

$$\frac{\dfrac{\psi(\iota x.\varphi(x))}{\exists x.\varphi(x)} \ (\iota E_1) \quad \begin{array}{c} \mathcal{D}_2 \\ \forall u \forall v.\varphi(u) \wedge \varphi(v) \to u = v \end{array} \quad \begin{array}{c} \mathcal{D}_3 \\ \forall w.\varphi(w) \to \chi(w) \end{array}}{\chi(\iota x.\varphi(x))} \ (\iota I)$$

(3.34)

To analyze the proof-theoretic meaning of nested definiteness, consider the following generic skeleton of a derivation of $\psi(\iota x.\chi(\iota y.\varphi(y), x))$ (with contexts omitted).

To fit the page, the derivation is displayed in several parts. For readability, some renaming of bound variables took place.

$$\mathcal{D}_1 : \quad \cfrac{\cfrac{\exists y.\varphi(y) \quad \forall u_1.\forall v_1.\varphi(u_1)\wedge\varphi(v_1)\rightarrow u_1 = v_1 \quad \forall w_1.\varphi(w_1)\rightarrow\chi(w_1, x)}{\chi(\iota y.\varphi(y), x)}\ (\iota I)}{\exists x.\chi(\iota y.\varphi(y), x)}\ (\exists I) \tag{3.35}$$

To fit the page, we use the following abbreviations.

$$\Gamma_1 =^{df.} \exists y.\varphi(y), \forall u_2.\forall v_2.\varphi(u_2)\wedge\varphi(v_2)\rightarrow u_2 = v_2, \forall w_2.\varphi(w_2)\rightarrow\chi(w_2, x)$$
$$\Gamma_2 =^{df.} \exists y.\varphi(y), \forall u_3.\forall v_3.\varphi(u_3)\wedge\varphi(v_3)\rightarrow u_3 = v_3, \forall w_3.\varphi(w_3)\rightarrow\chi(w_3, x)$$

$$\mathcal{D}_2 : \quad \cfrac{\cfrac{\cfrac{\cfrac{\cfrac{\Gamma_1}{\chi(\iota y.\varphi(y), u_2)}\ (\iota I) \quad \cfrac{\Gamma_2}{\chi(\iota y.\varphi(y), v_2)}\ (\iota I)}{\chi(\iota y.\varphi(y), u_2))\wedge\chi(\iota y.\varphi(y), v_2))}\ (\wedge I)}{\vdots}}{\cfrac{u_2 = v_2}{\chi(\iota y.\varphi(y), u_2)\wedge\chi(\iota y.\varphi(y), v_2)\rightarrow u_2 = v_2}\ (\rightarrow I)}}{\forall u_2\forall v_2.\chi(\iota y.\varphi(y), u_2)\wedge\chi(\iota y.\varphi(y), v_2)\rightarrow u_2 = v_2}\ (\forall I) \times 2 \tag{3.36}$$

$$\mathcal{D}_3 : \quad \cfrac{\cfrac{\cfrac{\cfrac{\exists y.\varphi(y) \quad \forall u_4.\forall v_4.\varphi(u_4)\wedge\varphi(v_4)\rightarrow u_4 = v_4 \quad \forall w_4.\varphi(w_4)\rightarrow\chi(w_4, x)}{\chi(\iota y.\varphi(y), u_4)}\ (\iota I)}{\vdots}}{\cfrac{\psi(u_4)}{\chi(\iota y.\varphi(y), u_4)\rightarrow\psi(u_4)}\ (\rightarrow I)}}{\forall u_4.\chi(\iota y.\varphi(y), u_4)\rightarrow\psi(u_4)}\ (\forall I) \tag{3.37}$$

$$\cfrac{\overset{\mathcal{D}_1}{\exists x.\chi(\iota y.\varphi(y), x)} \quad \overset{\mathcal{D}_2}{\forall u_2\forall v_2.\chi(\iota y.\varphi(y), u_2)\wedge\chi(\iota y.\varphi(y), v_2)\rightarrow u_2 = v_2} \quad \overset{\mathcal{D}_3}{\forall u.\chi(\iota y.\varphi(y), u_4)\rightarrow\psi(u_4)}}{\psi(\iota x.\chi(\iota y.\varphi(y), x))}\ (\iota I) \tag{3.38}$$

By inspecting the derivation of such a nested definiteness, we see that existence and uniqueness need to hold both for $\varphi(y)$ and for $\iota x.\chi(\iota y.\varphi(y), x)$, where the latter requires a unique x w.r.t. the unique y of the former. For (2.21), the rules require a unique boy, and a unique girl loved by *that* boy.

The proof-system for binary parallel definiteness is presented in Figure 2. The extension to n-ary parallelism should be obvious.

$$\frac{\begin{array}{c}\Gamma\vdash\exists x_1.\varphi_1(x_1,\overline{a}_1), \\ \Gamma\vdash\exists x_2.\varphi_2(x_2,\overline{a}_2)\end{array}\quad\begin{array}{c}\Gamma\vdash\forall u_1.\forall v_1.\varphi_1(u_1,\overline{a}_1)\wedge\varphi_1(v_1,\overline{a}_1)\rightarrow u_1=v_1, \\ \Gamma\vdash\forall u_2.\forall v_2.\varphi_2(u_2,\overline{a}_2)\wedge\varphi_2(v_2,\overline{a}_2)\rightarrow u_2=v_2 \\ \Gamma\vdash\forall w_1.\varphi_1(w_1,\overline{a}_1)\wedge\forall w_2.\varphi_2(w_2,\overline{a}_2)\rightarrow\psi(w_1,w_2,\overline{b})\end{array}}{\Gamma\vdash\psi(\iota x_1.\varphi_1(x_1,\overline{a}_1),\iota x_2.\varphi_2(x_2,\overline{a}_2),\overline{b}))}\ (\iota I)$$

$$\frac{\Gamma\vdash\psi(\iota x_1.\varphi_1(x_1,\overline{a}_1),\iota x_2.\varphi_2(x_2,\overline{a}_2),\overline{b})}{\Gamma\vdash\exists x_1.\varphi_1(x_1,\overline{a}_1)}\ (\iota E_{1,1})$$

$$\frac{\Gamma\vdash\psi(\iota x_1.\varphi_1(x_1,\overline{a}_1),\iota x_2.\varphi_2(x_2,\overline{a}_2),\overline{b})}{\Gamma\vdash\exists x_2.\varphi_2(x_2,\overline{a}_2)}\ (\iota E_{1,2})$$

$$\frac{\begin{array}{c}\Gamma\vdash\psi(\iota x_1.\varphi_1(x_1,\overline{a}_1),\iota x_2.\varphi_2(x_2,\overline{a}_2),\overline{b}) \\ \Gamma\vdash\varphi_1(u_1,\overline{a}_1) \\ \Gamma\vdash\varphi_2(u_2,\overline{a}_2) \\ \Gamma\vdash\varphi_1(v_1,\overline{a}_1) \\ \Gamma\vdash\varphi_2(v_2,\overline{a}_2)\end{array}}{\Gamma\vdash u_i=v_i}\ (\iota E_2),\quad u_i,v_i\ \text{fresh for }\Gamma$$

$$\frac{\Gamma\vdash\psi(\iota x_1.\varphi_1(x_1,\overline{a}_1),\iota x_2.\varphi_2(x_2,\overline{a}_2),\overline{b})}{\Gamma\vdash\forall z_1.\varphi_1(z_1,\overline{a}_1)\wedge\forall z_2.\varphi_2(z_2,\overline{a}_2)\rightarrow\psi(z_1,z_2,\overline{b})}\ (\iota E_3)$$

Figure 2: I/E-Rules for (binary) parallel definiteness

3.1   Imposing Presupposition

This section presents some ideas relating definiteness to *presupposition* in a proof-theoretic manner. A fuller proof-theoretic discussion, relating to a wider scope of phenomena related to presupposition, such as *projection, incorporation* and more, not just in case of definiteness, is deferred to a separate paper.

As a motivating example, consider the following. Suppose we want to prove (in our system) that

$$\text{Everyone bald owns no comb} \tag{3.39}$$

entails

$$\text{If the king of France is bald then he owns no comb} \tag{3.40}$$

This can be regimented as

$$\forall x.B(x){\rightarrow}C(x) \vdash B(\iota x.K(x)){\rightarrow}C(\iota x.K(x)) \tag{3.41}$$

where $C(x)$ can be read as **is comb-less** (abbreviating **has no comb**).

Presumably, the proof, using the standard I/E-rules for FOL, is as follows:

$$\cfrac{\cfrac{\cfrac{\forall x.B(x){\rightarrow}C(x)}{B(\iota x.K(x)){\rightarrow}C(\iota x.K(x))} \; (\forall E) \quad [B(\iota x.K(x))]_1}{C(\iota x.K(x))} \; (\rightarrow E)}{B(\iota x.K(x)){\rightarrow}C(\iota x.K(x))} \; (\rightarrow I^1) \tag{3.42}$$

As one can see, we *nowhere* introduce $B(\iota x.K(x))$ with a ιI-rule when:

- assuming $B(\iota x.K(x))$ and later discharging it, or

- deducing $B(\iota x.K(x))$ via $(\forall E)$!

Hence, existence and uniqueness were not assured in this derivation.

We modify the standard I/E-rules for FOL as follows[5].

Assumption: The standard assumption rule is

$$\cfrac{}{\Gamma, \varphi \vdash \varphi} \; (ass) \tag{3.43}$$

We modify it by enforcing the appropriate instances of $\mathbf{pr_1}$, $\mathbf{pr_2}$ (existence and uniqueness) to be added to the context, considered as *presuppositions*, *whenever φ has a ι-term as an argument*. This can be conceived as a side-condition on the standard assumption rule.

$$\cfrac{}{\Gamma, \mathbf{pr_1}, \mathbf{pr_2}, \varphi \vdash \varphi} \; (ass) \tag{3.44}$$

This modification affects both $(\rightarrow I)$ and $(\vee E)$, that introduce assumptions and discharge them. Thus, the regimentation of (3.40) now becomes

$$\forall x.B(x){\rightarrow}C(x), \mathbf{pr_1}, \mathbf{pr_2}, \vdash B(\iota x.K(x)){\rightarrow}C(\iota x.K(x)) \tag{3.45}$$

---

[5]This modification is inspired by the ideas in [6]. However, we do not have in FOL judgements of the form $\varphi : Prop$ that can serve as "handles" for imposing $\mathbf{pr_1}$, $\mathbf{pr_2}$ as presuppositions as done there, and we implement the same idea differently.

Elimination: When a ι-formula is a conclusion of a standard *FOL* elimination rule, is also no subject to the existence and uniqueness conditions. These have to be imposed by side-conditions. The standard $(\forall E)$-rule is

$$\frac{\Gamma \vdash \forall x. \varphi(x)}{\Gamma \vdash \varphi(t)} \; (\forall E) \qquad\qquad (3.46)$$

We modify the rule to

$$\frac{\Gamma \vdash \forall x. \varphi(x)}{\Gamma, \mathbf{pr_1}, \mathbf{pr_2} \vdash \varphi(t)} \; (\forall E) \qquad\qquad (3.47)$$

whenever t is a ι-term.

The other E-rules are modified in a similar way and we omit the details.

3.2 Reductions and Expansion

Recall that a *maximal formula* in a derivation is a formula serving both as a conclusion of an application of an I-rule as well as a major premise of an application of an E-rule. The presence of a maximal formula is considered a *detour* in the derivation. A reduction of a derivation with a maximal formula produces an equivalent derivation (with the same conclusion and the same (or less) assumptions). Such reductions are central to the property of *harmony* in proof-theoretic semantics [10, 1] and *normalization* [8].

Below are the three reductions for the singular ι-rules (contexts and parameters omitted). The notation for the parallel case is cluttered and omitted.

$$\cfrac{\cfrac{\begin{matrix} \mathcal{D}_1 & \mathcal{D}_2 & \mathcal{D}_3 \\ \exists x. \varphi(x) & \forall u. \forall v. \varphi(u) \wedge \varphi(v) \to u = v & \forall w. \varphi(w) \to \psi(w) \end{matrix}}{\psi(\iota x. \varphi(x))} \; (\iota I)}{\exists x. \varphi(x)} \; (\iota E_1) \qquad \rightsquigarrow_r \quad \begin{matrix} \mathcal{D}_1 \\ \exists x. \varphi(x) \end{matrix}$$

$$(3.48)$$

$$\dfrac{\dfrac{\begin{matrix}\mathcal{D}_1 & \mathcal{D}_2 & \mathcal{D}_3\\ \exists x.\varphi(x) & \forall u.\forall v.\varphi(u)\wedge\varphi(v){\rightarrow}u=v & \forall w.\varphi(w){\rightarrow}\psi(w)\end{matrix}}{\psi(\iota x.\varphi(x))}\ (\iota I)\quad \begin{matrix}\mathcal{D}_4 & \mathcal{D}_5\\ \varphi(u) & \varphi(v)\end{matrix}}{u=v}\ (\iota E_2)$$

$$\rightsquigarrow_r$$

$$\dfrac{\dfrac{\begin{matrix}\mathcal{D}_4 & \mathcal{D}_5\\ \varphi(u) & \varphi(v)\end{matrix}}{\varphi(u)\wedge\varphi(v)}\ (\wedge I)\qquad \dfrac{\dfrac{\mathcal{D}_2}{\forall u.\forall v.\varphi(u)\wedge\varphi(v){\rightarrow}u=v}}{\varphi(u)\wedge\varphi(v){\rightarrow}u=v}\ (\forall E)\times 2}{u=v}\ (\rightarrow E)$$

$$(3.49)$$

$$\dfrac{\dfrac{\begin{matrix}\mathcal{D}_1 & \mathcal{D}_2 & \mathcal{D}_3\\ \exists x.\varphi(x) & \forall u.\forall v.\varphi(u)\wedge\varphi(v){\rightarrow}u=v & \forall w.\varphi(w){\rightarrow}\psi(w)\end{matrix}}{\psi(\iota x.\varphi(x))}\ (\iota I)}{\forall w.\varphi(w){\rightarrow}\psi(w)}\ (\iota E_3)$$

$$\rightsquigarrow_r$$

$$\begin{matrix}\mathcal{D}_3\\ \forall w.\varphi(w){\rightarrow}\psi(w)\end{matrix}$$

$$(3.50)$$

An *expansion* of a derivation for some $\Gamma \vdash \varphi$ transforms it into an equivalent derivation in which φ is decomposed by applying E-rules and recomposed by applying I-rules. Expansions are central for the proof-theoretic semantics property of *stability* [10].

Below is the expansion establishing local completeness of the ι-rules.

$$\frac{\mathcal{D}}{\psi(\iota x.\varphi(x,\bar{a}),\bar{b})}$$

$$\leadsto_e$$

$$\cfrac{\cfrac{\mathcal{D}}{\psi(\iota x.\varphi(x))}\quad \cfrac{\cfrac{[\varphi(u)\wedge\varphi(v)]_2}{\varphi(u)}\;(\wedge E_1)\qquad \cfrac{[\varphi(u)\wedge\varphi(v)]_2}{\varphi(v)}\;(\wedge E_2)}{\cfrac{u=v}{\cfrac{\phi\cdot\sigma(u)\wedge\phi\cdot\sigma(v)\to u=v\quad u=v\to u=v}{\phi\cdot\sigma(u)\wedge\phi\cdot\sigma(v)\to u=v}\;(\to I^2)}\;(IA\times 2)}\;(\iota E_2)}{\psi(\iota x.\varphi(x,\bar{a}),\bar{b})}$$

$$\cfrac{\cfrac{\mathcal{D}}{\psi(\iota x.\varphi(x,\bar{a}),\bar{b})}}{\mathrm{E}\,x.\phi\cdot\sigma(\bar{a})}\;(\iota E_1)$$

$$\cfrac{\cfrac{\mathcal{D}}{\psi(\iota x.\varphi(x,\bar{a}),\bar{b})}}{\cfrac{\wedge m.\phi\cdot\sigma(\bar{a},w)\to\phi\cdot\psi(w,\bar{b})}{}\;(\iota E_3)}\;(\iota I)$$

(3.51)

326

4 Comparison With Other ND-Rules for Definiteness

We now compare the proposed I/E rules for definiteness to some other proposals found in the literature.

4.1 Hilbert and Bernays

Hilbert and Bernays [3] also suggest a contextual definition of 'ι', using the following I-rule (in our notation).

$$\frac{\Gamma \vdash \exists x.\varphi(x) \quad \Gamma \vdash \forall u.\forall v.\varphi(u) \land \varphi(v) \rightarrow u = v}{\Gamma \vdash \varphi(\iota x.\varphi(x))} \ (\iota I_{HB}) \tag{4.52}$$

The premises of the rule are the same as the respective two first premises of our rule (ιI), but the conclusion is weaker. It only allows a contextual inference where the main predicate (ψ in our rule) *coincides* with the predicate on which existence and uniqueness are imposed. It allows regimenting identity-expressing trivial sentences like

$$\text{The king of France is a king of France} \tag{4.53}$$

Hilbert and Bernays have no E-rule for ι. They do prove that the I-rule above is a conservative extension of the system without it.

We note that [3] (e.g., p. 391 of the 1st edition) does use more complicated expressions, involving nesting, such as:

$$\iota x.A(x, \iota y.B(y, \iota z.C(y, z))) \tag{4.54}$$

and

$$\iota x.\exists y A(\iota z.B(x, y, z)) \tag{4.55}$$

4.2 Kalish and Montague

In [4], Kalish and Montague more or less adopt the rule (ι_{HB}), but add another rule. The need of the other rule is due to the fact that they interpret $\iota x.\varphi(x)$ as a fixed element, the number zero, if existence and uniqueness do not obtain (improper description). The second rule, $(\iota_{KM}I)$ ensures, that all improper descriptions are equivalent.

$$\frac{\neg(\exists y \forall x.\varphi(x) \leftrightarrow x = y)}{\iota x.\varphi(x) = \iota y.\neg y = y} \ (\iota_{KM}I) \tag{4.56}$$

In addition to inheriting the limitation of (ι_{HB}), the approach of Kalish and Montague may lead to undesired results. For example, one can derive (the regimentation of): The natural number between 1 and 2 is 0.

4.3 Tennant

Tennant, in [12] (elaborating on [11]), proposes I/E-rules for a contextual proof-theoretic definition of ι. He does this within a uniform proof-theoretic definition of a class of subnectors he calls *abstraction operators*. He certainly shares our view of PTS as a meaning theory. However, the contexts he allows are only *identity statements* of the form $t = \iota x.\varphi(x)$ (where t is any term in the underlying object language). His rules are formulated in a *free logic* framework, where existential commitments are made explicit. He uses $\exists!t$, defined by $\exists!t =^{df.} \exists x.x = t$. The rules are presented below, in our notation.

$$\frac{\begin{array}{c}[\varphi(a)]_i, [\exists!a]_i \\ \vdots \\ a = t\end{array} \quad \exists!t \quad \begin{array}{c}[a=t]_j \\ \vdots \\ \varphi(a)\end{array}}{t = \iota x.\varphi(x)}\,(\iota_T)I^{i,j} \tag{4.57}$$

The E-rules derive each of the premises of $(\iota_T I)$ from the major premise $t = \iota x.\varphi(x)$.

This I-rule does not allow to infer directly a predication of the form $\psi(\iota x.\varphi(x))$. Instead, one has to infer $\psi(\iota x.\varphi(x)) \wedge \psi(t)$; for example,

$$\text{Louis is the (current) king of France and he is bald} \tag{4.58}$$

Representing nested definiteness is even more cumbersome.

4.4 Identity in the Matrix of ι-Terms

Identity can have two roles here:

1. In the premises of the (ιI)-rule.

2. As an instance of φ and especially of ψ.

We augment ND_ι with standard I/E-rules for identity (for example, [13]).

$$\frac{}{t = t}\,(= I) \qquad \frac{t = s \quad \varphi[x/t]}{\varphi[x/s]}\,(= E) \tag{4.59}$$

$(= I)$ and $(= E)$ are also known as Reflexivity and Replacement. Transitivity of $=$ is expressed as a special case of $(= E)$:

$$\frac{t = s \quad t = r}{s = r}\,(= E)$$

Nota bene, $(= I)$ and $(= E)$ are not in harmony. Consider the following two formulas, exemplifying the interplay between constants and ι-terms.

$$S(\iota x.x = d) \tag{4.60}$$

$$d = \iota x(x = d) \tag{4.61}$$

(Formulas of this form play a prominent role in the philosophical discussion of so-called "slingshot" arguments; see, e.g., [7].) The Russellian contextual analogues for (4.60) and (4.61) are (4.62) and (4.63), respectively.

$$\exists x.x = d \land \forall u \forall v.u = d \land v = d \rightarrow u = v \land \forall w.w = d \rightarrow S(w) \tag{4.62}$$

$$\exists x.x = b \land \forall u \forall v.u = b \land v = b \rightarrow u = v \land \forall w.w = b \rightarrow w = b \tag{4.63}$$

For purposes of illustration, consider

$$S(d) \tag{4.64}$$

(4.60) and (4.64) can be shown to be equivalent as follows:

$$\cfrac{\cfrac{\overline{d = d}\,(= I) \quad \cfrac{\cfrac{[S(\iota x.x = d)]_1}{\forall z.z = d \rightarrow Sz}\,(\iota E_3)}{d = d \rightarrow Sd}\,(\forall E)}{\cfrac{Sd}{S(\iota x.x = d) \rightarrow Sd}\,(\rightarrow I^1)}}{}\,(\rightarrow E) \tag{4.65}$$

and

$$\cfrac{\cfrac{\overline{d = d}\,(= I)}{\exists x.x = d}\,(\exists I) \quad \cfrac{\cfrac{\cfrac{[u = d \land v = d]_1}{u = d}\,(\land E_1) \quad \cfrac{[u = d \land v = d]_1}{v = d}\,(\land E_2)}{u = v}\,(= E)}{\cfrac{u = d \land v = d \rightarrow u = v}{\forall u.\forall v.u = d \land v = d \rightarrow u = v}\,(\forall I \times 2)}\,(\rightarrow I^1) \quad \cfrac{\cfrac{[Sd]_2}{d = d \rightarrow Sd}\,(\rightarrow I)}{\cfrac{\forall w.w = d \rightarrow Sw}{}\,(\forall I)}\,(\iota I)}{\cfrac{S(\iota x.x = d)}{Sd \rightarrow S(\iota x.x = d)}\,(\rightarrow I^2)} \tag{4.66}$$

(4.61) and $d = d$ can be shown to be equivalent in a similar way.

$$\cfrac{\cfrac{\overline{d = d}\,(= I) \quad \cfrac{\cfrac{[d = \iota x.x = d]_1}{\forall z.z = d \rightarrow d = z}\,(\iota E_3)}{d = d \rightarrow d = d}\,(\forall E)}{\cfrac{d = d}{d = \iota x.x = d \rightarrow d = d}\,(\rightarrow I^1)}}{}\,(\rightarrow E) \tag{4.67}$$

$$\frac{\dfrac{\dfrac{[u = d \land v = d]_1}{u = d} \, (\land E_1) \quad \dfrac{[u = d \land v = d]_1}{v = d} \, (\land E_2)}{u = v} \, (= E)}{\dfrac{\dfrac{d = d}{\exists x. x = d} \, (\exists I)}{\dfrac{u = d \land v = d \to u = v}{\dfrac{\forall u \forall v. u = d \land v = d \to u = v}{\dfrac{d = \iota x. x = d}{d = d \to d = \iota x (x = d)}} \, (\forall I \times 2)}} \quad \dfrac{\dfrac{[d = d]_2}{d = d \to d = d} \, (\to I)}{\dfrac{\forall w. w = d \to d = w}{} \, (\forall I)} \, (\iota I)$$

$$(4.68)$$

As these examples suggest, equivalence does not guarantee sameness of proof-theoretic meaning. Moreover, identity claims like (4.61) do not guarantee proof-theoretic synonymy of d and $\iota x. x = d$.

Also,

$$d = r \quad \text{and} \quad \iota x(x = d) = \iota x(x = r) \tag{4.69}$$

are equivalent:

$$\frac{\dfrac{d = r \quad d = \iota x(x = d)}{r = \iota x(x = d)} \, (= E) \quad r = \iota x(x = r)}{\iota x(x = d) = \iota x(x = r)} \, (= E) \tag{4.70}$$

$$\frac{\dfrac{\iota x(x = d) = \iota x(x = r) \quad \iota x(x = d) = d}{\iota x(x = r) = d} \, (= E) \quad \iota x(x = r) = r}{d = r} \, (= E) \tag{4.71}$$

We may combine ι-terms with identities in the matrix with parallel, nested, and parameterized definiteness. We illustrate this by means of (4.72). Natural language counterparts are cumbersome and evaded.

$$
\begin{aligned}
&Q(\iota x. x = d, \iota x. x = r) \\
&S(\iota x(x = d) \land Q(\iota y(y = r, x))) \\
&\exists y. S(\iota x. x = y)
\end{aligned}
\tag{4.72}
$$

5 Conclusions

In this paper we have presented an extension of first-order logic with the subnector term 'ι' (Russell) expressing definiteness in the form of existence and uniqueness. The extension is formulated in terms of a natural-deduction proof-system. By means of this proof-theoretic treatment of definiteness, there is a shift of focus from improper, non-referring ι-terms, to the grounds of assertion of sentences with ι-terms, including a treatment of presupposition. The ι-rules were shown to be harmonious, thereby qualifying as meaning-conferring.

This extension can save a basis for a more general study of PTS for subnectors in general, in contrast to the current focus of PTS (in logic) on sentential operators.

References

[1] Nissim Francez. *Proof-theoretic Semantics*. College Publications, London, 2015.

[2] Nissim Francez and Bartosz Wieckowski. A proof-theoretic semantics for contextual definiteness. In Enrico Moriconi and Laura Tesconi, editors, *Second Pisa Colloquium in Logic, Language and Epistemology*. ETS, Pisa, 2014.

[3] David Hilbert and Paul Bernays. *Grundlagen der Mathematik (vol. 1)*. Springer-Verlag, Berlin, Heidelberg, New York, 1934. 2nd edition, 1968.

[4] Donald Kalish and Richard Montague. *LOGIC techniques of formal reasoning*. Harcourt, Brace & World, New York Chicago San-Francisco, Atlanta, 1964.

[5] Steven T. Kuhn. Embedded definite descriptions: Russellian analysis and semantic puzzles. *Mind*, 109(435):443–454, 2000.

[6] Koji Mineshima. A presuppositional analysis of definite descriptions in proof theory. In Ken Satoh, Akihiro Inokuchi, Katashi Nagao, and Takahiro Kawamura, editors, *New Frontiers in Artificial Intelligence, JSAI 2007 Conference and Workshops, Miyazaki, Japan, June 18-22, 2007*, pages 214–227. Springer Verlag, Berlin, Heidelberg, 2008. Lecture Notes in Computer Science 4914.

[7] Stephen Neale. *Facing Facts*. Oxford University Press, 2001.

[8] Dag Prawitz. *Natural Deduction: A Proof-Theoretical Study*. Almqvist and Wicksell, Stockholm, 1965. Soft cover edition by Dover, 2006.

[9] Bertrand Russell. On denoting. *Mind*, 14(56):479–493, 1905.

[10] Peter Schroeder-Heister. Proof-theoretic semantics. In Edward N. Zalta, editor, *Stanford Encyclopaedia of Philosophy (SEP), http://plato.stanford.edu/*. The Metaphysics Research Lab, Center for the Study of Language and Information, Stanford University, Stanford, CA, 2011.

[11] Neil Tennant. *Anti-Realism and Logic*. Oxford University Press, Oxford, United Kingdom, 1987. Clarendon Library of Logic and Philosophy.

[12] Neil Tennant. A general theory of abstraction operators. *The philosophical quarterly*, 54(214):105–133, 2004.

[13] Anne S. Troelstra and Dirk van Dalen. *Constructivism in Mathematics, vol. I*. Elsevier, Amsterdam, 1988.

[14] Jan von Plato. A problem with normal form in natural deduction. *Math. Logic Quarterly*, 46:121–124, 2000.

[15] Jan von Plato. Natural deduction with general elimination rules. *Archive for Mathematical Logic*, 40:541–567, 2001.

[16] Alfred North Whitehead and Bertrand Russell. *Principia Mathematica*. Cambridge University Press, Cambridge, UK, 1910.

Received October 2015

Two Types of Indefinites: Hilbert & Russell

Norbert Gratzl[*]

Munich Center for Mathematical Philosophy, Ludwig-Maximilians-Universität München

N.Gratzl@lmu.de

Georg Schiemer[†]

University of Vienna, Universitätsstraße 7, 1090 Vienna

georg.schiemer@univie.ac.at

Abstract

This paper compares Hilbert's ϵ-terms and Russell's approach to indefinite descriptions, Russell's indefinites for short. Despite the fact that both accounts are usually taken to express indefinite descriptions, there is a number of dissimilarities. Specifically, it can be shown that Russell indefinites—expressed in terms of a logical ρ-operator—are not directly representable in terms of their corresponding ϵ-terms. Nevertheless, there are two possible translations of Russell indefinites into epsilon logic. The first one is given in a language with classical ϵ-terms. The second translation is based on a refined account of epsilon terms, namely *indexed* ϵ-terms. In what follows we briefly outline these approaches both syntactically and semantically and discuss their respective connections; in particular, we establish two equivalence results between the (indexed) epsilon calculus and the proposed ρ-term approach to Russell's indefinites.

Keywords: Indefinite descriptions, epsilon terms, choice semantics, Hilbert, Russell

We would like to thank Richard Zach, Hannes Leitgeb, and the participants of the *Epsilon 2015* conference in Montpellier for helpful discussions. We are also grateful to two anonymous referees for their comments on earlier drafts of this paper.

[*]Research on this paper was supported by the Alexander von Humboldt Foundation.

[†]Research on this paper was supported by the Austrian Science Fund (Grant number P 27718-G16) as well as by the German Research Foundation (Project "*Mathematics: Objectivity by Representation*").

1   Introduction

In linguistics and philosophy of language, one generally distinguishes between two ways to logically represent indefinite descriptions of the form

An A is a B.

The first approach goes back to Russell and views the expression "an A" as non-referential. Indefinites of this form are taken to function semantically like existential quantifiers (or variables bound by an existential quantifier). According to this *quantificational* account, the logical form of the above sentence is best captured by the following existentially quantified statement:

$$\exists x(A(x) \wedge B(x))$$

The second approach has roots in work by Hilbert and has recently been further developed by von Heusinger and Egli.[1] This is to view the expression "an A" as a constant term that denotes a particular object. Specifically, an indefinite phrase so understood can be presented logically by an epsilon term $\epsilon_x A(x)$. Informally speaking, this term picks out an *arbitrary* object that satisfies formula A if such an object exists. Accordingly, the indefinite description stated above is presented logically not in terms of a quantified statement, but in terms of the following statement:[2]

$$B(\epsilon_x A(x))$$

These two logical reconstructions of indefinite descriptions seem to be based on two different ways to understand indefinites. Let us dub them *Russell* and *Hilbert* indefinites. The central conceptual difference between them is usually taken to be the fact that unlike Hilbert's indefinites, Russell indefinites are not referring expressions or terms with a fixed reference. Nevertheless, as we want to show in this paper, there exists a natural way to represent Russell's ambiguous descriptions in terms of a logical language with a term-forming operator. Thus, in analogy to the representation of free choice indefinites in an epsilon-term logic, we will outline here a operator-based logic for the expression of Russel indefinites.

The central aim in this paper is to compare the representation of indefinites in term of epsilon logic with a Russellian approach to indefinite descriptions in terms of a logic based on a ρ-operator. As we will show, there is a number of dissimilarities between the two accounts. Specifically, it can be shown that Russell's ambiguous descriptions—expressed in terms of a logical ρ-operator—are not directly

---

[1]See, in particular, [16] and [15]

[2]See [16] for a detailed discussion of both approaches and for further references.

representable in terms of their corresponding ϵ-terms. Nevertheless, there are two possible translations of Russell indefinites into epsilon logic. The first one is based on an embedding of a language with ρ-terms into a classical language with ϵ-terms. The second translation is based on a refined account of epsilon terms, namely *indexed* ϵ-terms first introduced by von Heusinger.[3] In what follows we briefly outline these approaches to represent indefinites both syntactically and semantically and discuss their respective connections; in particular, we establish two equivalence results between the classical and indexed epsilon calculus on the one hand and the ρ-term approach to Russell's account of indefinite descriptions on the other hand.

The paper is organized as follows: Section 2 will introduce the extensional epsilon calculus EC as well as a suitable choice semantics for (closed) epsilon terms. Section 3 will then present a logic for ρ-terms based on Russell's remarks on indefinite descriptions. Section 4 will then give a closer comparison between the two logical representations of ambiguous descriptions. Specifically, we present a translation of the ρ-term presentation of "An A is a B" in classical epsilon logic (4.1) as well as in a language of indexed epsilon terms (4.2). Finally, section 5 will contain some concluding remarks and suggestions for future research.

2 Hilbert's ϵ-terms

A natural logical representation of indefinites (or indefinite descriptions) can be given in terms of epsilon terms, that is, terms formed with the help of an epsilon operator.[4] As understood by Hilbert, the ϵ-operator functions as a logical term-forming operator: given a first-order formula $A(x)$ with variable x occurring free in it, $\epsilon_x A(x)$ is a closed term in which all occurrences of x are bound. Informally speaking, this term refers to an arbitrary object satisfying the formula A if there exists such an object.[5]

Different epsilon calculi have been proposed in the literature since Hilbert to describe the logical behaviour of such terms. The *extensional* EC usually consists of two axiom schemes (in addition to the standard axioms and deduction rules of

---

[3]See, in particular, [16] and [9].

[4]Epsilon terms were originally introduced in Hilbert's proof-theoretic work on the foundations of mathematics in the 1920s. See, in particular, [14] and [17] for detailed historical discussions of the development of the epsilon calculus as well as of Hilbert's epsilon substitution method in his syntactic consistency proofs. Compare also [1] for a first systematic study of the epsilon calculus.

[5]The following discussion of epsilon logic follows closely the presentation given in [18]. See also [13] for a similar discussion of epsilon terms and their choice semantics.

first-order logic), namely

$$A(t) \rightarrow A(\epsilon_x A(x)) \qquad \text{(Critical formulas)}$$
$$\forall x(A(x) \leftrightarrow B(x)) \rightarrow \epsilon_x A(x) = \epsilon_x B(x) \qquad \text{(Extensionality)}$$

The second axiom expresses an extensionality principle for epsilon logic: if two formulas are equivalent, then their respective ϵ-representatives are identical.[6]

Hilbert's original motivation for the introduction of a calculus for epsilon terms was to show that one can explicitly define the first-order quantifiers in terms of epsilon terms in the following way:

$$\exists x A(x) :\leftrightarrow A(\epsilon_x A(x)) \qquad \text{(Def∃)}$$
$$\forall x A(x) :\leftrightarrow A(\epsilon_x \neg A(x)) \qquad \text{(Def∀)}$$

It is a well known fact that first-order predicate logic is embeddable in EC. This is based on a translation function $(.)^\epsilon$ that maps expressions of the a first-order language $\mathcal{L}$ to expressions of the language with epsilon-terms $\mathcal{L}_\epsilon$ (see [10]):

1. $e^\epsilon = e$, for e a variable or constant symbol

2. $P(t_1, \ldots, t_n)^\epsilon = P(t_1^\epsilon, \ldots, t_n^\epsilon)$

3. $f(t_1, \ldots, t_n)^\epsilon = f(t_1^\epsilon, \ldots, t_n^\epsilon)$

4. $(\neg A)^\epsilon = \neg A^\epsilon$

5. $(A \wedge B)^\epsilon = A^\epsilon \wedge B^\epsilon$

6. $(A \vee B)^\epsilon = A^\epsilon \vee B^\epsilon$

7. $(\exists x(A(x)))^\epsilon = A^\epsilon(\epsilon_x A(x)^\epsilon)$

8. $(\forall x(A(x)))^\epsilon = A^\epsilon(\epsilon_x \neg A(x)^\epsilon)$

As a consequence of this, any first-order formula can be represented as a quantifier-free formula in $\mathcal{L}_\epsilon$, a result which was of central importance in Hilbert's proof theoretic work, in particular, in his two ϵ-theorems.[7]

---

[6]It should be noted here that the extensionality axiom was already mentioned in Hilbert's work, but not used in the proofs of his famous epsilon theorems. The axiom is discussed again in Asser's study of the epsilon calculus [2] as well as in [6]. See [18] and [8] for modern presentations of intensional and extensional epsilon calculi.

[7]Compare again [17] and [10] for further details.

Turning to the semantic interpretation of (extensional) EC, we saw that epsilon terms of the form $\epsilon_x A(x)$ were understood by Hilbert and subsequent logicians to function as referring indefinite expressions, i.e. as terms that pick out *any* object which satisfies the formula $A(x)$ under the condition that there are such objects. Compare, for instance, Hilbert & Bernays' informal description of the semantic interpretation of such terms in the second volume of *Grundlagen der Mathematik* (1939):

> Syntactically, [the ϵ-symbol] provides a function of a variable predicate, which–besides the argument to which the variable bound by the ϵ-symbol refers–may contain free variables as arguments ("parameters"). The value of this function for a given predicate A (for fixed values of the parameters) is an object of the universe for which–according to the semantical translation of the formula (ϵ_0)–the predicate A holds, provided that A holds for any object of the universe at all. [5, p.12]

A natural model-theoretic formalization of this understanding of epsilon terms is given today in terms of a choice-functional semantics. A choice semantics for the extensional EC can be characterized as follows:[8] an interpretation $\mathfrak{M}$ of the language $\mathcal{L}_\epsilon$ has the form $\langle D, I \rangle$ with D a domain and I an interpretation function for the signature of $\mathcal{L}_\epsilon$. We further hold that $s : Var \to D$ is an assignment function on $\mathfrak{M}$. The ϵ-operator is interpreted by an extensional choice function of the form $\delta : \wp(D) \to D$ such that, for any $X \subseteq D$:

$$\delta(X) = \begin{cases} x \in X, & \text{if } X \neq \emptyset; \\ x \in D & \text{otherwise.} \end{cases}$$

Such a choice function assigns a "representative" object to any non-empty subset of D. It gives an arbitrary object from domain D in case the set X is empty.

Based on this notion of extensional choice functions, one can give a choice-functional semantics for EC. Valuation rules for terms of $\mathcal{L}_\epsilon$ not containing epsilon terms are specified as for standard first-order logic. In addition, the semantic evaluation of ϵ-terms is specified relative to a structure $\mathfrak{M}$, assignment function s, and a choice function δ on $\mathfrak{M}$ based on the following valuation rule:[9]

$$val^{\mathfrak{M}, \delta, s}(\epsilon_x A(x)) = \delta(\{d \in D \mid \mathfrak{M}, s[x/d] \models A(x)\}).$$

---

[8]Early formulations of a choice-functional semantics for the epsilon calculus were given in [2] and in [6]. See also [8] and [18] for modern presentations of a choice semantics for extensional EC. Both [6] and [18] contain a proof of the completeness of EC with respect to this semantics.

[9]The following rule applies only to closed epsilon terms. For a more comprehensive discussion of the semantics of extensional and intensional epsilon logics, including valuation rules for open epsilon terms see [8].

The semantic value of a given term $\epsilon_x A(x)$ is thus the object that the choice function δ picks out from the truth-set of formula A. If the set defined by A is empty, then δ picks out any object in D otherwise. Based on this, the semantic notions of satisfaction of formulas of $\mathcal{L}_\epsilon$ can then be specified in the usual way and in direct analogy with first-order logic.

This choice-functional semantics for EC can be seen as a way to make precise the particular *indefinite* character of epsilon terms. As is argued in [13], the reason to view such terms as indefinites lies precisely in their semantic character, more specifically, in the kind of "arbitrary reference" usually associated with such terms. This mode of reference (typical also for instantial terms in logic and mathematical reasoning) has recently been described by Magidor and Breckenridge in the following way:

> *Arbitrary Reference* (AR): It is possible to fix the reference of an expression arbitrarily. When we do so, the expression receives its ordinary kind of semantic-value, though we do not and cannot know which value in particular it receives. [3, p.378]

This kind of undetermined reference is also characterstic for Hilbert's understanding of epsilon terms as indefinite expressions. Thus, in Hilbert's account of indefinite phrases, indefiniteness is explained best in the sense that such phrases refer *arbitrarily* to objects. Moreover, one can view the valuation rule for ϵ-terms stated above as a way to make precise this very notion of arbitrary reference. We can thus paraphrase the epsilon-term representation $B(\epsilon_x(A(x))$ of the indefinite description stated in the introduction as "An *arbitrary* A is a B." With this in mind, let us now turn to Russell's account of indefinite descriptions.

3   Russell ρ-terms

We want to motivate our presentation of Russell's account of indefinite descriptions with the following quote:

> The definition is as follows: The statement that an object having the property ϕ has the property ψ means: The joint assertion of ϕx and ψx is not always false. So far as logic goes, this is the same proposition as might be expressed by some ϕ's are ψ's; but rhetorically there is a difference, because in the one case there is a suggestion of singularity, and in the other case of plurality. [12, p.171]

Given this quote, an ambiguous description in this sense is the occurrence of an indefinite phrase "an A" in a context B, viz. "an A is a B".[10] As we saw in the Introduction, the standard formalization of this in first-order logic is:

$$\exists x(A(x) \wedge B(x))$$

where both A and B are unary predicates or formulas.[11] Russell's indefinites can alternatively be expressed in terms of a term-forming operator that is in several ways similar to Hilbert's ϵ-operator. For a given formula A with x occurring free in it, let $\rho_x A(x)$ be a term standing for "an x, such that x has A". This ρ-operator can then be defined in the following way (relative to some context):[12]

$$B(\rho_x A(x)) :\leftrightarrow \exists x(A(x) \wedge B(x)) \qquad \text{(Def } \rho)$$

In this paper, we assume that for Russell indefinite descriptions can function (at least on the surface) as singular terms – as the following quote shows:

> The identity in 'Socrates is a man' is identity between an object named (accepting 'Socrates' as a name, subject to qualifications explained later) and an object ambiguously described. [12, p.172]

Again, nowadays we are more inclined to view a sentence as 'Socrates is a man' as an atomic sentence in which 'is a man' is predicated from (a singular term) 'Socrates'.

With respect to (Def ρ) Russell's famous theory of definite description can be seen as an extension of his account of indefinite descriptions by adding a uniqueness condition to the existential condition already present in (Def ρ). A contextual definition of an indefinite description can also be constructed from the definite description "*the A is a B*", expressed by $\exists x(A(x) \wedge \forall y(A(y) \rightarrow x = y) \wedge B(x))$, simply by dropping the uniqueness clause.[13]

For reasons belonging to Russell's particular approach to proper names, neither definite nor indefinite descriptions belong to the class of proper singular terms.

---

[10]For a closer discussion of the relation between ambiguous and definite descriptions, see also Russell's classical paper [12].

[11]A problem with this existential reading of indefinite descriptions in the formalization of natural language discourse is that $A(x)$ and $B(x)$ are treated symmetrically in $\exists x(A(x) \wedge B(x))$. In the natural language sentence above, this is not necessarily the case. We would like to thank one of the reviewers for bringing this point to our attention.

[12]It should noted here that this operator-based interpretation of indefinite descriptions was not given by Russel himself, but has been developed by the second author of the present article.

[13]Compare Russell's own presentation of his theory of descriptions in [11].

Instead, his stance on both types of descriptions is that they are incomplete symbols, i.e. that the meaning of the (in)definite description is constituted by some context.

In analogy to Russell's considerations for definite descriptions ρ-terms are not given a direct interpretation; rather these terms are interpreted in a *contextual* way, or are defined contextually. If we follow Russell in his understanding of both definite and indefinite descriptions, then there is no need of extending the semantical framework of first order predicate logic since the semantical conditions for formulas containing a ρ-term (or a definite description) can be directly read off the righthand side of (Def ρ).

However, if we wish to to give ρ-terms special semantical considerations, we outline an approach for doing so. This approach is presented rather informally here: given a model $\mathfrak{M}$, let $\mathcal{A} \subseteq dom(\mathfrak{M})$ be the set of objects defined by formula A, and let $\mathcal{B} \subseteq dom(\mathfrak{M})$ be the set defined by formula B. Intuitively speaking, the operator ρ picks out one element in $\mathcal{A}$ that is also in $\mathcal{B}$ (assuming that their intersection is non-empty). The central conceptual idea underlying this Russellian account of indefinites is a kind of semantic *context dependency*, that is the fact that the specification of an A-representative picked out by the operator depends on the particular sentential context in which formula A occurs. In terms of the informal semantics underlying the ρ-operator, this point is given by the constraint that the selection of a particular ρ-representative of set $\mathcal{A}$ is specified only relative to a given 'context' set $\mathcal{B}$ in which the ρ-representative also occurs. Thus, in a slogan, one can say that the reference of a given term is a function of its particular sentential context. Clearly, this only works if neither $\mathcal{A}$ nor $\mathcal{B}$ is empty. In the case that $\mathcal{A}$ is empty, one could think of a solution familiar from the treatment of definite descriptions as done by Carnap. In this case we would require a *chosen object* which is by fiat in the extension of every predicate.

4 Russell and Hilbert indefinites

The relationship between Hilbert's and Russell's accounts of indefinite descriptions can be studied in a precise way by comparing the two underlying logics and their respective term-forming operators. In particular, it can be shown that the two logical representations of "An A is a B" in terms of an epsilon and a rho operator do not coincide.[14] Thus, given two first-order formulas A, B, we can show that:

$$B(\epsilon_x A(x)) \nleftrightarrow B(\rho_x A(x))$$

Proof sketch: To see that the left-to-right implication does not hold, consider a model $\mathfrak{M}$ where $\mathcal{A} = \emptyset$ as well as a choice-function δ interpreting the ϵ-operator

---

[14]The following discussion follows closely the presentation of this result in [13].

such that $\delta(\mathcal{A}) \in \mathcal{B}$. Relative to $\mathfrak{M}$ and δ, the antecedent will turn out true. However, the consequent will be false given that $\mathcal{A} \cap \mathcal{B} = \emptyset$. It follows from this that $\exists x(A(x) \wedge B(x))$ is false and therefore also the right-hand side of the above equivalence. In order to show the right-to-left direction to be non-valid, consider a model where $\mathcal{A} \cap \mathcal{B} \neq \emptyset$ and $\mathcal{A} \not\subseteq \mathcal{B}$. Consider a choice function interpreting the ϵ-operator such that $\delta(\mathcal{A}) = x \notin \mathcal{B}$. The right hand side formula will clearly be true in this model. Nevertheless, the epsilon formula on the left-hand side will be false relative to the particular choice function δ.[15]

This shows that *Hilbert* and *Russell* indefinites are not identical. The main conceptual reason for this fact lies in their different semantic nature. As we saw, Hilbert indefinites are characterized by a specific mode of reference, that is, by the fact that the terms representing such indefinites refer *arbitrarily* to objects in the domain. The indefinite nature of such terms is thus best explained in terms of their arbitrary reference. By contrast, indefiniteness in Russell's sense primarily means *non-uniqueness* of reference (in opposition to his account of definite descriptions). Moreover, another central semantic feature of ρ-terms is, as we saw, the fact that their reference is not specified in isolation, but *contextually*, that is, relative to a given sentential context. This semantic *context dependency* is clearly missing in the choice semantic treatment of the extensional ϵ-logic presented in Section 2. As we saw above, the semantic value of an epsilon term is specified in a model relative to a specific choice function. Given such a choice-functional interpretation, the semantic value of $\epsilon_x A(x)$ remains *stable* under changes of sentential contexts in which the term might occur.

The question remains whether Russell's and Hilbert's accounts of indefinite descriptions, if expressed by means of rho-terms and epsilon-terms respectively, are inter-translatable. The answer to this is positive. In fact, we will present two different ways in which Russell's account of indefinite descriptions can be expressed in a language containing an epsilon operator.

4.1 Russell indefinites in EC

The first way to translate the Russellian account of definition descriptions into epsilon logic is based on the classical language of epsilon terms outlined in section 2. Recall that first-order predicate logic can be embedded into EC based on a translation function specified above. Given that RC is embeddable in predicate logic, it follows that RC must also be interpretable in (classical) EC. In particular, we can translate the ρ-term representation of the indefinite description "An A is a B" into

---

[15]Compare, again, [13] for a discussion of this result.

a quantifier-free statement of $\mathcal{L}_\epsilon$ of the following from:

$$B(\rho_x A(x)) \leftrightarrow (\exists x (A(x) \wedge B(x)))^\epsilon$$
$$\leftrightarrow A(\underbrace{\epsilon_x(A(x) \wedge B(x))}_{e1}) \wedge B(\underbrace{\epsilon_x(A(x) \wedge B(x))}_{e1})$$

Intuitively speaking, the context dependency of the Russell indefinite is reflected here on the right-hand side by the epsilon term built from the conjunction of formulas $A(x)$ and $B(x)$. One could say that context dependency of the semantic interpretation of the ρ-term in the sentential context $B(\rho_x A(x))$ is *internalized* here in a more complex ϵ-term. Notice, moreover, the defining formula on the right-hand side contains two identical epsilon terms $\epsilon_x(A(x) \wedge (B(x))$.[16] The fact that the term occurs in both conjuncts on the right-hand side is central for the translation of Russell indefinite descriptions into EC. Put differently, it can easily be shown that the weaker equivalence statement

$$B(\rho_x A(x)) \leftrightarrow B(\epsilon_x(A(x) \wedge B(x)))$$

is not generally true. To see this, consider a model in which $\mathcal{A} \cap \mathcal{B} = \emptyset$ and a choice function δ interpreting the ϵ-operator such that $\delta(\mathcal{A} \cap \mathcal{B}) \in \mathcal{B}$. (The choice function picks out an arbitrary member of the model domain here that happens to be in $\mathcal{B}$.). The right-hand side of the formula would then be true, but the left-hand side would clearly be false (by definition of the ρ-operator). Counterexamples of this form are explicitly ruled out in the above stronger equivalence statement by the fact that any possible semantic value of the term $\epsilon_x(A(x) \wedge B(x))$ is forced to be a member of *both* $\mathcal{A}$ and $\mathcal{B}$ if the right-hand side of formula is to be true.

4.2 Russell indefinites in indexed EC

While the translation of Russell's indefinite descriptions in the classical language of extensional EC is somewhat cumbersome, it turns out that they can be formulated in a natural extension of it, namely in a language of *indexed* epsilon terms. Indexed epsilon terms have been subject to recent investigation, both in the logical and semantic literature.[17]

Roughly, the language $\mathcal{L}_{\epsilon_i}$ of an indexed epsilon-calculus (IEC) contains ϵ-operators $\epsilon_i x$ indexed by context variables $i, j, \ldots$ or context constants $c, c', \ldots$.

---

[16]Thus, for this simple example of indefinite descriptions we do not have to concern us here with nested epsilon terms and the *rank* and *degree* of epsilon terms. See [10] for further details on this topic.

[17]Such ϵ-terms were first discussed in work by Egli and von Heusinger (e.g. [4], [16]) and subsequently (more systematically) by Mints & Sarenac in [9].

TWO TYPES OF INDEFINITES

If $A(x)$ is a formula with a free variable x, and i a context variable, then $\epsilon_i x A(x)$ is a term of the language. The context variables occurring in such a term can also be bound by existential and universal quantifiers and allow the formulation of sentences such as $\exists i \exists j B(\epsilon_i x A(x), \epsilon_j x A(x))$.[18] Intuitively speaking, the context-indices of an epsilon-symbol represent different contexts (or situations) in which the ϵ-term may occur. The epsilon operator thus picks out one particular A-representative relative to a particular context i, and possibly a different object relative to another context j. This context variability can be expressed semantically in terms of indexed choice functions that are intended to represent such contexts.

The semantic interpretation of $\mathcal{L}_{\epsilon_i}$ differs from that of $\mathcal{L}_\epsilon$ only in one point.[19] In the case of classical epsilon terms, choice functions interpreting the ϵ-operator are considered as *external* to a model. In the case of indexed epsilon terms, a family of possible choice functions is incorporated in the model. A *choice structure* interpreting $\mathcal{L}_{\epsilon_i}$ is thus a triple $\mathfrak{M} = \langle D, I, \mathbb{F} \rangle$ where D and I are interpreted as before and $\mathbb{F}$ is a non-empty set of choice functions. As pointed out in [9], one can understand the context variables of an ϵ-operator as ranging over this collection of choice functions. An assignment function s on $\mathfrak{M}$ then maps individual variables to elements in the domain D and context variables to elements in $\mathbb{F}$. A valuation rule for indexed ϵ-terms can then be specified analogously to the above case:[20]

$$val^{\mathfrak{M},s}(\epsilon_i x A(x)) = s(i)(\{d \in D \mid \mathfrak{M}, s[x/d] \models A(x)\}),$$

where $s(i) \in \mathbb{F}$.

A number of axioms have been introduced in [9] to describe the logical behaviour of indexed epsilon terms. These contain, in particular, the following variants of the axioms of classical EC:

$A(t) \rightarrow A(\epsilon_a x A(x))$, with a a context term.	(Critical formulas)
$\forall x (A(x) \leftrightarrow B(x)) \rightarrow \forall i (\epsilon_i x A(x) = \epsilon_i x B(x))$	(Extensionality)
$\varphi[i/a] \rightarrow \exists i \varphi$, with a and i context terms.	(EI for context variables)

The second axiom of extensionality states that equivalent formulas have the same ϵ-representative in all possible contexts. The third axiom states that if a formula contains an epsilon term with a context constant a, then a can be substituted by an

---

[18] Quantifiers can be defined in the language of indexed epsilon terms in a number of equivalent ways. For instance, the existential quantifier can be specified by $\exists x A(x) :\leftrightarrow \exists i A(\epsilon_i x A(x))$ or by $\exists x A(x) :\leftrightarrow \forall i A(\epsilon_i x A(x))$. See [9, p.619].

[19] The following discussion is based on the presentation given in [9].

[20] See again [9] for a more detailed presentation.

existentially bound context variable i.[21]

Russell's indefinite descriptions turn out to be embeddable in the language of IEC. In particular, it can be shown that the following equivalence holds:[22]

$$B(\rho_x A(x)) \leftrightarrow \exists x(A(x)) \wedge \exists i B(\epsilon_i x A(x))$$

The right hand side of the formula states that (i) set $\mathcal{A}$ is nonempty and (ii) there exists at least one context in which the element picked out from $\mathcal{A}$ by the corresponding choice function also lives in set $\mathcal{B}$. This is precisely the claim also expressed on the left hand side of the formula. Thus, both sides are true if and only if there exists an element in the intersection of $\mathcal{A}$ and $\mathcal{B}$. A proof of this theorem can be given in a combined indexed epsilon and rho-calculus:

Proof sketch: Consider first the left-to-right direction: assume $B(\rho_x A(x))$. Then, by definition (Def ρ), it follows that $\exists x(A(x) \wedge B(x))$ and therefore also $\exists x A(x)$. A theorem in IEC is $\exists x(A(x) \wedge B(x)) \rightarrow \exists i B(\epsilon_i x A(x))$ (see [9, p.622]) Together, these results give us $B(\rho_x A(x)) \rightarrow \exists x(A(x)) \wedge \exists i B(\epsilon_i x A(x))$.

The other direction follows directly from another theorem in IEC, namely $\exists x(A(x)) \wedge \exists i B(\epsilon_i x(A(x))) \rightarrow \exists x(A(x) \wedge B(x))$ and (Def ρ) (see again [9, p.622]).

This result shows that Russell's account of indefinite descriptions (and the *contextual* principle of indefinites implicit in it) can also be represented in terms of Hilbert's epsilon terms if one allows the generalization of the language of EC to include context indices. The main reason for this is that the extended language of IEC allows one to capture also *syntactically* (that is, by means of context variables and context quantifiers) the kind of semantic context sensitivity that is already expressible *metatheoretically* for standard EC in terms of the quantification over choice functions.[23]

5   Conclusion

What did we achieve in this paper? We presented two accounts of indefinite descriptions. The first one was based on Hilbert's ϵ-terms, the second one was Russell's

---

[21] Given this choice semantics, Mints & Sarenac also present a Henkin-style proof of the completeness of IEC. It should be noted here that Hans Leiß (LMU Munich) has recently argued in his talk at EPSILON2015 in Montpellier that the completeness proof in [9] may contain a gap. He has also suggested that a completeness theorem can be proven if context equality in IEC is dropped. See [7] for further details.

[22] The stronger and, arguably, more intuitive equivalence $B(\rho_x A(x)) \leftrightarrow \exists i B(\epsilon_i x A(x))$ does not hold. One can think of a model with $\mathcal{A} = \emptyset$ and an assignment of choice function δ to variable i such that $\delta(\mathcal{A}) \in \mathcal{B}$. It follows that $\exists i B(\epsilon_i x A(x))$ is true and $B(\rho_x A(x))$ is false.

[23] This point has first been stressed by von Heusinger with respect to indefinite noun phrases. As he puts it, IEC allows a "a uniform representation of indefinite [noun phrases] by means of indexed epsilon terms" [16, 261].

account of indefinite descriptions formulated here in terms of what we coined ρ-terms. Both accounts can be seen as formal investigations of indefinites. There are, however, significant differences between them. In the case of Hilbert's epsilon terms, the indeterminacy comes from the inherent understanding of ϵ-terms themselves; a fact that relates directly to the semantic conditions outlined in section 2. In the semantic framework that we have outlined here (i.e. a semantics based on choice functions), the reference of an ϵ-term is always ensured even if the predicate on which the ϵ-term depends on is empty. We then related this choice-functional treatment of epsilon logic to a notion from the philosophy of language, namely arbitrary reference; we claimed that this type of undetermined reference to objects is characteristic for Hilbert's understanding of ϵ-terms as indefinite expressions.

The second approach described in the paper was in the spirit of Russell's conception of indefinite descriptions. The semantic indeterminacy of Russell indefinites can be qualified in two ways: (a) as we have said in section 3, Russell's account of ambiguous descriptions can be seen as a restriction of his view of definite descriptions (simply be dropping the uniqueness condition). Interesting here is the fact that indeterminacy enters by its formal interpretation of formulas containing ρ-terms as existentially quantified sentences. However, there is more, i.e. (b) there is also a context dependency which is expressed by the fact that indefinite descriptions have to occur in a certain context, otherwise they would be meaningless – we are following here Russell's doctrine of *incomplete symbols*. Semantically speaking, there is no real need to extend the usual framework of first order predicate logic. Nonetheless, we have discussed a possibly promising alternative semantical route briefly. A more considerate model-theoretic study of ρ-terms will be subject to future research.

We have then addressed the question concerning interrelations between the two approaches. As was pointed out, there is no direct connection between these two calculi. If we, however, employ the $(.)^\epsilon$-translation (outlined in section 2) to the righthand side of (Def. ρ) we have found that there is a connection of two formal approaches after all. Another interconnection between Hilbert and Russell indefinites has been established by exploiting the fact that Russell's indefinites always rely on some (sentential) context. This was the main thought behind why we have chosen to turn to an indexed ϵ-calculus for the representation of Russell's ambiguous descriptions. The indexed ϵ-calculus has the syntactical means to express both the context dependency of Russell's indefinites (by hand of an index) and furthermore allows us to quantify over indices which is possible in Hilbert's ϵ-calculus only on the semantic side.

The present paper contains a philosophical outline of a rather formal investigation of Russell and Hilbert indefinites. A more rigorous presentation of the material covered here will have to be given in a future paper.

References

[1] W. Ackermann. Begründung des "tertium non datur" mittels der Hilbertschen Theorie der Widerspruchsfreiheit. *Mathematische Annalen*, 93:1–36, 1924.

[2] G. Asser. Theorie der logischen Auswahlfunktionen. *Zeitschrift für mathematische Logik und Grundlagen der Mathematik*, 3:30–68, 1957.

[3] W. Breckenridge and O. Magidor. Arbitrary reference. *Philosophical Studies*, 158(377-400), 2012.

[4] U. Egli and K. von Heusinger. The epsilon operator and E-type pronouns. In U. et al. Egli, editor, *Lexical Knowledge in the Organization of Language*, pages 121–141. Amsterdam: Benjamins, 1995.

[5] D. Hilbert and P. Bernays. *Grundlagen der Mathematik. Vol. II*. Berlin; Heidelberg; New York: Springer, 2nd edition, [1939] 1970.

[6] A. C. Leisenring. *Mathematical Logic and Hilbert's Epsilon-Symbol*. London: Macdonald, 1969.

[7] H. Leiß. Equality of contexts in the indexed ε-calculus. Presentation at *Epsilon2015*, Montpellier, 13 June, 2015.

[8] W. P. M. Meyer Viol. *Instantial Logic. An Investigation into Reasoning with Instances*. ILLC Dissertation Series 1995–11. Ph.D. thesis, University of Utrecht, 1995.

[9] G. M. Mints and D. Sarenac. Completeness of an indexed epsilon-calculus. *Archive for Mathematical Logic*, 42:617–625, 2003.

[10] G. Moser and R. Zach. The epsilon calculus and Herbrand complexity. *Studia Logica*, 82(1):133–155, 2006.

[11] B. Russell. On denoting. *Mind*, 14:479–493, 1905.

[12] B. Russell. *Introduction to Mathematical Philosophy*. London: George Allen and Unwin; New York: The Macmillan Company, 1919.

[13] G. Schiemer and N. Gratzl. The epsilon reconstruction of theories and scientific structuralism. *Erkenntnis*, doi 10.1007/s10670-015-9747-9, 2015.

[14] W. Sieg. Hilbert's programs: 1917-1922. *Bulletin of Symbolic Logic*, 5:1–44, 1999.

[15] Egli U. (eds.) von Heusinger, K., editor. *Reference and Anaphoric Relations*. Dordrecht: Kluwer, 2000.

[16] K. von Heusinger. The reference of indefinites. In K. von Heusinger and U. Egli, editors, *Reference and Anaphoric Relations*, pages 265–284. Dordrecht: Kluwer, 2000.

[17] R. Zach. The practice of finitism. epsilon calculus and consistency proofs in Hilbert's program. *Synthese*, 137:211–259, 2003.

[18] R. Zach. Lectures on the epsilon calculus. *Lecture notes*, 2009.

 Received 30 November 2015

ON EQUALITY OF CONTEXTS AND COMPLETENESS OF THE INDEXED ϵ-CALCULUS

HANS LEIß

Centrum für Informations- und Sprachverarbeitung
Ludwig-Maximilians-Universität München
Oettingenstr.67, 80539 München
`leiss@cis.uni-muenchen.de`

Abstract

G.E. Mints and D. Sarenac [1] (2003) provided a completeness theorem for the "indexed" ϵ-calculus with respect to the class of "choice"-structures. These are generalized second-order structures with a non-empty set of choice functions. The language L_{ϵ_i} extends the first-order language L by quantification over choice functions, an equality predicate $\doteq$ for choice functions, and "indexed ϵ"-terms $\epsilon_i x \alpha$ interpreted as the object chosen by choice function i from the set of objects x satisfying α.

We point out a gap in the completeness proof of [1] that cannot be fixed: with an equality predicate between choice functions, it is impossible to extend every consistent set of formulas to a maximal consistent one. In particular, the theory saying that a and b are different choice functions, but choose the same object for each definable set, is consistent, but has no model of the kind constructed in the completeness proof of Mints and Sarenac.

When the equality between choice functions is removed, a modification of Mints/Sarenac's construction indeed provides a model for each consistent set of sentences. This works even if ϵ-terms $\epsilon i \alpha$ for contexts are admitted, provided one adds weak extensionality axioms for this choice function ϵ for contexts.

However, the modified system seems less adequate for the applications to the semantics of natural language, which gave rise to the introduction of the indexed ϵ-calculus.

Keywords: Indexed epsilon calculus, completeness theorem, context equality

1 Introduction

U. Egli and K. von Heusinger [4, 3, 2] have introduced indexed ϵ-terms to interpret definite and indefinite descriptions using choice-functions depending on a given context: the term $\epsilon_i x \alpha$ denotes the object satisfying α which is the most salient (prominent) such with respect to context i. To model the dynamics of discourse, they assume a salience hierarchy among discourse entities and express changes of this hierarchy through updates of context-dependent choice functions. A choice function is here the abstraction of a "context"; it orders each set N of objects and thereby interprets noun phrases like "an N", "another N", "the third N" etc. in a context-dependent way. The evaluation of an indefinite noun phrase "an N" in context i leads to an updated context j that differs from i only in the ordering it gives to N: the element chosen by i is now, in j, the first, most salient one of N.

G. Mints and D. Sarenac [1] propose a language $L\epsilon$ and a calculus $S\epsilon_i$, the "indexed ϵ-calculus", with context quantifiers $\exists i, \forall i$ ranging over a set of choice functions for the universe. They give a completeness proof in Henkin-style with respect to weak (functional) second-order structures.

We draw attention to the fact that in [1], on the one hand, contexts are considered to be *indices of* choice functions, but on the other hand, context terms are interpreted as choice functions, not as indices of such, and the language $L\epsilon$ has an equality predicate for context terms. A *choice function* for a set A of objects is a function $f : \mathcal{P}(A) \to A$ that assigns an element $f(X) \in X$ to each non-empty $X \subseteq A$ in its domain. The completeness proof of Mints and Sarenac extends a consistent set Γ of $L\epsilon$-sentences to a maximal consistent set Δ in $L\epsilon^+$, the extension of $L\epsilon$ by infinitely many Henkin-constants, and then builds a term model $\mathcal{A}$ from equivalence classes of closed terms of $L\epsilon^+$. The set A of individuals of $\mathcal{A}$ consists of the equivalence classes $[t] := \{\, s \mid t \doteq s \in \Delta \,\}$ of object terms t, while its set $\mathbb{F}$ of functions is a set of choice functions $\Phi_{[a]} : \mathcal{P}(A) \to A$ for A indexed by Δ-equivalence classes $[a] = \{\, b \mid a \doteq b \in \Delta \,\}$ of context terms a. By construction, two elements of $\mathbb{F}$ are equal if and only if they assign the same elements to the same definable sets of individuals.

Thus, according to the term model construction, not only does any consistent set of $L\epsilon$-sentences have a model, but it also has a model where different choice functions differ on some definable set. But this cannot be true. Consider the set

$$\Gamma := \{a \not\doteq b\} \cup \{\, \forall i \forall \vec{y}(\epsilon_a x \alpha \doteq \epsilon_b x \alpha) \mid \alpha(x, \vec{y}, \vec{i}) \text{ an } L\epsilon\text{-formula} \,\}.$$

Suppose Γ is consistent. Then the term model construction provides a model $\mathcal{A}$ where the choice functions assigned to a and b are equal, since they agree on the definable subsets; on the other hand, they must be different, since $a \not\doteq b \in \Gamma$. So Γ must be inconsistent. Then there are finitely many formulas α_k such that

$$\vdash \bigwedge_k \forall \vec{y} \forall \vec{i}(\epsilon_a x \alpha_k \doteq \epsilon_b x \alpha_k) \to a \doteq b.$$

By the equational axioms and proof rules of the calculus $S\epsilon_i$, this improves to an equivalence. Hence, the equality of choice functions is definable, which is also unplausible. This raises some doubts on the completeness proof. Our goal is to clarify the situation.

Omitting the motivating background of applications in linguistics, we present syntax, semantics and proof system of the indexed ϵ-calculus in section 2 and sketch the Mints/Sarenac completeness proof in section 3. We then explain in section 4 where their proof has a gap and provide a counterexample. Section 5 shows that, with suitable modification, the completeness proof works if the equality between context terms is dropped from the language. In the concluding section 6 we indicate why the equality of context terms would, however, be essential for giving a first-order version of the higher-order linguistic theories of dynamic interpretation of discourse that motivated the introduction of the indexed ϵ-calculus.

2 The indexed ϵ-calculus

We present the indexed ϵ-calculus in a notation close to that of Mints/Sarenac [1].

2.1 Syntax and semantics of $L\epsilon$

The language $L\epsilon$ has *context terms* a, *object terms* t, and *formulas* α as defined by the grammar

$$
\begin{aligned}
a &:= i \mid c \\
t &:= x \mid d \mid f(t_1, \ldots, t_n) \mid \epsilon_a x \alpha \\
\alpha &:= \perp \mid t_1 \doteq t_2 \mid a_1 \doteq a_2 \mid P(t_1, \ldots, t_n) \mid \neg \alpha \mid (\alpha_1 \to \alpha_2) \mid \exists i \alpha
\end{aligned}
$$

where x ranges over countably many object variables, i over countably many context variables; c stands for context constants, d for object constants, f for functions and P for predicates on objects. Object quantifiers are given as abbreviations: $\exists x \alpha := \exists i \alpha[x/\epsilon_i x \alpha]$ with fresh i. We omit the standard definitions of free variables and substitutions $[x/t]$ and $[i/a]$ of free variables by terms; let it suffice to mention that $\epsilon_i x \alpha$ binds x, but not i.

A *choice function for a set A* is a function $f : \mathcal{P}(A) \to A$ such that $f(S) \in S$ for every non-empty set $S \subseteq A$. By L we mean the sublanguage of $L\epsilon$ without context terms (and hence without object terms of the form $\epsilon_a x \alpha$). A *choice structure* for $L\epsilon$ is a two-sorted structure

$$
\mathcal{A} = (A, P^{\mathcal{A}}, \ldots, f^{\mathcal{A}}, \ldots, d^{\mathcal{A}}, \ldots; \mathbb{F}, c^{\mathcal{A}}, \ldots)
$$

349

where $(A, P^{\mathcal{A}}, \ldots, f^{\mathcal{A}}, \ldots, d^{\mathcal{A}}, \ldots)$ is an ordinary first-order structure for L and $\mathbb{F}$ is a non-empty set of choice functions for A, with $c^{\mathcal{A}} \in \mathbb{F}$ for every context constant c. Notice that $\mathcal{A}$ is a *weak* functional second order structure: $\mathbb{F} \subseteq \mathcal{P}(A) \to A$ need not be the full set of all choice functions for A, let alone all functions from $\mathcal{P}(A)$ to A.

An *assignment* s *over* $\mathcal{A}$, $s : \mathit{Var} \to \mathcal{A}$, assigns to each object variable x an element $s(x) \in A$ and to every context variable i a choice function $s(i) \in \mathbb{F}$. The *update* of s at x by $b \in A$ is written as $s(x|b)$, similarly $s(i|f)$ for $f \in \mathbb{F}$. The *value* $s(a) \in \mathbb{F}$, $s(t) \in A$ and $s(\alpha) \in \{0,1\}$ of a context term a, an object term t and a formula α is defined via

$$
\begin{aligned}
s(c) &:= c^{\mathcal{A}}, \\
s(d) &:= d^{\mathcal{A}}, \\
s(f(t_1, \ldots, t_n)) &:= f^{\mathcal{A}}(s(t_1), \ldots, s(t_n)), \\
s(\epsilon_a x \alpha) &:= s(a)(\{b \in A \mid s(x|b)(\alpha) = 1\}), \\
s(P(t_1, \ldots, t_n)) &:= \begin{cases} 1, & \text{if } (s(t_1), \ldots, s(t_n)) \in P^{\mathcal{A}}, \\ 0, & \text{else,} \end{cases} \\
s(\exists i \alpha) &:= \max\{ s(i|f)(\alpha) \mid f \in \mathbb{F} \},
\end{aligned}
$$

etc. for the remaining clauses.

By simultaneous induction on the structure of contexts, terms and formulas, one can show that the values do not depend on the assignment at variables which are not free in the argument term or formula:

Lemma 2.1 (Coincidence). *Let s_1, s_2 be assignments over $\mathcal{A}$.*

1. *If $s_1 =_{free(a)} s_2$, then $s_1(a) = s_2(a)$.*

2. *If $s_1 =_{free(t)} s_2$, then $s(t_1) = s_2(t)$.*

3. *If $s_1 =_{free(\alpha)} s_2$, then $s_1(\alpha) = s_2(\alpha)$.*

From this, a further induction on the structure of terms and formulas provides the substitution property, relating syntactic substitution with updates of the assignment:

Lemma 2.2 (Substitution). *For object terms t, u, context terms a, b, formulas α, and assignments $s : \mathit{Var} \to \mathcal{A}$,*

$$
\begin{aligned}
s(t[x/u]) &= s(x|s(u))(t) & s(a[i/b]) &= s(i|s(b))(a) \\
s(\alpha[x/u]) &= s(x|s(u))(\alpha) & s(t[i/b]) &= s(i|s(b))(t) \\
& & s(\alpha[i/b]) &= s(i|s(b))(\alpha).
\end{aligned}
$$

2.2 Deductive system $S\epsilon_i$

We restrict ourselves to the simpler system $S\epsilon_i$ of [1] and ignore the extension $S\epsilon_i^{fin}$ with axioms asserting the existence of choice functions with defined values on finitely many definable sets.

Axioms: 1. All substitution instances of propositional tautologies.

2. Equality axioms:
   for object terms t, s, context terms a, b and atomic formulas α,

$$t \doteq t \qquad a \doteq a$$
$$t \doteq s \rightarrow (\alpha[x/t] \leftrightarrow \alpha[x/s]) \qquad a \doteq b \rightarrow (\alpha[i/a] \leftrightarrow \alpha[i/b])$$

3. Critical formulas: $\alpha[x/t] \rightarrow \alpha[x/\epsilon_a x \alpha]$, if t is substitutable for x in α.

4. Extensionality axioms for choice functions:

$$\forall x(\alpha \leftrightarrow \beta) \rightarrow \forall i(\epsilon_i x \alpha \doteq \epsilon_i x \beta).$$

5. Quantifier axioms for choice functions:

$$\alpha[i/a] \rightarrow \exists i \alpha$$

Rules:

$$\frac{\alpha, \quad (\alpha \rightarrow \beta)}{\beta}, \qquad \frac{\alpha[i/j] \rightarrow \beta}{\exists i \alpha \rightarrow \beta}, \quad j \notin \mathit{free}(\exists i \alpha \rightarrow \beta).$$

In the extensionality axioms, it is tacitly assumed that $i \notin \mathit{free}(\alpha \leftrightarrow \beta)$, as otherwise the axiom would amount to

$$\exists i \forall x(\alpha \leftrightarrow \beta) \rightarrow \forall i(\epsilon_i x \alpha \doteq \epsilon_i x \beta),$$

which is certainly too strong.

Mints and Sarenac claim the following soundness and completeness theorems:

Theorem 2.1 ([1], 5.1, 5.2). *$S\epsilon_i$ is sound: if $\vdash \alpha$, then $\models \alpha$. $S\epsilon_i$ is complete: if $\models \alpha$, then $\vdash \alpha$.*

The proof given in [1] treats the more general case of $T \vdash \alpha$ and $T \models \alpha$ for arbitrary sets T of formulas, showing that a consistent set Γ ($= T \cup \{\neg\alpha\}$) of formulas has a model $\mathcal{A}$. In the following section, we sketch the completeness proof and then point out where it has a gap.

3 Sketch of the completeness proof for $S\epsilon_i$

The proof of theorem 2.1 in [1] is a modification of Henkin's completeness proof for first-order predicate logic as presented in [5]. It consist of an informal first step, which extends a consistent set Γ of sentences to a maximal consistent set $\Delta \supseteq \Gamma$ of sentences with additional constants, and a more explict second step, which defines a term model of Δ. The informal first step can be summarized as:

Lemma 3.1 (?). *Every consistent set Γ of $L\epsilon$-sentences can be extended to a maximal consistent set Δ of $L\epsilon^+$-sentences, where $L\epsilon^+$ is the extension of $L\epsilon$ by infinitely many new context constants.*

The authors of [1] just say: "We extend Γ to a maximal consistent set Δ containing witnesses from a set C of new constants: for any context variable i, and any formula α, there is a constant $c \in C$ such that

$$(\exists i \alpha \rightarrow \alpha[i/c]) \in \Delta."$$

Henkin constants for objects are not needed, as terms $\epsilon_c x \alpha$ can be used instead.

Lemma 3.2 ([1], 5.1). *Δ is deductively closed, i.e. if $\Delta \vdash \alpha$, then $\alpha \in \Delta$.*

Then define the term model $\mathcal{A}$ for Δ as follows: Let A and I be the set of equivalence classes

$$[t] := \{s \mid t \doteq s \in \Delta\} \quad \text{respectively} \quad [a] := \{b \mid a \doteq b \in \Delta\}$$

of closed object resp. context terms of $L\epsilon^+$. Define an L-structure

$$(A, P^{\mathcal{A}}, \dots, f^{\mathcal{A}}, \dots, d^{\mathcal{A}}, \dots)$$

from Δ as usual. Then fix a choice function f for A, define

$$\Phi_{[a]}(S) \quad := \quad \begin{cases} [\epsilon_a x \alpha], & \text{if } S = \{[t] \mid \alpha[x/t] \in \Delta\} \text{ for some formula } \alpha(x), \\ f(S), & \text{otherwise,} \end{cases}$$

and put $\mathcal{A} = (A, P^{\mathcal{A}}, \dots, f^{\mathcal{A}}, \dots, d^{\mathcal{A}}, \dots; \mathbb{F}, c^{\mathcal{A}}, \dots)$ using $\mathbb{F} := \{\Phi_{[a]} \mid [a] \in I\}$ and $c^{\mathcal{A}} := \Phi_{[c]}$.

The maximality of Δ and the axioms on equality, extensionality and critical formulas imply that $\mathcal{A}$ is well-defined and a choice structure for $L\epsilon$. (Details are given in the proof of the corresponding lemma 5.4 in section 5 below.)

The final step in the proof then is that by simultaneous induction on terms and formulas (again, details are as for the corresponding lemma 5.5 below) one gets:

Lemma 3.3. *(cf.[1], 5.8) Let s be a variable assignment for $\mathcal{A}$, such that $s(x_j) = [t_j]$ and $s(i_k) = \Phi_{[a_k]}$. Then*

1. $s(b) = \Phi_{[b[\vec{i}/\vec{a}]]}$, *for each context term $b(\vec{i})$.*

2. $s(u) = [u[\vec{x}/\vec{t}, \vec{i}/\vec{a}]]$, *for each object term $u(\vec{x}, \vec{i})$.*

3. $s(\alpha) = 1$ *if and only if $\alpha[\vec{x}/\vec{t}, \vec{i}/\vec{a}] \in \Delta$, for each formula $\alpha(\vec{x}, \vec{i})$.*

By the last clause, it follows that $\mathcal{A}$ is a model of Δ, hence of Γ.

4 Incompleteness of $S\epsilon_i$

Although the calculus $S\epsilon_i$ is called the *indexed ϵ-calculus*, the choice structures $\mathcal{A}$ do not come with a space I of indices as *names* of choice functions; rather, the indices or context terms a are interpreted directly as choice functions for A. In the constructed term model $\mathcal{A}$, different indices $[a] \neq [b]$ correspond to different choice functions, because if $[a] \neq [b]$, then $a \doteq b \notin \Delta$, hence $a \neq b \in \Delta$ since Δ is maximal, hence $s(i \neq j) = 1$ for assignments s with $s(i) = \Phi_{[a]}$, $s(j) = \Phi_{[b]}$ by Lemma 3.3, 3., hence $\Phi_{[a]} \neq \Phi_{[b]}$ (a case skipped over in [1]). It follows that

$$a \doteq b \in \Delta \iff \Phi_{[a]} = \Phi_{[b]}. \tag{1}$$

Moreover, two choice functions of $\mathbb{F}$ are equal if they agree on the definable subsets of A, because on undefinable subsets they agree with the fixed choice function f. Let Def_A be the set of all object sets

$$\{ d \in A \mid s(x|d)(\alpha) = 1 \}$$

which are definable by $L\epsilon$-formulas $\alpha(x, \vec{y}, \vec{i})$, using parameters from A and $\mathbb{F}$ via assignments s. By Lemma 3.3, 3. these are the sets

$$S = \{ [t] \mid \alpha[x/t] \in \Delta \} \tag{2}$$

for $L\epsilon^+$-formulas $\alpha(x)$. Two functions $\Phi_{[a]}, \Phi_{[b]}$ agree on (2) iff $\epsilon_a x \alpha \doteq \epsilon_b x \alpha \in \Delta$, so that by (1),

$$a \doteq b \in \Delta \iff \{ \epsilon_a x \alpha \doteq \epsilon_b x \alpha \mid \alpha(x) \text{ an } L\epsilon^+\text{-formula} \} \subseteq \Delta. \tag{3}$$

The problem now is:

Problem 4.1. *How is (3) achieved in the construction of Δ from an initial consistent set Γ? Is there really a proof of Lemma 3.1?*

The familiar construction for first-order logic builds an increasing chain $\Gamma =:$ $\Delta_0 \subseteq \Delta_1 \subseteq \ldots$ of consistent sets $\Delta_n \subseteq \Delta_{n+1}$ whose union Δ is maximal consistent. Using an enumeration $\{\varphi_n \mid n \in \mathbb{N}\}$ of all formulas, with suitable repetitions, Δ_{n+1} is $\Delta_n \cup \{\varphi_n\}$ or $\Delta_n \cup \{\neg\varphi_n\}$, depending on whether the first is consistent or not. But in the case of $L\epsilon^+$, this apparently does not work. Suppose φ_n is $a \doteq b$, and $\Delta_n \cup \{a \doteq b\}$ is inconsistent. Then $\Delta_n \vdash a \not\doteq b$. Therefore, $\Delta \supseteq \Delta_n$ must contain some formula $\epsilon_a x \alpha \not\doteq \epsilon_b x \alpha$. So why is there some formula $\alpha(x) \in L\epsilon^+$ such that $\Delta_{n+1} := \Delta_n \cup \{\epsilon_a x \alpha \not\doteq \epsilon_b x \alpha\}$ is consistent? Δ_n need not already contain such an inequation between ϵ-terms, as it might just contain the inequation $a \not\doteq b$.

Notice that in $S\epsilon_i$, equality for objects and equality for contexts have the same properties, except for the additional extensionality axioms for choice functions. But clearly, equality of choice functions $f, g : \mathcal{P}(A) \to A$ is a second-order notion:

$$f = g \iff \forall B \in \mathcal{P}(A)(f(A) = g(A)).$$

This makes it unplausible that $S\epsilon_i$ can be complete, if $\doteq$ is meant to be identity of choice functions. Below we construct a counterexample against the completeness part of theorem 2.1.

In section 5 we will check that a completeness theorem holds for an indexed ϵ-calculus *without* $\doteq$ for context terms. It seems plausible that one can have a completeness result for an indexed ϵ-calculus with respect to generalized three-sorted choice structures $(A, \mathcal{D}, \mathbb{F}, \ldots)$ where $\mathcal{D} \subseteq \mathcal{P}(A)$, $\mathbb{F} \subseteq \mathcal{P}(A) \to A$ is a set of choice functions for A, and equality $\doteq$ on $\mathbb{F}$ is axiomatized by

$$i \doteq j \leftrightarrow \forall P(\epsilon_i x P(x) \doteq \epsilon_j x P(x))$$

with set quantifiers ranging over $\mathcal{D}$, comprehension axioms and $\exists/\forall$-rules for set quantifiers. In the term model construction, to satisfy $a \not\doteq b$ one would add a new set constant D and force a and b to disagree on D, where D would turn out not to be a definable set of objects of the term model. However, we havn't checked the details.

4.1 Counterexample against the completeness of $S\epsilon_i$

As sketched in the introduction, the completeness theorem for $S\epsilon_i$ is intuitively challenged by the theory claiming the existence of two different choice functions that yet agree on all definable sets. So, let

$$\Gamma(a, b) := \{a \not\doteq b\} \cup \{\forall \vec{y} \forall \vec{i}(\epsilon_a x \alpha \doteq \epsilon_b x \alpha) \mid \alpha(x, \vec{y}, \vec{i}) \text{ an } L\epsilon\text{-formula}\},$$

where a and b are two context variables[1] that are not used as bound variables below.

---

[1] We do not consider a, b new constants in order not to change L and the notion of $L\epsilon$-definability.

First note that the consistency of $\Gamma(a, b)$ contradicts the model construction of Mints/Sarenac. If $\Gamma(a, b)$ is consistent, (by lemma 3.1) there is a maximal consistent extension $\Delta \supseteq \Gamma(a, b)$. In the completion process, universal quantifiers in $\Gamma(a, b)$ are instantiated by all possible parameters, so that

$$\{ \epsilon_a x \alpha \doteq \epsilon_b x \alpha \mid \alpha(x) \text{ an } L\epsilon^+\text{-formula} \} \subseteq \Delta.$$

Hence, by (3), $a \doteq b \in \Delta$, and by the term model construction, a and b get the same interpretation, $\Phi_{[a]} = \Phi_{[b]}$. But then the term model satisfies $a \doteq b$, while $a \not\doteq b \in \Delta$.

Lemma 4.1. $\Gamma(a, b)$ *is consistent.*

Proof. Let $\mathcal{A}$ be a choice structure for $L\epsilon$ with a countably infinite set A of objects and a countable set $\mathbb{F}$ of choice functions for A. Pick some $f \in \mathbb{F}$ and use it to interpret a. If there is $g \in \mathbb{F}$ such that $(\mathcal{A}, f, g)$ is a model of $\Gamma(a, b)$, we are done.

Otherwise, since $Def_{\mathcal{A}}$ is countable, there is some set $B \subseteq A$ such that $B \notin Def_{\mathcal{A}}$. Let g be a choice function for A that differs from f just on the set B. Let $\mathcal{A}'$ be the choice structure $\mathcal{A}$ with function space $\mathbb{F}' := \mathbb{F} \cup \{g\}$ instead of $\mathbb{F}$. If $B \notin Def_{\mathcal{A}'}$, we can interpret b by g and have a model $(\mathcal{A}', f, g)$ of $\Gamma(a, b)$, since $f(S) = g(S)$ for *all* subsets of A except B, in particular, for all $L\epsilon$-definable object sets S.[2]

To show $B \notin Def_{\mathcal{A}'}$, we now show $Def_{\mathcal{A}'} \subseteq Def_{\mathcal{A}}$. In particular, B is not $L\epsilon$-definable with f and g as parameters. (The reverse inclusion $Def_{\mathcal{A}} \subseteq Def_{\mathcal{A}'}$ is clear: in $\mathcal{A}'$, the parameter g can be used to restrict quantification over $\mathbb{F}'$ to quantification over $\mathbb{F}$.)

Claim: For any assignment s' over $\mathcal{A}'$ that maps a, b to f, g there is a translation $\cdot^{s'}$ of $L\epsilon$-terms and -formulas such that for all object terms t and formulas α:

$$s'(t) = s(\overline{t}^{s'}) \quad \text{and} \quad s'(\alpha) = s(\overline{\alpha}^{s'}), \tag{4}$$

where s is the assignment over $\mathcal{A}$ such that

$$s(x) := s'(x) \quad \text{and} \quad s(j) := \begin{cases} f, & \text{if } s'(j) = g \\ s'(j) & \text{else} \end{cases}$$

for object variables x and context variables j.

The translation $\cdot^{s'}$ depends on s' (but not on what s' does on object variables), since context variables mapped to g by s' need a special treatment. The idea is to replace equations between context terms involving value g to atomic formulas $\top$

---

[2]But notice that since formulas may contain quantifiers ranging over $\mathbb{F}$, the meaning of a defining formula changes when $\mathbb{F}$ is extended. Likewise, $\mathcal{A}$ and $\mathcal{A}'$ may not satisfy the same $L\epsilon$-sentences.

or $\bot$, replace quantification over $\mathbb{F}'$ to quantification over $\mathbb{F}$ and a special case for witness g, and replace a choice with g from a definable set in $\mathcal{A}'$ by a choice with f from the set equivalently defined without referring to g.

For object terms t, the translation $\overline{t}^{s'}$ is given by the clauses

- $\overline{x}^{s'} = x$,

- $\overline{d}^{s'} = d$,

- $\overline{h(t_1, \ldots, t_n)}^{s'} = h(\overline{t_1}^{s'}, \ldots, \overline{t_n}^{s'})$,

- $\overline{\epsilon_i x \alpha}^{s'} = \epsilon_i x \overline{\alpha}^{s'}$ for context variables i,

- $\overline{\epsilon_c x \alpha}^{s'} = \epsilon_c x \overline{\alpha}^{s'}$ for context constants c.

For formulas α, the translation $\overline{\alpha}^{s'}$ is given by the clauses

- $\overline{t_1 \dot{=} t_2}^{s'} = \overline{t_1}^{s'} \dot{=} \overline{t_2}^{s'}$,

- $\overline{a_1 \dot{=} a_2}^{s'} = \begin{cases} a_1 \dot{=} a_2, & \text{if none of } a_1, a_2 \text{ is a variable mapped to } g \text{ by } s', \\ \bot, & \text{if exactly one of } a_1, a_2 \text{ is a variable mapped to } g \text{ by } s', \\ \top, & \text{else.} \end{cases}$

- $\overline{P(t_1, \ldots, t_n)}^{s'} = P(\overline{t_1}^{s'}, \ldots, \overline{t_n}^{s'})$,

- $\overline{(\alpha_1 \vee \alpha_2)}^{s'} = (\overline{\alpha_1}^{s'} \vee \overline{\alpha_2}^{s'})$,

- $\overline{\neg \alpha}^{s'} = \neg \overline{\alpha}^{s'}$,

- $\overline{\exists j \alpha}^{s'} = (\overline{\alpha}^{s'(j|g)}[j/a] \vee \exists j \overline{\alpha}^{s'})$,

using the connective $\vee$ rather than $\rightarrow$ to make the result more readable.

It is routine to prove (4) by induction on the nesting depth of ϵ-operators in terms and formulas. We only show the slightly subtle cases:

1. If t is $\epsilon_j x \alpha$ and the claim is true for α and $s'(x|d)$, then

$$\begin{aligned} s'(\epsilon_j x \alpha) &= s'(j)(\{ d \mid s'(x|d)(\alpha) = 1 \}) \\ &= s'(j)(\{ d \mid s(x|d)(\overline{\alpha}^{s'(x|d)}) = 1 \}) \\ &= s'(j)(\{ d \mid s(x|d)(\overline{\alpha}^{s'}) = 1 \}) \\ &= s(j)(\{ d \mid s(x|d)(\overline{\alpha}^{s'}) = 1 \}) \\ &= s(\epsilon_j x \overline{\alpha}^{s'}) \\ &= s(\overline{\epsilon_j x \alpha}^{s'}). \end{aligned}$$

For the middle step, notice that if $s'(j) = g$, then $s(j) = f$ chooses the same element from the set $\{\, d \mid s(x|d)(\overline{\alpha}^{s'}) = 1 \,\} \in Def_{\mathcal{A}}$.

2. If α is $a_1 \doteq a_2$ and these are context variables with $s'(a_1) = g \neq s'(a_2)$, then

$$s'(a_1 \doteq a_2) = 0 = s(\bot) = s(\overline{a_1 \doteq a_2}^{s'}),$$

and likewise in the other cases of context equations.

3. If α is $\exists j\beta$ and the claim is true for β and $s'(j|g)$ resp. $s'(j|h)$, then

$$
\begin{aligned}
s'(\exists j\beta) = 1 \quad &\Longleftrightarrow \quad \text{for some } h \in \mathbb{F}' \; s'(j|h)(\beta) = 1 \\
&\Longleftrightarrow \quad s'(j|g)(\beta) = 1 \text{ or for some } h \in \mathbb{F} \; s'(j|h)(\beta) = 1 \\
&\Longleftrightarrow \quad s(j|f)(\overline{\beta}^{s'(j|g)}) = 1 \text{ or for some } h \in \mathbb{F} \; s(j|h)(\overline{\beta}^{s'(j|h)}) = 1 \\
&\Longleftrightarrow \quad s(j|f)(\overline{\beta}^{s'(j|g)}) = 1 \text{ or for some } h \in \mathbb{F} \; s(j|h)(\overline{\beta}^{s'}) = 1 \\
&\Longleftrightarrow \quad s(j|f)(\overline{\beta}^{s'(j|g)}) = 1 \text{ or } s(\exists j \overline{\beta}^{s'}) = 1 \\
&\Longleftrightarrow \quad s(\overline{\beta}^{s'(j|g)}[j/a]) = 1 \text{ or } s(\exists j \overline{\beta}^{s'}) = 1 \\
&\Longleftrightarrow \quad s(\overline{\beta}^{s'(j|g)}[j/a] \vee \exists j \overline{\beta}^{s'}) = 1 \\
&\Longleftrightarrow \quad s(\overline{\exists j\beta}^{s'}) = 1.
\end{aligned}
$$

Notice that we may assume $s'(j) \in \mathbb{F}$ and hence $\overline{\beta}^{s'(j|h)} = \overline{\beta}^{s'}$. Moreover, $\overline{\beta}^{s'(j|g)}$ may have free occurrences of j as ϵ-index, but since $s'(a) = f = s(a)$, we have $s(j|f)(\overline{\beta}^{s'(j|g)}) = s(\overline{\beta}^{s'(j|g)}[j/a])$ by the substitution lemma 2.2.

It follows from (4) that if an object set is defined in $\mathcal{A}'$ as $B' = \{\, d \mid s'(x|d)(\alpha) = 1 \,\}$, then it is defined in $\mathcal{A}$ as $B' = \{\, d \mid s(x|d)(\overline{\alpha}^{s'}) = 1 \,\}$. Hence, from $B \notin Def_{\mathcal{A}}$ we get $B \notin Def_{\mathcal{A}'}$. $\qquad\square$

As remarked above, lemma 4.1 refutes the claim of Mints and Sarenac that every consistent set of $L\epsilon$-sentences could be extended to a maximal consistent set of $L\epsilon^+$-sentences (lemma 3.1). Since $\Gamma(a, b)$ is infinite, it does not immediatly refute the completeness claim for single sentences, theorem 2.1.

5 Completeness of the indexed ϵ-calculus without equality for contexts

We now show that for a consistent set of sentences that don't use the $\doteq$-predicate for context terms, a modification of the construction by Mints and Sarenac indeed provides a term model.

5.1 Syntax and semantics of L'_{ϵ_i}

The language L'_{ϵ_i} we use in the following differs from Mints/Sarenac's language $L\epsilon$ in that it doesn't have equations $a_1 \doteq a_2$ between context terms among the formulas.

We define *context terms* a, *object terms* t, and *formulas* α by the grammar

$$
\begin{aligned}
a & := i \mid c \\
t & := x \mid d \mid h(t_1, \ldots, t_n) \mid \epsilon_i x \alpha \\
\alpha & := \bot \mid t_1 \doteq t_2 \mid P(t_1, \ldots, t_n) \mid \neg\alpha \mid (\alpha_1 \to \alpha_2) \mid \exists i\alpha,
\end{aligned}
$$

where i ranges over an infinite set of *context variables*, x over an infinite set of *object variables*. Abbreviation: $\exists x\alpha := \alpha[x/\epsilon_c x\alpha]$ for some fixed context constant c.

The semantics for L'_{ϵ_i} is the same as for $L\epsilon$, i.e. an L'_ϵ-structure

$$
(A, P^{\mathcal{A}}, \ldots, h^{\mathcal{A}}, \ldots, d^{\mathcal{A}}, \ldots; \mathbb{F}, c^{\mathcal{A}}, \ldots)
$$

is a first-order structure $(A, P^{\mathcal{A}}, \ldots, h^{\mathcal{A}}, \ldots, d^{\mathcal{A}}, \ldots)$ for L, extended by a nonempty set $\mathbb{F} \subseteq \mathcal{P}(A) \to A$ of choice-functions for A and an interpretation of context constants c by choice-functions $c^{\mathcal{A}} \in \mathbb{F}$.

The coincidence and substitution lemmas hold as for $L\epsilon$.

5.2 The indexed ϵ-calculus S'_{ϵ_i} without context equality

We use the same axioms and rules as Mints/Seranac, except that (a) equational axioms and rules for equality between context terms are omitted and (b) the extensionality axioms for choice functions are strengthened.

Axioms: 1. All substitution instances into propositional tautologies.

2. Equality axioms for object terms:

$$
t \doteq t, \qquad t \doteq s \to (\alpha[x/t] \leftrightarrow \alpha[x/s])
$$

3. Critical formulas: $\alpha[x/t] \to \alpha[x/\epsilon_a x\alpha]$

4. Extensionality axioms for choice functions:

$$
\forall x(\alpha \leftrightarrow \beta) \to \epsilon_a x\alpha \doteq \epsilon_a x\beta
$$

5. Quantifier axioms: $\alpha[i/a] \to \exists i\alpha$

Rules:

$$
\frac{\alpha, \quad (\alpha \to \beta)}{\beta}, \qquad \frac{\alpha[i/j] \to \beta}{\exists i\alpha \to \beta}, \quad j \notin \mathit{free}(\exists i\alpha \to \beta).
$$

In the above extensionality axioms, the context term a may occur in $(\alpha \leftrightarrow \beta)$. These axioms could equivalently be stated as

$$\forall i(\forall x(\alpha \leftrightarrow \beta) \rightarrow \epsilon_i x \alpha \doteq \epsilon_i x \beta)$$

where i may be free in α, β. They imply the extensionality axiom of Mints/Sarenac,

$$\forall x(\alpha \leftrightarrow \beta) \rightarrow \forall i(\epsilon_i x \alpha \doteq \epsilon_i x \beta),$$

in which $i \notin \mathit{free}(\alpha, \beta)$. In von Heusinger's use of indexed ϵ-terms to interpret definite noun phrases, *the N* in context i is expressed as $\epsilon_i x N(x)$, while *the other N* or *the second N* in context i is expressed by $\epsilon_i y(N(y) \wedge y \not\doteq \epsilon_i x N(x))$. Hence, as *the second N* and *the first N* should be treated similarly, the extensionality of a choice function must not be restricted to properties independent of the choice function.

5.3   Completeness theorem for S'_{ϵ_i}

Without $\doteq$ for context terms in the language, the analog of lemma 3.1 holds. Let C be an infinite set of new context constants and $L'_{\epsilon_i}(C)$ the extension of L'_{ϵ_i} by C.

Lemma 5.1. *Every consistent set Γ of L'_{ϵ_i}-sentences has a maximal consistent extension $\Delta \supseteq \Gamma$ in the language $L'_{\epsilon_i}(C)$.*

Proof. Since the problematic case where Γ contains an inequation between context terms no longer occurs, the proof familiar from the standard case of L-sentences works. Let $\{\varphi_n \mid n \in \mathbb{N}\}$ be an enumeration of all $L'_{\epsilon_i}(C)$-sentences, repeating every sentence infinitely often. Put $\Delta_0 := \Gamma$, Δ_{n+1} as follows, depending on φ_n:

- φ_n is of the form $\bot$, $t_1 \doteq t_2$, $P(t_1, \ldots, t_k)$, $\neg\alpha$ or $(\alpha_1 \rightarrow \alpha_2)$:

$$\Delta_{n+1} := \begin{cases} \Delta_n \cup \{\varphi_n\}, & \text{if this is consistent,} \\ \Delta_n \cup \{\neg\varphi_n\}, & \text{else.} \end{cases}$$

- φ_n is $\exists i\alpha$: choose $c \in C$ not occurring in formulas of $\Delta'_n := \Delta_n \cup \{\varphi_n\}$; put

$$\Delta_{n+1} := \begin{cases} \Delta_n, & \text{if } \Delta'_n \text{ is inconsistent,} \\ \Delta'_n, & \text{else, if } \alpha[i/a] \in \Delta_n \text{ for some closed term } a, \\ \Delta'_n \cup \{\alpha[i/c]\}, & \text{else.} \end{cases}$$

By induction, each Δ_n and hence $\Delta := \bigcup\{\Delta_n \mid n \in \mathbb{N}\}$ is consistent. Since each $L'_{\epsilon_i}(C)$-sentence is considered infinitely often, Δ is maximal consistent. $\square$

Lemma 5.2. *Every maximal consistent set Δ of $L'_{\epsilon_i}(C)$-sentences has a model $\mathcal{A}$.*

We first define $\mathcal{A}$ and show that it is a choice structure for Δ in the subsequent lemmas.

Define the relations $s =_\Delta t$ between closed object terms s, t and $a \approx_\Delta b$ between closed context terms a, b of $L'_{\epsilon_i}(C)$ by

$$s =_\Delta t \quad :\Longleftrightarrow \quad s \doteq t \in \Delta,$$
$$a \approx_\Delta b \quad :\Longleftrightarrow \quad \{\, \epsilon_a x \beta \doteq \epsilon_b x \beta \mid \beta(x) \text{ an } L'_{\epsilon_i}(C)\text{-formula} \,\} \subseteq \Delta.$$

By the equational axioms and the fact that Δ is deductively closed, $=_\Delta$ is a congruence relation on the closed object terms, and $\approx_\Delta$ is an equivalence relation on the closed context terms.

Lemma 5.3. *If $a \approx_\Delta b$, then*

1. *$t[i/a] \doteq t[i/b] \in \Delta$, for each object term $t(i)$, and*

2. *$\alpha[i/a] \in \Delta$ iff $\alpha[i/b] \in \Delta$, for each formula $\alpha(i)$ of $L'_{\epsilon_i}(C)$.*

Proof. By simultaneous induction over terms and formulas. Consider 1. for the term $\epsilon_i x \beta$. We have $\forall x(\beta[i/a] \leftrightarrow \beta[i/b]) \in \Delta$, for otherwise, since Δ is maximal consistent, $\exists x \neg(\beta[i/a] \leftrightarrow \beta[i/b]) \in \Delta$, and then there is a closed term t such that $\beta[i/a][x/t] \in \Delta$ iff $\beta[i/b][x/t] \notin \Delta$, contradicting the induction hypothesis for the formula $\beta[x/t]$. Since Δ is closed under the extensionality axioms, $\epsilon_a x \beta[i/a] \doteq \epsilon_a x \beta[i/b] \in \Delta$, and because $a \approx_\Delta b$, we have $\epsilon_a x \beta[i/b] \doteq \epsilon_b x \beta[i/b] \in \Delta$. Since Δ is closed under the equality axioms, it follows that $(\epsilon_i x \beta)[i/a] \doteq (\epsilon_i x \beta)[i/b] \in \Delta$.

To show 2. for atomic formulas, use 1. and the equality axioms. $\square$

Let $[t]$ be the $=_\Delta$-congruence class of t and $[a]$ be the $\approx_\Delta$-equivalence class of a. Define a model $\mathcal{A}$ for Δ as follows. The universe A of objects of $\mathcal{A}$ is

$$A := \{\, [t] \mid t \text{ a closed object term} \,\}.$$

The object constants d, function symbols h and relation symbols P of L are interpreted, as usual, by

$$d^{\mathcal{A}} \quad := \quad [d],$$
$$h^{\mathcal{A}}([t_1], \ldots, [t_n]) \quad := \quad [h(t_1, \ldots, t_n)],$$
$$P^{\mathcal{A}}([t_1], \ldots, [t_n]) \quad := \quad \begin{cases} 1, & \text{if } P(t_1, \ldots, t_n) \in \Delta, \\ 0, & \text{else.} \end{cases}$$

Since Δ is maximal consistent and deductively closed, $(A, P^{\mathcal{A}}, \ldots, h^{\mathcal{A}}, \ldots, d^{\mathcal{A}}, \ldots)$ is a well-defined first-order structure for L.

To add a universe $\mathbb{F}$ of choice functions for A, let

$$I := \{ [a] \mid a \text{ a closed context term} \}$$

be the set of $\approx_\Delta$-equivalence classes of closed context terms and $f : \mathcal{P}(A) \to A$ be a fixed choice function for A. For $[a] \in I$, put

$$\Phi_{[a]}(S) := \begin{cases} [\epsilon_a x\alpha], & \text{if } S = \{[t] \mid \alpha[x/t] \in \Delta\} \text{ for some formula } \alpha(x), \\ f(S), & \text{otherwise.} \end{cases}$$

Then take

$$\mathbb{F} := \{ \Phi_{[a]} \mid [a] \in I \}, \qquad c^{\mathcal{A}} := \Phi_{[c]},$$

and put

$$\mathcal{A} := (A, P^{\mathcal{A}}, \ldots, h^{\mathcal{A}}, \ldots, d^{\mathcal{A}}, \ldots; \mathbb{F}, c^{\mathcal{A}}, \ldots).$$

Lemma 5.4. $\mathcal{A}$ is a well-defined L'_{ϵ_i}-structure. In particular,

1. $\Phi_{[a]}(S)$ does not dependent on the representative of $[a]$ or the defining formula of S,

2. $\Phi_{[a]}$ is a choice function for A.

Proof. 1. Suppose $a \approx_\Delta b$, and $S_\alpha = S_\beta$ for $S_\alpha := \{ [t] \mid \alpha[x/t] \in \Delta \}$, $S_\beta := \{ [t] \mid \beta[x/t] \in \Delta \}$. Then $\forall x(\alpha \leftrightarrow \beta) \in \Delta$, because otherwise, since Δ is maximal consistent, $\neg\forall x(\alpha \leftrightarrow \beta) \in \Delta$, hence $\exists x\neg(\alpha \leftrightarrow \beta) \in \Delta$, hence for some t, $\{\neg\alpha[x/t], \beta[x/t]\} \subseteq \Delta$ or $\{\alpha[x/t], \neg\beta[x/t]\} \subseteq \Delta$, which contradicts $S_\alpha = S_\beta$. By the extensionality axiom (in its strong form above), it follows that

$$\{\epsilon_a x\alpha \doteq \epsilon_a x\beta, \epsilon_b x\alpha \doteq \epsilon_b x\beta\} \subseteq \Delta.$$

From $a \approx_\Delta b$, we know $\epsilon_a x\beta \doteq \epsilon_b x\beta \in \Delta$. With the equality axioms for object terms and as Δ is deductively closed, we get $\epsilon_a x\alpha \doteq \epsilon_b x\beta \in \Delta$, hence $\Phi_{[a]}(S_\alpha) = [\epsilon_a x\alpha] = [\epsilon_b x\beta] = \Phi_{[b]}(S_\beta)$. If $S \subseteq A$ is not definable, then $\Phi_{[a]}(S) = f(S) = \Phi_{[b]}(S)$. So $\Phi_{[a]} = \Phi_{[b]}$ is a well-defined function on $\mathcal{P}(A)$.

2. If $\emptyset \neq S$ is undefinable, then $\Phi_{[a]}(S) = f(S) \in S$. If $\emptyset \neq S = \{ [t] \mid \alpha[x/t] \in \Delta \}$, there is a term t with $\alpha[x/t] \in \Delta$. By the axiom of critical formulas, $\alpha[x/\epsilon_a x\alpha] \in \Delta$, hence $\Phi_{[a]}(S) = [\epsilon_a x\alpha] \in S$. So $\Phi_{[a]}$ is a choice function for A. Hence, the term model $\mathcal{A}$ of Δ is a well-defined choice structure. $\square$

It remains to be seen that $\mathcal{A}$ is a model of Δ, hence of Γ.

Lemma 5.5. *Let s be an assignment over $\mathcal{A}$ such that $s(x_j) = [t_j]$ for $j < |\vec{x}|$ and $s(i_k) = \Phi_{[a_k]}$ for $k < |\vec{i}|$. Then*

1. $s(b) = \Phi_{[b[\vec{i}/\vec{a}]]}$ *for each context term $b(\vec{i})$.*

2. $s(u) = [u[\vec{x}/\vec{t}, \vec{i}/\vec{a}]]$, *for each object term $u(\vec{x}, \vec{i})$,*

3. $s(\alpha) = 1$ *if and only if $\alpha[\vec{x}/\vec{t}, \vec{i}/\vec{a}] \in \Delta$, for each formula $\alpha(\vec{x}, \vec{i})$.*

Proof. By induction on the complexity of terms and formulas, 1.-3. are shown simultaneously, for all assignments over $\mathcal{A}$.

1. $s(i_k) = \Phi_{[a_k]} = \Phi_{[i_k[\vec{i}/\vec{a}]]}$, and $s(c) = c^{\mathcal{A}} = \Phi_{[c]} = \Phi_{[c[\vec{i}/\vec{a}]]}$.

2. We only consider the most interesting case:

$$
\begin{aligned}
s(\epsilon_a x \alpha) &= s(a)(\{\, [t] \mid s(x|[t])(\alpha) = 1 \,\}) \\
&= \Phi_{[a[\vec{i}/\vec{a}]]}(\{\, [t] \mid \alpha[\vec{x}/\vec{t}, \vec{i}/\vec{a}, x/t] \in \Delta \,\}) \quad \text{(1. for s, 3. for $s(x|[t])$)} \\
&= \Phi_{[a[\vec{i}/\vec{a}]]}(\{\, [t] \mid \alpha[\vec{x}/\vec{t}, \vec{i}/\vec{a}][x/t] \in \Delta \,\}) \quad \text{(since $\vec{t}, \vec{a}$ are closed)} \\
&= [\epsilon_{a[\vec{i}/\vec{a}]} x \alpha[\vec{x}/\vec{t}, \vec{i}/\vec{a}]] \\
&= [(\epsilon_a x \alpha)[\vec{x}/\vec{t}, \vec{i}/\vec{a}]].
\end{aligned}
$$

3. If α is an equation $t \doteq u$, then

$$
\begin{aligned}
s(t \doteq u) = 1 &\iff s(t) = s(u) \\
&\iff [t[\vec{x}/\vec{t}, \vec{i}/\vec{a}]] = [u[\vec{x}/\vec{t}, \vec{i}/\vec{a}]] \quad \text{(by 2. for t, u and s)} \\
&\iff t[\vec{x}/\vec{t}, \vec{i}/\vec{a}] \doteq u[\vec{x}/\vec{t}, \vec{i}/\vec{a}] \in \Delta \\
&\iff (t \doteq u)[\vec{x}/\vec{t}, \vec{i}/\vec{a}] \in \Delta.
\end{aligned}
$$

Similarly, if α is an atomic formula $P(u_1, \ldots, u_n)$. By induction, the claim holds for the negation and implication of formulas. Finally, if α is an existential formula, $\exists i \beta$, we have

$$
\begin{aligned}
s(\exists i \beta) = 1 &\iff \text{for some } f \in \mathbb{F}, \ s(i|f)(\beta) = 1 \\
&\iff \text{for some } [a] \in I, \ s(i|\Phi_{[a]})(\beta) = 1 \\
&\iff \text{for some } [a] \in I, \ \beta[\vec{x}/\vec{t}, \vec{i}/\vec{a}, i/a] \in \Delta \quad \text{(3. for β, $s(i|\Phi_{[a]})$)} \\
&\iff \text{for some } [a] \in I, \ \beta[\vec{x}/\vec{t}, \vec{i}/\vec{a}][i/a] \in \Delta \\
&\iff (\exists i.\beta[\vec{x}/\vec{t}, \vec{i}/\vec{a}]) \in \Delta \quad \text{(by quantifier and Henkin axioms)} \\
&\iff (\exists i \beta)[\vec{x}/\vec{t}, \vec{i}/\vec{a}] \in \Delta.
\end{aligned}
$$

It follows that $\mathcal{A}$ is a model of Δ. $\qquad\qquad\qquad\qquad\qquad\qquad\qquad$ $\square$

5.4 Adding a choice function ϵ for contexts

While definite and indefinite noun phrases are represented in predicate logic using quantifiers over individuals, their representation by indexed ϵ-terms uses quantifiers over contexts resp. choice functions. But Hilbert originally introduced ϵ-terms in order to *replace* quantifiers altogether. Applying this to quantifiers over choice functions, as discussed by von Heusinger [3], section 3.7, one is lead to ϵ-terms $\epsilon i \alpha$ to choose a context i such that α. The introduction of Mints/Sarenac [1] suggests the investigation of this extension.

We now sketch how to adapt the completeness theorem for L'_{ϵ_i} to obtain a completeness theorem for $L'_{\epsilon_i} + \epsilon i$, the extension of L'_{ϵ_i} by ϵ-terms for contexts.

In syntax, we change the grammar for context terms to

$$ a := i \mid c \mid \epsilon i \alpha. $$

For the semantics, we use structures $(A, P^{\mathcal{A}}, \ldots; \mathbb{F}, c^{\mathcal{A}}, \ldots; \epsilon^{\mathcal{A}})$ where $(A, P^{\mathcal{A}}, \ldots)$ is a first-order L-structure, $\mathbb{F} \subseteq \mathcal{P}(A) \to A$ a set of choice functions for A with $c^{\mathcal{A}} \in \mathbb{F}$, and $\epsilon^{\mathcal{A}} : \mathcal{P}(\mathbb{F}) \to \mathbb{F}$ a fixed choice function for $\mathbb{F}$.

We add axioms for a weak form of extensionality for the context choice ϵ:

$$ \forall i (\alpha \leftrightarrow \beta) \to \epsilon_{\epsilon i \alpha} x \varphi \doteq \epsilon_{\epsilon i \beta} x \varphi, \qquad \text{if } x \notin \mathit{free}(\alpha \leftrightarrow \beta), i \notin \mathit{free}(\varphi) \qquad (5) $$

Notice that these *do not use a $\doteq$-predicate for contexts*.

Since there are no new formulas, the extension of a consistent set Γ of sentences to a maximal consistent set Δ does not need to be changed. In the term model construction for Δ, we define a choice function $\Phi_\epsilon : \mathcal{P}(\mathbb{F}) \to \mathbb{F}$ for $\mathbb{F}$ by

$$ \Phi_\epsilon(S) := \begin{cases} \Phi_{[\epsilon i \alpha]}, & \text{if } S = \{\, \Phi_{[a]} \mid \alpha[i/a] \in \Delta, [a] \in I \,\} \text{ for some formula } \alpha(i), \\ G(S), & \text{else.} \end{cases} $$

This is well-defined, i.e. independent of the defining formula for S: if

$$ S_\alpha := \{\, \Phi_{[a]} \mid \alpha[i/a] \in \Delta, [a] \in I \,\} = \{\, \Phi_{[a]} \mid \beta[i/a] \in \Delta, [a] \in I \,\} =: S_\beta, $$

then $\forall i (\alpha \leftrightarrow \beta) \in \Delta$: otherwise, $\exists i \neg (\alpha \leftrightarrow \beta) \in \Delta$, hence we have $\neg (\alpha \leftrightarrow \beta)[i/a] \in \Delta$ for some context constant a, hence $\alpha[i/a] \notin \Delta$ iff $\beta[i/a] \in \Delta$, which contradicts $S_\alpha = S_\beta$. From $\forall i (\alpha \leftrightarrow \beta) \in \Delta$ and weak extensionality (5) for ϵ, we get

$$ \{\, \epsilon_{\epsilon i \alpha} x \varphi \doteq \epsilon_{\epsilon i \beta} x \varphi \mid \varphi(x) \text{ an } L'_{\epsilon_i}(C)\text{-formula} \,\} \subseteq \Delta, $$

i.e. $\epsilon i\alpha \approx_\Delta \epsilon i\beta$ resp. $[\epsilon i\alpha] = [\epsilon i\beta]$, hence $\Phi_{[\epsilon i\alpha]} = \Phi_{[\epsilon i\beta]}$. Therefore, $\Phi_\epsilon(S) = \Phi_{[\epsilon i\alpha]}$ is independent of the choice of the defining formula $\alpha(i)$ for S.

It follows that 1. of lemma 5.5 extends to such choice context-terms:

$$
\begin{aligned}
s(\epsilon i\alpha) \quad &:= \quad \Phi_\epsilon(\{\, f \mid s(i|f)(\alpha) = 1 \,\}) \\
&= \quad \Phi_\epsilon(\{\, \Phi_{[a]} \mid [a] \in I, s(i|\Phi_{[a]})(\alpha) = 1 \,\}) \\
&= \quad \Phi_\epsilon(\{\, \Phi_{[a]} \mid [a] \in I, \alpha[\vec{x}/\vec{t}, \vec{i}/\vec{a}, i/a] \in \Delta \,\}) \\
&= \quad \Phi_\epsilon(\{\, \Phi_{[a]} \mid [a] \in I, \alpha[\vec{x}/\vec{t}, \vec{i}/\vec{a}][i/a] \in \Delta \,\}) \\
&= \quad \Phi_{[\epsilon i\alpha[\vec{x}/\vec{t},\vec{i}/\vec{a}]]} \\
&= \quad \Phi_{[(\epsilon i\alpha)[\vec{x}/\vec{t},\vec{i}/\vec{a}]]}.
\end{aligned}
$$

We here need a somewhat stronger induction hypothesis than in the lemma, since the context terms $\epsilon i\alpha$ may have free individual variables.

6   Conclusion

The intention of Mints and Sarenac was to provide "a managable version" of U. Egli's and K. von Heusinger's higher-order theories of context change in dynamic interpretation of texts. It seems that the gap in their completeness proof for S_{ϵ_i} is due to a mere oversight, since true equality for choice functions is a second-order notion.

To model the dynamics of discourse, von Heusinger uses a second-order "salience update operation ρ" where the choice function $\rho(\Phi, S)$ agrees with the choice function Φ except on the set S. Thus, it seems useful to extend $L\epsilon$ by the restriction of (a relational version of) ρ to definable sets, the *update relation*

$$
i \approx_{x,\alpha} j : \Longleftrightarrow \bigwedge_\beta (\neg \forall x(\alpha \leftrightarrow \beta) \to \epsilon_i x\beta \doteq \epsilon_j x\beta),
$$

which says that i and j agree on all definable sets except $\{\, x \mid \alpha \,\}$. But even the simpler relation $i \approx j$ of equality of choice-functions on *all* definable sets appears to go beyond first-order, and must not be mixed with true equality $\doteq$ of functions, as the calculus S_{ϵ_i} would have it. Though the completeness theorem works if $\doteq$ for contexts is dropped from the language, as shown for S'_{ϵ_i}, it seems unplausible that such a system really gives a basis for a "managable version" of von Heusinger's theory of dynamic interpretation of discourse, where $\approx_{x,\alpha}$ resp. $\approx$ are essential.

Acknowledgement

I thank Christian Meyer for a discussion on the counterexample and two referees for their constructive comments.

References

[1] G.E. Mints and Darko Sarenac. Completeness of indexed ϵ-calculus. *Archive for Mathematical Logic*, Vol. 42.7, pages 617–625, 2003.

[2] Klaus von Heusinger. Reference and salience. In A. von Stechow F.Hamm, J.Kolb, editors, *The Blaubeuren Papers. Proc. Workshop on Recent Developments in the Theory of Natural Language Semantics*, SfS-Report 08-95, pages 149–172, 1995.

[3] Klaus von Heusinger. Der Epsilon-Operator in der Analyse natürlicher Sprache. Teil I: Grundlagen. Arbeitspapier 59. Fachbereich Sprachwissenschaft Universität Konstanz, 1993.

[4] Urs Egli and Klaus von Heusinger. Definite Kennzeichnungen als Epsilon-Audrücke. In G. Lüdi, C.-A. Zuber, editors, *Akten des 4. regionalen Linguistentreffens*. Romanisches Seminar, Universität Basel, pages 105–115, 1993.

[5] H.B. Enderton. A Mathematical Introduction to Logic. Academic Press, 1972.

Received 3 December 2015

NOUN PHRASES IN JAPANESE AND EPSILON-IOTA-TAU CALCULI

SUMIYO NISHIGUCHI
Tokyo University of Science
1-3 Kagurazaka, Shinjuku, Tokyo 162-8601, Japan
`nishiguchi@rs.tus.ac.jp`

Abstract

The existence of determiners in Japanese, a language that does not possess explicit determiners, has so far been argued mainly from a syntactic perspective. This study presents a semantic account of the existence of determiners in Japanese. I propose that Japanese NPs are the properties of type $<e,t>$ in general, and that the null determiners that correspond to epsilon, iota and tau operators accompany bare NPs and contribute to indefiniteness, definiteness, and genericity, respectively.

Keywords: null, determiner, (in)definite, generic

1 Determiners in Japanese?

Unlike in Indo-European languages, explicit determiners, like *a, an, the*, and *some* in English do not exist in Japanese. Furthermore, neither plurality nor definiteness is marked morphologically. Consider the example in (1).

(1) Otoko-no hito-ga   haitte-ki-ta.
    male-POSS person-NOM enter-come-PAST
    "A man came in."

In order to emphasize singularity, a numeral *hitori* "1-person" should be added to the noun *otoko* "man," as in (2).

(2) 1-ri-no   otoko-no hito-ga   haitte-ki-ta.
    1-CL-POSS male-POSS person-NOM enter-come-PAST
    "One man entered."

I would like to thank the reviewers and the audience at Epsilon 2015 for useful suggestions.

Although not equivalent to *the*, as a means to express definiteness, pronominal modifiers or demonstratives, expressed by the *so*-series in Japanese, are added to serve deictic purposes, as in (3).

A reminder on Japanese demonstratives

Demonstratives can be bound pronouns of focused noun phrases, such as *only/also/even NP* as shown by the Weak Crossover Effect of focus particles [14] as in (3) and (4).

(3) So-ko-no   jugyoin-ga   supa-o   uttae-ta.
    that-place-POSS employee-NOM supermarket-ACC sue-PAST
    "Their employee sued the supermarket."

(4) *So-ko-no   jugyoin-ga   supa-{nomi/mo/sae}-o   uttae-ta.
    that-place-POSS employee-NOM supermarket-{only/also/even}-ACC sue-PAST
    "Their employee sued the supermarket."

Demonstratives can form donkey anaphora as in (5) whereas bare NPs do not, as in (6).

(5) a. Inu$_i$-o   kat-teiru   hito-wa   so$_i$-itsu-o   nader-u.
       dog-ACC keep-PROG person-TOP that-one-ACC pet-PRES
       "Everyone who keeps a dog pets it."

    b. Baiorin$_i$-o mot-teiru   hito-wa   so$_i$-re-o   migak-u.
       violin-ACC keep-PROG person-TOP it-thing-ACC polish-PRES
       "Everyone who has a violin polishes it."

(6) a. Inu$_i$-o   kat-teiru   hito-wa   inu$_{?i/j}$-o nader-u.
       dog-ACC keep-PROG person-TOP do-ACC   pet-PRES
       "Everyone who keeps a dog pets it."

    b. Baiorin$_i$-o mot-teiru   hito-wa   baiorin$_{?i/j}$-o migak-u.
       violin-ACC keep-PROG person-TOP violin-ACC   polish-PRES
       "Everyone who has a violin polishes it."

(7) So-no   otoko-no   hito-ga   haitte-ki-ta.
    that-POSS male-POSS person-NOM enter-come-PAST
    "That/the man entered."

368

The absence of explicit determiners has raised both syntactic and semantic questions. In terms of syntactic theory, the DP (determiner phrase) Hypothesis has been advocated in the works of [29] and [10], according to which every language has some or other form of D [10], without exception. The DP Hypothesis [1] suggests that determiners are the head of the DP, rather than of the noun phrase (NP).

(8)

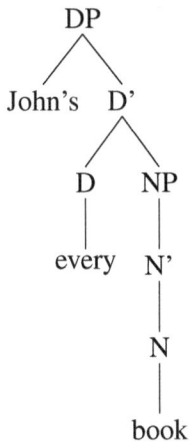

(9)

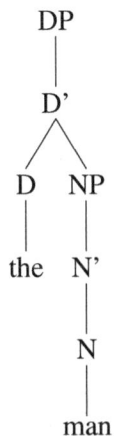

[10] assumes that D should play a role in Logical Form (LF), even though it is not pronounced in languages without overt determiners.

On the other hand, from the semantic point of view, the application of Generalized Quantifier theory [2] to a determiner-free language like Japanese has become an issue [31,

26]. In English, a determiner like *a* combines with the set expression *man* in (10) and produces the quantifier *a man*.

(10) a. $[\![a \quad man]\!] = \lambda P.\text{man}' \cap P \neq \emptyset$

b. $[\![a]\!] = \lambda P.\lambda Q.P \cap Q \neq \emptyset$

[15] pointed out that the GQ theory does not straightforwardly apply to Japanese because Japanese NPs do not always correspond to quantifiers. Rather, predicates such as adjectives constitute quantifiers in Japanese, as in (11).

(11) Tokyo-wa   gakusei-ga   oi
     Tokyo-NOM student-NOM many
     "There are many students in Tokyo."

[26] further argues that predicative adjectives are quantifiers in Japanese, and that the generalized quantifier theory applies to Japanese language. According to his analysis, D(E), *oi* "many" is a determiner in L(GQ)-language as in (12).

(12) (Tokyo-wa)(gakusei-ga)oi(E)

As the number of arguments of an adjectival quantifier is not limited to two, [24] proposed polymorphic predicative quantifiers as in (13).

(13) $[\![many]\!]_{<et^n, t>} = \lambda P_1, P_2, P_3, ..., P_n. |P_1 \cap P_2 \cap, ..., \cap P_n| \geq |P_n \cdot c|$

In section 2, I will consider bare NPs without determiners, and show that Japanese NPs are inherently of the predicate type, in which a determiner type null operator is attached to bare NPs. Section 3 will explain that the null determiners represent (in)definiteness or genericity, and iota, epsilon, and tau operators correspond to these determiners.

2   Bare NPs with Null Operators

In this section, I consider the semantic types of bare nouns in Japanese. I suggest the existence of null operators in determiner position that take bare NPs as arguments [22]. Common nouns in English can be viewed as a set of individuals whose semantic values may be functions from individuals to truth-values, cf. [11]. [27] assigns bare plural NPs in predicate positions type <e,t>.

(14) $[\![dog]\!] = \lambda x \in D_e.\text{dog}'(x)$

However, [6] assumes the Nominal Mapping Parameter across languages and claims that bare/common nouns in Chinese-type languages without plural marking, are mass/kind-denoting individuals of type <e>. [18] further points out that common nouns with the [+human] feature can be followed by a plural marker, e.g., *gakusei-tachi* "student-pl," therefore, they can be interpreted as countable common nouns whose extensions are sets of atoms. Without plural markers, they stay kind or mass of type <e>.

(15) a. $[\![hon\text{"book"}]\!] = x \in D_e\text{: } x \sqsubseteq \text{book'}$

("$\alpha \sqsubseteq \beta$" means that α is a component or a member of β, and book' is the plural kind that comprises all of the atomic members of the book kind)

b. $[\![gakusei\text{"student"}]\!] = \lambda x.\text{student'}(x)$

Nevertheless, the plural marker *tachi* "PL" is attached not only to human denoting nouns in Japanese. According to a corpus search[1], animals such as *raion* "lion," *pengin* "penguin," *neko* "cat," *inu* "dog," *uma* "horse," *dobutsu* "animal," and *kemono* "beast" as well as *kani* "crab," *kin* "germ," *kusa* "grass," and *sakura* "cherry blossom" can also take a plural marker. Therefore, the [+human] feature is not sufficient to distinguish common nouns which can be mapped to <e,t> type by means of the plural marker.

Hence, I rather interpret nouns in Japanese to be consistently either the set of individuals or the function from an entity to truth-values. The lexical entry for the NP *inu* "dog" in sentence (16) is given in (17a).

The example in (16) is one with the stage-level predicate *oyogu* 'swim':

(16) Inu-ga   oyogu.
      dog(s)-NOM swim
      "A dog swims / Dogs swim."

(17) a. $[\![inu\text{"dog"}]\!] = \lambda x \in D_e.\text{dog'}(x)$

b. $[\![oyogu\text{"swim"}]\!] = \lambda x.\text{swim'}(x)$

In order to avoid type-mismatch between the subject DP *inu* "dog" and the predicate *oyogu* "swim," the DP should be either <e> or <et,t>, as the predicate *oyogu* "swim" is a one-place holder or a property of type <e,t>, as in (17). Ascribing the noun *inu* as type <e,t> then necessitates the phonetically empty (/ε/) D of type <et,e> or <et,<et,t>> given in (18), which combines with the NP and returns the individual or the set of properties.

---

[1] *Chunagon*, BCCWJ-NT, search result of "common noun + *tachi*"

(18) a. $[\![D/\epsilon/]\!] = \lambda f \in D_{<e,t>}.$ $[\lambda g \in D_{<e,t>}.$ there is some $x \in D_e$ such that $f(x)= g(x)=1]$

  b. $[\![D/\epsilon/\_inu\text{``}dog\text{''}]\!] = [\lambda g \in D_{<e,t>}$. there is some $x \in D_e$ such that $f(x)=$ dog'$(x)=1]$

It is comparable with the analysis of bare singular nouns in Brazilian Portuguese in [30]. Even though Brazilian Portuguese allows overt determiners, bare singulars appear in argument positions. Therefore, bare singular nouns are claimed to be DPs with empty determiners and no number.[2]

3   Epsilon, Iota, and Tau Operators

3.1   (In)definiteness and Genericity

With regard to the compositional calculation of meaning, I assume that the ϵ (epsilon), ι (iota), and τ (tau) operators lower the types of common nouns into type $<e>$. The use of the ϵ operator follows its use for Japanese nouns in [4, 23, 25].

Even without the overt determiner, bare NPs in Japanese show indefiniteness, definiteness, or genericity. *Inu* "dog" in (19a) signifies a specific and familiar dog to discourse participants, so that the definiteness can be translated using the *iota* operator [28]. In contrast, the dog in (19b) is indefinite, unspecific and discourse new. (19c) is a statement about dogs as a kind or species in general. Generic statements are only expressed by the sentences with the topic marker *-wa* [19]. In view of the genericity, we will use the τ operator that corresponds to D. (19d) is another generic statement about department stores as a kind.

(19) a. Inu-ga   pakutto   ho-ni   kamitsui-ta.
        dog-NOM ONOMATOPEA cheek-GOAL bite-PAST
        "The dog bit the cheek."

        $\iota x.\text{dog}(x) \wedge \text{bite(s)}(x)$
        Attributed to C.W.Nicole, *C.W.Nicole-no Kuromihe Nikki*, Translated by Kazuyo Takeuchi, Kodansha, Tokyo

  b. Toku-de inu-ga   hoe-ta.
        far-LOC dog-NOM bark-PAST
        "A dog barked afar."
        $\epsilon x.\text{dog}(x) \wedge \text{bark-afar}(x)$
        Attributed to Hiromi Kawakami, *Kamisama*, Chuokoronshinsha, Tokyo

---

[2]I thank an anonymous reviewer for drawing my attention to bare singulars in Brazilian Portuguese.

c. Inu-wa   hito-ni   tsuki, neko-wa ie-ni   tsuku.
dog-TOP people-DAT attach cat-TOP house-DAT attach
"Dogs are attached to people and cats to their house."

τx,y.dog(x) $\wedge$ human(y) $\rightarrow$ attached-to(y)(x)

d. Depaato-wa   saizu-ga hofu-desu.
department.store-TOP size-NOM rich-HON
"Department stores have wide range of (clothing) sizes."

τx.department.store(x) $\rightarrow$ have-various-sizes(x)

The ϵ and τ calculi were used in [12, 13]. The ϵ operator is the dual of the τ operator, even though it has been noted that [3]'s τ in fact corresponds to ϵ.

As shown by [8], τ terms have universal force despite returning individuals of type <e>. Both τ and ϵ terms substitute for the argument x of type <e>, as shown in (20).

(20) Characteristic axioms for ϵ and τ. Let $\phi \in$ Wff:

(ϵ) $\exists$x.ϕ $\Rightarrow$ ϕ [x/ϵx.ϕ]

(τ) ϕ[x/τx.ϕ] $\Rightarrow$ $\forall$x.ϕ

From a linguistic point of view, lowering the type of quantifying NPs to individuals of type <e> by means of ι, ϵ, and τ calculi eases semantic calculation. The individuals of type <e> also adequately represents kind-denoting bare NPs without determiners.

3.2   Possessive DPs with Multiple Operators and Existential Presupposition

Let us consider the representation of more complex DPs with another DP embedded. Multiple Ds are present in a single possessive or genitive DP. For example, in *boshi-no fujin* "the lady with a hat", there is no overt determiner that modifies either *boshi* "hat" or *fujin* "lady." However, the context specifies that *the lady with a hat* is a definite description whereas *hat* is indefinite. The use of the epsilon and iota operators in the places of the null determiners eases calculation of the entire DP.

(21) a. *boshi* "hat": ϵx.hat'(x): some x satisfying hat'(x), if there is one

b. *fujin* "lady": ιy[lady'(y)]: the unique x satisfying lady'(x), if there is such a thing

c. *no* "POSS" : λX λY ιy[Y(y) $\wedge$ R(ϵx.X)(y)]

d. *boshi-no fujin* "the lady with a hat": ιy[lady'(y) $\wedge$ λe[manner(e) = with'(ϵx.hat')(y)]

The use of the epsilon operator is plausible in (21) because *boshi* "hat" is not a definite description—the existence of *boshi* "hat" is not presupposed. The existence of the hat is canceled by the negation in (22), even though negation is known to be a presupposition hole [16]. Therefore, there is no existential presupposition of the existence of the hat, and the use of the epsilon operator is appropriate.

(22) a. Kono-heya-ni-wa   boshi-no hito-wa   i-**nai**.
          this-room-LOC-TOP hat-POSS person-TOP exist-NEG
          "There is no one with a hat in this room."

  b. $\neg\exists x,y.person(x)\wedge R(x,y)\wedge hat(y)\wedge here(x)$

The indefinite DP *hon* "book" is embedded under the question in (23a), and the reply in (23b) suggests that the existence of the book is denied.

(23) a. Hon-o   yomi-mashi-ta-ka?
          book-ACC read-HON-Q
          "Did you read a book?

  b. Zenzen.
      at all
      "Not at all."

(24) provides the Combinatory Categorial Grammar notation for (23a) (Steedman 2000). The inference rule "Lex" derives lexical entries for each phrase. *Hon* "book" is a NP whose semantic entry is the epsilon term. *Hon-o* "book-ACC" combines with the predicate *yomimashita* "read" and then with the null subject, which is the hearer. The sentence now combines with the question marker and receives the interrogative interpretation.

(24)

$$\frac{\cfrac{hon-o}{NP : \varepsilon x.book'}\ \text{Lex} \quad \cfrac{yomimashita}{NP\backslash(NP\backslash S) : \lambda w, x, y.read'(x)(y)(w)}\ \text{Lex}}{\cfrac{NP : h}{NP\backslash S : \lambda w, y.read'(\varepsilon x.book')(y)(w)}\ \text{Lex}}\ <$$

$$\cfrac{S : \lambda w.read'(\varepsilon x.book')(h)(w) \quad \cfrac{ka}{S\backslash(S/S) : \lambda p_{<st>}.\lambda w_s.q_{<st>}[q = p \vee q = \neg p](w)}\ \text{Lex}}{S\backslash S : \lambda w_{<s>}.\lambda q_{<st>}.q_{<st>}[q = read'(\varepsilon x.book')(h) \vee q = \neg read'(\varepsilon x.book')(h)](w)}\ <$$

(s: speaker, h: hearer, w, w': possible worlds, R: epistemic accessibility relation between possible worlds)

Paying attention to the internal structure of the DP, the ι and ϵ operators reside in D and provide definiteness for the DP *hito* "person" in (26) and indefiniteness for *boshi* "hat" in (27). The entire DP is a definite description whose head noun is *hito* "person," as shown in (28).

(25) boshi-no hito
      hat-POSS person
      "the hat person"

(26)

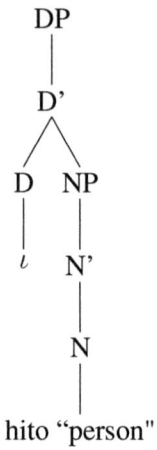

hito "person"

(27)

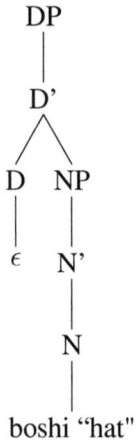

boshi "hat"

(28)

```
            DP
            |
            D'
           /  \
          D    NP
          |   /  \
          ι  DP    D'
             |    /  \
             D'  D    NP
            / \  |    |
           D  NP poss hito
           |   |
           ε   N'
               |
               N
               |
          boshi "hat"
```

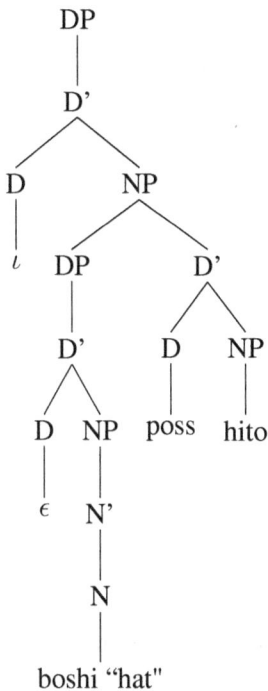

3.3 Tau-Operator for Genericity

As discussed in section 3.1, a generic statement is expressed without any determiner in Japanese. The bare NP has a generic reading in (29), by which *inu* "dog" can refer to dogs in general as a kind, and walking is the habitual property that predicates dog species. The bare NP *inu* "dog" can also be a specific definite dog, which has the habit of walking [19].

(29) Inu-wa   aruku.
dog-TOP walk
"Dogs (as a kind) walk (as a habit)/ The (specific) dog walks (as a habit)."

As used by [17, 21] and others, the generic operator Gen has been widely used in the linguistics literature. Gen is a generic quantifier that unselectively quantifies over individuals and events [9]. With the generic subject reading of (29), given in (30a), in general, if you are a dog, you walk, or participate in the event of walking. Gen quantifies over individuals and also events. In contrast, when the bare NP is definite, the specific dog has the generic property of events of walking as in (30b). Generic operator over events is no different from the adverb of quantification such as *often*, cf. [20].

(30) a. Gen x,e[dog(x) & agent(e) = x → walk(e)]

   b. Gen e[agent(e) = ιx.dog(x) → walk(e)]

In (30a), dogs are usually the agents of the habitual event of walking. This generic quantifier is different from a universal quantifier in that exceptions are allowed: some dogs who have had traffic accidents may not be able to walk, but this does not falsify the generic statement [5]. Therefore, generic quantification does not correspond to universal quantification in (33a) but to quantification by *most* as in (33b).

(31) #Ashi-o kossetsushita-inu-wa aruku.
     leg-ACC break-dog-TOP   walk
     "Dogs with broken legs walk (in general)"

(32) All dogs bark and a dog wearing a muzzle is a dog. ⇒ A dog wearing a muzzle barks.

(33) a. $\forall$x[dog'(x) → walk'(x)]

   b. $\forall$x[x$\in$dog'$\wedge$ |x| $\geq$ 1/2|dog'| → bark'(x)]

In French, generic DPs may contain a definite or indefinite determiner.[3] Similar to generic statements in English and Japanese, the existence of exceptional non-studious students does not falsify the sentence in (34). If the generic statement corresponds to the universal quantifier, the statement should be truth-conditionally false.

(34) a. Les étudiants étudient.
        the students study
        "The students study."

   b. L'étudient étudie.
        the student study
        "The student study."

   c. Un étudiant étudie.
        a   student study
        "A student studies."

   d. Des étudiants étudient.
        some students study
        "Students study."

---

[3]I thank Christian Retore for drawing my attention to French generic NPs.

Therefore, the exact representation of a generic statement should be that in (35) in line with [7].

(35) $\forall x.P(x) \wedge R(x) \to Q(x)$

There is a contextually supplied R that selects only the relevant majority of the set member of P. Thus, exceptional members of P which are not Q do not falsify the statement.

Thus, my use of the tau operator in this paper corresponds to (35) in the exact sense.

4 Conclusion

A language without determiners express (in)definiteness and genericity with a null determiner that corresponds to iota, epsilon, and tau operators. Even without overt determiners, indefinite description is identified with the absence of existential presupposition. Noun phrases in Japanese are uniformly regarded as the properties of type $<e,t>$.

References

[1] Steven Paul Abney. *The English Noun Phrase in its Sentential Aspect*. PhD thesis, MIT, 1987.

[2] Jon Barwise and Robin Cooper. Generalized quantifiers and natural language. *Linguistics and Philosophy*, 4:159–219, 1981.

[3] Nicolas Bourbaki. *Théorie des Ensembles*. Hermann, Paris, 1956.

[4] R. Cann, Ruth Kempson, and L. Marten. The dynamics of language: an introduction. volume 35 of *Syntax and Semantics*. Academic Press, Amsterdam, San Diego, 2005.

[5] Gregory N. Carlson. A unified analysis of the English bare plural. *Linguistics and Philosophy*, 1:413–457, 1977.

[6] Gennaro Chierchia. Plurality of mass nouns and the notion of "semantics parameter". In Susan Rothstein, editor, *Events and Grammar*, pages 53–103. Kluwer Academic Publishers, London, 1998.

[7] Renaat Declerck. The origins of genericity. *Linguistics*, 29:79–102, 1991.

[8] David Devidi. Intuitionistic ϵ and τ calculi. *Mathematical Logic Quarterly*, 41:523–546, 1995.

[9] Donka Farkas and Henriëtte de Swart. Article choice in plural generics. *Lingua*, 117:1657–1676, 2007.

[10] Naoki Fukui. The principles-and-parameters approach: A comparative syntax of English and Japanese,. In Masayoshi Shibatani and Theodora Bynon, editors, *Approaches to Language Typology*, pages 327–372. Clarendon Press, Oxford, 1995.

[11] Irene Heim and Angelika Kratzer. *Semantics in Generative Grammar*. Blackwell, Oxford, 1998.

[12] David Hilbert and Paul Bernays. *Grundlagen der Mathematik*, volume 1. Springer, Berlin, 1934.

[13] David Hilbert and Paul Bernays. *Grundlagen der Mathematik*, volume 2. Springer, Berlin, 1939,1970.

[14] Hajime Hoji. *Logical Form Constraints and Configurational Structures in Japanese*. PhD thesis, University of Washington, 1985.

[15] M. Imani. On quantification in Japanese. *English Linguistics*, 7:87–104, 1990.

[16] Lauri Kartunnen. Presuppositions of compound sentences. *Linguistic Inquiry*, 4:169–193, 1973.

[17] Manfred Krifka, Francis Jeffry Pelletier, Gregory N. Carlson, Alice ter Meulen, Gennaro Chierchia, and Godehard Link. Genericity: An introduction. In *The Generic Book*. The University of Chicago Press, Chicago, 1995.

[18] Takeo Kurafuji. Plural morphemes, definiteness and the notion of semantic parameter. In *GLOW in Asia*, pages 361–378. 2002.

[19] S.-Y. Kuroda. *Japanese Syntax and Semantics*. Springer, Berlin, 1992.

[20] David Lewis. *Adverbs of Quantification*. Cambridge University Press, Cambridge, 1998.

[21] Alda Mari, Claire Beyssade, and Fabio Del Prete. Introduction. In *Genericity*. Oxford University Press, Oxford, 2012.

[22] Sumiyo Nishiguchi. Logical properties of Japanese kakari joshi. *Osaka University Papers in English Linguistics*, 7:115–133, 2003.

[23] Sumiyo Nishiguchi. Extended qualia-based lexical knowledge for disambiguation of Japanese postposition no. In Thomas Icard, editor, *Proceedings of the 14th Student Session of the European Summer School for Logic, Language, and Information*, pages 137–147. 2009.

[24] Sumiyo Nishiguchi. Quantifiers in Japanese. In Peter. Bosch et al, editor, *Logic, Language and Computation: TbiLLC 2007*, volume 5422 of *LNCS*. Springer, 2009.

[25] Sumiyo Nishiguchi. Ccg of questions and focus. In *Proceedings of JSAI2011*, 2011.

[26] Norihiro Ogata. Japanese particles and generalized quantifiers. *Proceedings of the 4th Summer Conference 1990 Tokyo Linguistics Forum*, 1991.

[27] Barbara Partee. Noun phrase interpretation and type-shifting principles. In Jeroen Groenendijk et al, editor, *Studies in Discourse Representation Theory and the Theory of Generalized Quantifiers*, pages 115–143. Foris Publications, Dordrecht, 1986.

[28] Bertrand Russell. On denoting. *Mind*, 14(56):479–493, 1905.

[29] Mamoru Saito and Keiko Murasugi. *N'*-deletion in Japanese. In *UConn Working Papers in Linguistics*. University of Conneticut, 1989.

[30] Christina Schmitt and Alan Munn. Against the nominal mapping parameter: Bare nouns in Brazilian Portuguese. In *NELS 29*. 1999.

[31] Kin-ichiro Shirai. Japanese noun-phrases and particles *wa* and *ga*. In Jeroen Groenendijk et al., editor, *Foundations of Pragmatics and Lexical Semantics*, pages 63–80. Foris Publications, Dordrecht, 1987.

Received

HILBERT'S ϵ-OPERATOR IN DOCTRINES

FABIO PASQUALI
University of Aix-Marseille
pasquali@dima.unige.it

Abstract

We study doctrines with Hilbert's ϵ-operator and their applications to two examples of interest, with a particular focus on the one based on naive set theory.

Keywords: Categorical logic, doctrines.

1 Introduction

In this paper we introduce the notion of doctrine with Hilbert's ϵ-operator and we study some properties of this class of doctrines, especially in connection with naive set theory. Our aim is to give an introductory account to the theory of doctrines with the ϵ-operator and at the same time to present some new results.

Section 2 is a brief introduction to the basic facts of the theory of doctrine, that does not require any background in category theory. We discuss two archetypal examples: the first is the doctrine that arises from the contravariant powerset functor, the other is the doctrine that arises from the Lindenbaum-Tarski algebra of any given first order language. The material collected in section 2 is well known and expert readers may skip directly to section 3.

Section 3 contains the definition of doctrine with the ϵ-operator, which is taken from [3]. We show that the Lindenbaum-Tarski algebra of a first order language $\mathcal{L}$ gives rise to a doctrine which has the ϵ-operator if and only if $\mathcal{L}$ is Hilbert's epsilon calculus. We show also that the contravariant powerset functor on the category of sets is a doctrine which has the ϵ-operator, provided that we use the axiom of choice

The author thanks professors Christian Retoré, Vito Michele Abrusci and Giuseppe Rosolini for many useful discussions and advises. This work has been funded by the French-Chinese project ANR-11-IS02-0002 and NSFC 61161130530 Locali.

and the excluded middle at the meta-level.

It was Lawvere who first recognized that quantifiers are adjoints [2]. More specifically the existential quantifiers behave like left adjoints and the universal quantifiers behave like right adjoints (to certain substitution functors whose description is left implicit in this introduction). As in this paper we focus on Hilbert's ϵ-operator in the intuitionistic framework, we are particularly interested in existential quantifiers. The notion of left adjoint alone is not enough to completely encompass the notion of quantification. This can be resumed by the motto: *quantifiers are adjoints, but adjoints are not quantifiers*. In fact, in the general case, left adjoints, as opposed to existential quantifiers, fail in validating two fundamental properties that are known as the Beck-Chevalley condition and Frobenius reciprocity which are crucial in describing the interplay between existential quantification and substitution (the former) and between existential quantification and binary conjunctions (the latter). Left adjoints that validate both the Beck-Chevalley condition and Frobenius reciprocity are the appropriate categorical tool to express the existential quantification and in section 4 we prove that both the conditions are derivable in the presence of the ϵ-operator.

In section 5 we present an application of the theory of doctrines with the ϵ-operator to naive intuitionistic set theory. As the ϵ-operator is connected to the axiom of choice, in view of the known result of Diaconescu [1], a care is required in order to allow this form choice to hold without being classic. We achieve this by dropping the powerset axiom. Under this restriction we show that the category of sets gives rise to a doctrine with the ϵ-operator if and only if every function defined on a subset, i.e. a function $f \colon A \longrightarrow B$, where $A \subseteq X$ (for some set X), can be extended to a function defined on the whole set, i.e. there is $\overline{f} \colon X \longrightarrow B$ which agrees with f on A, i.e. the restriction of $\overline{f}$ to A is f.

2   Preliminaries

In this section we introduce the notion of doctrine. These are the main mathematical tool with which we will be concerned in this article. Before giving the formal definition of doctrine, we look at to relevant and known situations and we will continuously refer at these throughout the whole paper. The first deals with sets in the naive sense, while the second deals with first order languages.

2.1   Powersets

Suppose A and B are sets. We use the standard notation

$$f \colon A \longrightarrow B$$

to denote that f is a function with domain A and codomain B. If there is a function $g \colon B \longrightarrow C$ then there is a third function from A to C, i.e. the composition of f and g, that we denote by

$$gf \colon A \longrightarrow C$$

The composition is associative: $(fg)k = f(gk)$. And identity arrows are neutral elements: $id_B f = f$ and $f id_A = f$. In other words, sets and functions form a category.

For every pair of sets A and B we denote by $A \times B$ their cartesian product and by π_A and π_B the projections from $A \times B$ to each of the factors. In other words, sets and functions form a category with binary products.

As usual we say that X is a subset of B if for every $x \in X$ it is that $x \in B$ and we denote this by

$$X \hookrightarrow B \qquad \text{or equivalently} \qquad X \subseteq B$$

Both arrows are called inclusion. Thus a subset is the domain of an inclusion. For every set A we have an assignment

$$A \mapsto \mathbb{P}(A)$$

where we denote by $\mathbb{P}(A)$ the totality of inclusions with codomain A or, equivalently, the totality of subsets of A. We are tempted to say that $\mathbb{P}(A)$ is the set of subsets of A. For the moment, the fact that $\mathbb{P}(A)$ is a set or not is irrelevant and we will freely call such a collection set. But a careful distinction will play a crucial role in the last section.

For every function $f \colon X \longrightarrow Y$ we have an assignment

$$f \mapsto f^{-1} \colon \mathbb{P}(Y) \longrightarrow \mathbb{P}(X)$$

where for a subset U of Y it is

$$f^{-1}(U) = \{x \in X \mid f(x) \in U\}$$

The assignment f^{-1} is monotone, i.e. if $U \subseteq V$, then $f^{-1}(U) \subseteq f^{-1}(V)$.

For every projection, say $\pi_A \colon A \times B \longrightarrow A$ (actually for every function, but this generality is not required in this example), we have an assignment

$$\mathrm{Im}_{\pi_A} \colon \mathbb{P}(A \times B) \longrightarrow \mathbb{P}(A)$$

where for $R \subseteq A \times B$ it is

$$\mathrm{Im}_{\pi_A}(R) = \{a \in A \mid \exists b \in B \ (a,b) \in R\}$$

The assignment Im_{π_A} is monotone, i.e. if $R \subseteq S$, then $\mathrm{Im}_{\pi_A}(R) \subseteq \mathrm{Im}_{\pi_A}(S)$.

The functions π_A^{-1} and Im_{π_A}, the so-called inverse image function and direct image function, have the following property: for every $S \in \mathbb{P}(A)$ and every $R \in \mathbb{P}(A \times B)$ it holds

$$\mathrm{Im}_{\pi_A}(R) \subseteq S \quad \text{if and only if} \quad R \subseteq \pi_A^{-1}(S)$$

Using a more categorical language, we say that Im_{π_A} is left adjoint to π_A^{-1}.

2.2 First order languages

We shall refer at doctrines with the ϵ-operator as epsilon doctrines or, more concisely, ϵ-doctrines.

Suppose $\mathcal{L}$ is a first order language. Suppose also that V is a countable infinite set of variables. We write $\vec{x}$ to abbreviate a finite list of distinct variables, i.e.

$$\vec{x} = (x_1, x_2, ..., x_n)$$

We shall call $\vec{x}$ context. Suppose that for another context $\vec{y} = (y_1, y_2, ..., y_k)$ there are k terms

$$t_1(\vec{x})$$

$$t_2(\vec{x})$$

$$\vdots$$

$$t_k(\vec{x})$$

we shall denote simultaneous substitution with an arrow

$$[\vec{t}/\vec{y}] \colon \vec{x} \longrightarrow \vec{y}$$

If there is a second arrow $[\vec{q}/\vec{z}]\colon \vec{y} \longrightarrow \vec{z}$, where $\vec{z} = (z_1, z_2, ..., z_h)$, then there is a third arrow

$$\vec{x} \xrightarrow{\quad [q_1[\vec{t}/\vec{y}]/z_1, ..., q_h[\vec{t}/\vec{y}]/z_h] \quad} \vec{z}$$

to which we refer as composition of substitutions. Composition is associative and the identical substitution, i.e. the substitution of the form $[\vec{x}/\vec{x}]$ is a neutral element.

It is worth to note that, according with our definitions and notation, an arrow $f\colon \vec{x} \longrightarrow \vec{y}$ exists if and only if there are k terms in the free variables $\vec{x}$.

What we proved so far is that contexts are the objects of a category whose arrows are given by terms substitutions.

For every pair of contexts $\vec{x}$ and $\vec{y}$ we write

$$\vec{x} \times \vec{y}$$

to denote a chosen list $\vec{w}$ of as many distinct variables as the sum of the number of variables in $\vec{x}$ and of that in $\vec{y}$, such as

$$(w_1, w_2, ..., w_n, w_{n+1}, ..., w_{n+k})$$

We denote by $\pi_{\vec{x}}\colon \vec{x} \times \vec{y} \longrightarrow \vec{x}$ the substitution of the variables in $\vec{x}$ with the first n in $\vec{w}$ and and by $\pi_{\vec{y}}\colon \vec{x} \times \vec{y} \longrightarrow \vec{y}$ the substitution of the variables in $\vec{y}$ with the last k variables in $\vec{w}$, i.e.

$$\pi_X = [w_1/x_1, w_2/x_2, ..., w_n/x_n]$$

$$\pi_Y = [w_{n+1}/y_1, w_{n+2}/y_2, ..., w_{n+k}/y_k]$$

We shall call these arrows projections. For every context $\vec{z} = (z_1, z_2, ..., z_j)$, and every pair of arrows $[\vec{t}/\vec{x}]\colon \vec{z} \longrightarrow \vec{x}$ and $[\vec{t'}/\vec{y}]\colon \vec{z} \longrightarrow \vec{y}$, the arrow $\langle [\vec{t}/\vec{x}], [\vec{t'}/\vec{y}] \rangle \colon \vec{z} \longrightarrow \vec{w}$ defined by the following simultaneous substitutions

$$[t_1/w_1, t_2/w_2, ..., t_n/w_n, t'_1/w_{n+1}, t'_2/w_{n+2}, ..., t'_k/w_{n+k}]$$

is the unique arrow such that

$$\pi_X \circ \langle [\vec{t}/\vec{x}], [\vec{t'}/\vec{y}] \rangle = [\vec{t}/\vec{x}]$$

$$\pi_Y \circ \langle [\vec{t}/\vec{x}], [\vec{t'}/\vec{y}] \rangle = [\vec{t'}/\vec{y}]$$

In other words, the category of contexts and terms has binary products.

The easiest case happens when variables in $\vec{x}$ and variables in $\vec{y}$ are distinct, then we can put

$$\vec{x} \times \vec{y} = (x_1, x_2, ..., x_n, y_1, y_2, ..., y_k)$$

For every context $\vec{x}$ we have an assignment

$$\vec{x} \mapsto LT(\vec{x})$$

where $LT(\vec{x})$ is the Lindenbaum-Tarski order of well-formed formulas of $\mathcal{L}$. I.e. an element ϕ of $LT(\vec{x})$ is a well-formed formula of $\mathcal{L}$ with free variables in $\vec{x}$ and the order is given by implication: ϕ is less than or equal to ψ if and only if $\phi \Rightarrow \psi$ is derivable.

For every arrow $[\vec{t}/\vec{y}]\colon \vec{x} \longrightarrow \vec{y}$ we have an assignment

$$(-)[\vec{t}/\vec{y}]\colon LT(\vec{y}) \longrightarrow LT(\vec{x})$$

which maps a formula ϕ with free variables $\vec{y}$ into the formula $\phi[\vec{t}/\vec{y}]$ whose free variables are in $\vec{x}$. The assignment is monotone, i.e. for ϕ, ψ in $LT(\vec{x})$ if $\phi \Rightarrow \psi$ is derivable, then $\phi[\vec{t}/\vec{y}] \Rightarrow \psi[\vec{t}/\vec{y}]$ is derivable.

For every projection $\pi_{\vec{x}}\colon \vec{x} \times \vec{y} \longrightarrow \vec{x}$ we have an assignment

$$\exists \vec{y}(-)\colon LT(\vec{x} \times \vec{y}) \longrightarrow LT(\vec{x})$$

which maps a formula ϕ with free variables in $\vec{x} \times \vec{y}$ into the formula $\exists_{y_1, y_2, .., y_k} \phi$ with free variables in $\vec{x}$.

For every ρ in $LT(\vec{x} \times \vec{y})$ and every σ in $LT(\vec{x})$ it holds that

$$\exists \vec{y} \rho \Rightarrow \sigma \quad \text{if and only if} \quad \rho \Rightarrow \sigma[\pi_{\vec{x}}]$$

Using a categorical language we can say that the existential quantification is left adjoint to substitution along projections of contexts.

In any first order language the two following rules

$$\frac{\rho \Rightarrow \sigma[\pi_{\vec{x}}]}{\exists \vec{y} \rho \Rightarrow \sigma} \exists^L_{in} \qquad \frac{\exists \vec{y} \rho \Rightarrow \sigma}{\rho \Rightarrow \sigma[\pi_{\vec{x}}]} \exists^L_{out}$$

with the obvious constraint on free variables, are derivable. We can actually do one step further.

Proposition 2.1. (Lawvere) The rules $\exists_{in}^L$ and $\exists_{out}^L$ are derivable if and only if the rules $\exists I$ and $\exists E$ are derivable.

Proof. This is in [2] and [4]. $\qquad\qquad\qquad\qquad\qquad\qquad\qquad\qquad\qquad\qquad$ □

For uniformity sake, it will be convenient to deal with collections $LT(\vec{x})$ which are posets rather then preorders. So we shall consider the poset reflection of the preorders above. I.e. an element of $LT(\vec{x})$ is an equivalence class $[\phi]$ of well formed formulas of $\mathcal{L}$ with no more free variables than those in $\vec{x}$, where ϕ is equivalent to ψ if and only if $\phi \Leftrightarrow \psi$ is derivable.

The situations described in 2.1 and in 2.2 share many common features. They both present a category and two assignments, one maps each object of the category to a partially ordered set and the other maps (contravarianly) each arrow of the category to a monotone function between partially ordered sets and this monotone function has a left adjoint.

2.3 Doctrines

The similarities outlined at the end of the previous subsection suggest that both the situations in 2.1 and in 2.2 should be special cases of a single abstract structure. This is provided by the notion of doctrine.

Recall that a **category** $\mathcal{C}$ is a collection of objects and a collection of arrows between objects, such that arrows compose associatively and for every object A there is an arrow $id_A : A \longrightarrow A$ which is the neutral element of the composition.

For categories $\mathcal{C}$ and $\mathcal{E}$ a **functor** $F:\mathcal{C} \longrightarrow \mathcal{E}$ is an assignment which maps objects and arrows of $\mathcal{C}$ into objects and arrows of $\mathcal{E}$ in such a way that composition and identities are preserved, i.e. $F(id_A) = id_{F(A)}$ and $F(fg) = F(f)F(g)$.

We write $G:\mathcal{C}^{op} \longrightarrow \mathcal{E}$ to denote an assignment which maps objects and arrows of $\mathcal{C}$ into objects and arrows of $\mathcal{E}$ in such a way that identities are preserved and if $f:A \longrightarrow B$ in $\mathcal{C}$ then $G(f):G(B) \longrightarrow G(A)$ in $\mathcal{E}$ and $G(fg) = G(g)G(f)$. G is said **contravariant** functor from $\mathcal{C}$ to $\mathcal{E}$.

A category $\mathcal{C}$ has **binary products** if for every pair of objects A and B in $\mathcal{C}$ there is a diagram

$$A \xleftarrow{\ \pi_A\ } A \times B \xrightarrow{\ \pi_B\ } B$$

where π_A and π_B are called **projections** and such that for every object Z and every pair of arrows $f\colon Z \longrightarrow A$ and $g\colon Z \longrightarrow B$ there is a unique function $\langle f, g \rangle\colon Z \longrightarrow A \times B$ with

$$\pi_A \langle f, g \rangle = f$$

$$\pi_B \langle f, g \rangle = g$$

If there are two arrows $t\colon A \longrightarrow X$ and $k\colon B \longrightarrow Y$, we denote by $t \times k$ the induced arrow

$$\langle t\pi_A, k\pi_B \rangle = t \times k\colon A \times B \longrightarrow X \times Y$$

The category **Sets** whose objects are sets and whose arrows are functions is a category with binary products. Where the product of the sets A and B is their cartesian product.

If $\mathcal{L}$ is a first order language and V is an infinite countable set of variables, the category $\mathbb{C}^L$ whose objects are contexts and arrows are terms (as we have described in 2.2) is a category with binary products.

We denote by **Pos** the category whose objects are partially ordered sets and arrows are monotone functions.

Definition 2.2. A **doctrine** is a pair $(\mathcal{C}, P)$ where $\mathcal{C}$ is a category with binary products and P is a contravariant functor

$$P\colon \mathcal{C}^{op} \longrightarrow \mathbf{Pos}$$

such that for every projection $\pi_A\colon A \times B \longrightarrow A$ in $\mathcal{C}$ the monotone function

$$P(f)\colon P(A) \longrightarrow P(A \times B)$$

has a left adjoint, i.e. a monotone function

$$\Sigma_{\pi_A}\colon P(A \times B) \longrightarrow P(A)$$

such that for every $s \in P(A)$ and every $r \in P(A \times B)$ it holds that

$$\Sigma_{\pi_A}(r) \leq s \quad \text{if and only if} \quad r \leq P(\pi_A)(s)$$

Example 2.3. The situation described in subsection 2.1 gives rise to a doctrine $(\mathbf{Sets}, \mathbb{P})$. For a set A the poset $\mathbb{P}(A)$ is the collection of subsets of A ordered by inclusion. For a function f and and for a projection $\pi_A\colon A \times B \longrightarrow A$ it is

$$\mathbb{P}(f) = f^{-1} \quad \text{and} \quad \Sigma_{\pi_A} = \mathrm{Im}_{\pi_A}$$

Example 2.4. Suppose $\mathcal{L}$ is a first order language. Suppose that V is an infinite countable set of variables. The situation described in subsection 2.2 gives rise to a doctrine $(\mathbb{C}^{\mathcal{L}}, LT^{\mathcal{L}})$. For an arrow $[\vec{t}/\vec{y}]$ it is

$$LT([\vec{t}/\vec{y}])(-) = (-)[\vec{t}/\vec{y}]$$

and for a projection $\pi_{\vec{x}} : \vec{x} \times \vec{y} \longrightarrow \vec{x}$ it is

$$\Sigma_{\pi_{\vec{x}}} = \exists \vec{y} \, (-)$$

In many applications of interest it is necessary to work with doctrines with a richer structure with respect to the one introduced in definition 2.2. In this paper we will mainly consider the following simple one.

Definition 2.5. A doctrine $(\mathcal{C}, P)$ is said **primary** if for every A in $\mathcal{C}$ the poset $P(A)$ has a binary meet operation $\wedge$ and for every arrow $f : X \longrightarrow A$ and every α, β in $P(A)$ it is

$$P(f)(\alpha \wedge \beta) = P(f)(\alpha) \wedge P(f)(\beta)$$

In other words posets of the form $P(A)$ are meet-semilattices and functions of the form $P(f)$ are homomorphisms of meet-semilattices.

Example 2.6. The doctrine $(\mathbf{Sets}, \mathbb{P})$ introduced in example 2.3 is clearly primary. For every set A the intersection is a meet operation on the collection of its subsets and intersection commutes with inverse images.

Example 2.7. The doctrine $(\mathbb{C}^{\mathcal{L}}, LT^{\mathcal{L}})$ introduced in example 2.4 is primary as conjunctions are known to be meets with respect to the order given by the implication. Thus for two wellformed formulas ϕ and ψ with free variables in $\vec{x}$ define

$$[\phi] \wedge [\psi] = [\phi \ \& \ \psi]$$

Conjunctions commute with substitution: $(\phi \ \& \ \psi)[\vec{t}/\vec{x}] \leftrightarrow (\phi[\vec{t}/\vec{x}]) \ \& \ (\psi[\vec{t}/\vec{x}])$

From the example above we can draft the following correspondence: in a doctrine $(\mathcal{C}, P)$ we can think of

- objects of $\mathcal{C}$ as contexts

- arrows of $\mathcal{C}$ as terms

- elements in $P(A)$ as formulas with free variables in the context A

- functions $P(f)$ as substitutions of terms in formulas

- left adjoints along projections as the existential quantifiers.

After proposition 2.1 it might appear obvious that left adjoints are all what we need to express existential quantification in the framework of doctrines. This is not completely true, for at least two reasons. These are the so called Beck-Chevalley condition and Frobenius reciprocity.

The Beck-Chevalley condition

Consider the following fact. Suppose $R(x, y)$ is a well formed formula and t a term such that y is not among the free variables of t then the two assignments

$$R(x, y) \mapsto R(x, y)[t/x] = R(t, y) \mapsto \exists y \ R(t, y)$$

$$R(x, y) \mapsto \exists y \ R(x, y) \mapsto \exists y \ R(x, y)[t/x] = \exists y \ R(t, y)$$

end up to the same (and therefore equivalent) formulas.

In this respect, doctrines have the feature that for a commutative square of the form

$$
\begin{array}{ccc}
A \times Y & \xrightarrow{\pi_A} & A \\
{\scriptstyle t \times id_Y} \downarrow & & \downarrow {\scriptstyle t} \\
X \times Y & \xrightarrow[\pi_Y]{} & X
\end{array}
$$

the condition of left adjoint naturally brings to the standard inequality

$$P(t)\Sigma_{\pi_A} \leq \Sigma_{\pi_A} P(id_X \times t)$$

which is not necessarily an equality. When in a doctrine the inequality above is an equality, we say that left adjoints satisfy the Beck-Chevalley condition.

Frobenius Reciprocity

Consider the following fact. Suppose $R(x, y)$ is a well formed formula in free variables x and y. Suppose $P(x)$ is a well formed formula in the free variable x. The following is derivable

$$\exists y \ (R(x, y) \wedge P(x)) \leftrightarrow \exists y \ (R(x, y)) \wedge P(x)$$

In this respect, doctrines have the feature that for every pair of objects X and Y of $\mathcal{C}$, for every $r \in P(X \times Y)$ and every $p \in P(X)$ the condition of left adjoint brings to the following standard inequality

$$\Sigma_{\pi_X}(r \wedge P(\pi_X)(p)) \leq \Sigma_{\pi_X}(r) \wedge p$$

which is not necessarily an equality.

When the inequality above is an equality we say that left adjoints satisfy the Frobenius Reciprocity.

The validity of the Back-Chevalley condition and the validity of Frobenius reciprocity are so important to motivate the following.

Definition 2.8. A primary doctrine where left adjoints of the form Σ_π satisfy the Beck-Chevalley condition and Frobenius reciprocity is said **existential**.

Example 2.9. It is clear from what we said above that the doctrine $(\mathbb{C}^{\mathcal{L}}, LT^{\mathcal{L}})$ of example 2.4 is existential.

Example 2.10. Is is an easy exercise to show that for every binary relation $R \subseteq X \times Y$ and for every subset $P \subseteq X$ it is

$$\mathrm{Im}_{\pi_X}(R \cap \pi_X^{-1}(P)) = \mathrm{Im}_{\pi_X}(R) \cap P$$

In other words the doctrine $(\mathbf{Sets}, \mathbb{P})$ is existential, as the Beck-Chevalley condition trivially holds in this doctrine.

3 The Epsilon operator in doctrines

In this section we introduce doctrines with the ϵ-operator and we relate our definition with doctrines in examples 2.3 and 2.4. In particular we show that for a first order language $\mathcal{L}$ the doctrine $(\mathbb{C}^{\mathcal{L}}, LT^{\mathcal{L}})$ has the ϵ-operator if and only if $\mathcal{L}$ is Hilbert's ϵ-calculus. Moreover we show that, using the axiom of choice and the excluded middle, also the doctrine $(\mathbf{Sets}, \mathbb{P})$ has the ϵ-operator.

Following the correspondence that we traced in the previous section, i.e. if $(\mathcal{C}, P)$ is a doctrine, we look at arrows of $\mathcal{C}$ as terms and at elements in the image of P as formulas, then we expect that $(\mathcal{C}, P)$ has the ϵ-operator if it has an operator that associates to any formulas ϕ an arrow ϵ_ϕ of $\mathcal{C}$, satisfying certain properties. In order to have the doctrine described in 2.3 among our examples we need to avoid the codomain of ϵ_ϕ to empty (unless also the domain is empty). Thus we want a notion of emptyness for objects of categories. We adopt the following.

Empty object: an object 0 of a category $\mathcal{C}$ is said to be **empty** if and only if every arrow to 0 is an isomorphism. I.e. 0 in $\mathcal{C}$ is empty if and only if for every arrow $f: A \longrightarrow 0$ there is an arrow $g: 0 \longrightarrow A$ such that

$$fg = id_0 \quad \text{and} \quad gf = id_A$$

We have now all the ingredients to introduce the ϵ-operator in doctrines.

Definition 3.1. A doctrine $(\mathcal{C}, P)$ has the ϵ-operator if for every pair of objects X and Y of $\mathcal{C}$ and for every ϕ in $P(X \times Y)$ if Y is not empty then there is an arrow $\epsilon_\phi: X \longrightarrow Y$ such that

$$\Sigma_{\pi_X}(\phi) = P(\langle id_X, \epsilon_\phi \rangle)(\phi)$$

We want now to comment on this definition of doctrine with ϵ-operator in relation to the two examples introduced in the previous section.

First order languages

Suppose $\mathcal{L}$ is a first order language and V a countable set of variables. Consider the doctrine $(\mathbb{C}^{\mathcal{L}}, LT^{\mathcal{L}})$ introduced in example 2.4. First note that the category of contexts $\mathbb{C}^{\mathcal{L}}$ has no empty objects, so the constraint on Y in definition 3.1 is vacuous in this example.

We have that $(\mathbb{C}^{\mathcal{L}}, LT^{\mathcal{L}})$ is an epsilon doctrine if and only if $\mathcal{L}$ is Hilbert's epsilon calculus. In fact suppose $\mathcal{L}$ is Hilbert's ϵ-calculus and suppose that ϕ is in $LT(\vec{x} \times \vec{y})$, i.e. ϕ is a well-formed formula with free variables in $\vec{x} \times \vec{y} = \vec{x} \times (y_1, y_2, ..., y_k)$. Then we have epsilon terms such that

$$\phi_1 \leftrightarrow \exists y_k \phi \leftrightarrow \phi[\epsilon_{y_k, \phi}/y_k]$$

$$\phi_2 \leftrightarrow \exists y_{k-1} \phi_1 \leftrightarrow \phi_1[\epsilon_{y_{k-1}, \phi_1}/y_{k-1}]$$

$$\vdots$$

$$\phi_k \leftrightarrow \exists y_1 \phi_{k-1} \leftrightarrow \phi_{k-1}[\epsilon_{y_1, \phi_{k-1}}/y_1]$$

Therefore there is an arrow of $\mathbb{C}^{\mathcal{L}}$

$$[\epsilon_{y_1, \phi_{k-1}}/y_1, \epsilon_{y_2, \phi_{k-2}}/y_2, ..., \epsilon_{y_k, \phi}/y_1]: \vec{x} \longrightarrow \vec{y}$$

which is the desired one.

Conversely if $(\mathbb{C}^{\mathcal{L}}, LT^{\mathcal{L}})$ is a doctrine with the ϵ-operator and ϕ is a formula with free variables $x_1, ..., x_n$ and y, then ϕ is a formula in $LT((x_1, ..., x_n) \times (y))$. Therefore we have an arrow

$$\epsilon_\phi \colon (x_1, ..., x_n) \longrightarrow (y)$$

such that

$$\exists y \; \phi = \Sigma_{\pi_{(y)}}(\phi) \leftrightarrow LT(\langle id_{\vec{x}}, \epsilon_\phi \rangle)(\phi) = \phi[x_1/x_1, x_2/x_2, ..., \epsilon_\phi(x_1, ..., x_n)/y]$$

From which the claim.

Powersets

We now turn on our second example. The doctrine $(\mathbf{Sets}, \mathbb{P})$ introduced in example 2.3 is an ϵ-doctrine. Suppose $\phi \subseteq X \times Y$ where Y is not empty. Let a be an element of Y. Consider the function $\epsilon_\phi \colon X \longrightarrow Y$ defined by the following assignment

$$\epsilon_\phi(x) = \begin{cases} y & \text{if } \{y \in Y \mid (x,y) \in \phi\} \neq \emptyset \text{ and } y \text{ is any of its elements} \\ \\ a & \text{otherwise} \end{cases}.$$

Then it is

$$\begin{aligned} \langle id_X, \epsilon_\phi \rangle^{-1}\phi &= \{x \in X \mid (x, \epsilon_\phi(x)) \in \phi\} \\ &= \{x \in X \mid \exists y \in Y \; (x,y) \in \phi\} \\ &= \mathrm{Im}_{\pi_X}\phi \end{aligned}$$

from which the claim.

Note that in building the function ϵ_ϕ we allow ourself to pick an element from each set of a family of non-empty sets, namely those sets of the form $\{y \in Y \mid (x,y) \in \phi\}$. We also allow ourself to define functions by cases. In other words we used both the axiom of choice and the law of excluded middle. We shall return on this point in the last section.

4 Existential doctrines and the ϵ-operator

In section 2 we pointed out that the the condition of being left adjoints is not enough to prove the validity of both the Beck-Chevalley condition and the Frobenius reciprocity. In this section we show both this conditions become provable in doctrines with the ϵ-operator.

Lemma 4.1. Suppose $(\mathcal{C}, P)$ is a doctrine with the ϵ-operator. For every projection $\pi_X \colon X \times Y \longrightarrow Y$ where Y is not an empty object, for every $\psi \in P(X \times Y)$ and every $h \colon X \longrightarrow Y$, it is

$$P(\langle id_X, h \rangle)(\psi) \le P(\langle id_X, \epsilon_\psi \rangle)(\psi)$$

Proof. Since Σ_{π_X} is left adjoint to $P(\pi_X)$ it is

$$\psi \le P(\pi_X)\Sigma_{\pi_X}(\psi) = P(\pi_X)P(\langle id_X, \epsilon_\psi \rangle)(\psi)$$

Apply $P(\langle id_X, h \rangle)$ to both sides of the inequality to have

$$P(\langle id_X, h \rangle)(\psi) \le P(\langle id_X, h \rangle)P(\pi_X)P(\langle id_X, \epsilon_\psi \rangle)(\psi)$$

In the right hand side we have $P(\langle id_X, h \rangle)P(\pi_X) = P(\pi_X \langle id_X, h \rangle) = P(id_X)$ from which the claim. $\qquad\square$

For the rest of the section, unless specified otherwise, we shall consider only doctrines $(\mathcal{C}, P)$ in which the category $\mathcal{C}$ does not have an empty object. We will turn on this latter case later.

Proposition 4.2. Every doctrine with the ϵ-operator validates the Beck-Chevalley condition.

Proof. We need to prove that for every projection $\pi_X \colon X \times Y \longrightarrow X$ the left adjoint Σ_{π_X} satisfying the Beck-Chevalley condition.

For every ψ in $P(X \times Y)$ the ϵ-operator generates an arrow $\epsilon_\psi \colon X \longrightarrow Y$ with

$$\Sigma_{\pi_X}(\psi) = P(\langle id_X, \epsilon_\psi \rangle)(\psi)$$

Consider an arrow $t \colon A \longrightarrow X$ and the composition $\epsilon_\psi t \colon A \longrightarrow Y$. By 4.1 it is

$$P(\langle id_A, \epsilon_\psi t \rangle)P((t \times id_Y))(\psi) \le P(\langle id_A, \epsilon_{P(t \times id_Y)(\psi)} \rangle)P(t \times id_Y)(\psi)$$

since the following diagram

$$
\begin{array}{ccc}
A & \xrightarrow{\quad t \quad} & X \\
{\scriptstyle \langle id_A, \epsilon_\psi t \rangle} \downarrow & {\scriptstyle \langle t, \epsilon_\psi t \rangle} \searrow & \downarrow {\scriptstyle \langle id_X, \epsilon_\psi \rangle} \\
A \times Y & \xrightarrow[\quad t \times id_Y \quad]{} & X \times Y
\end{array}
$$

commutes, i.e. $(t \times id_Y)\langle id_A, \epsilon_\psi t \rangle = \langle t, \epsilon_\psi t \rangle = \langle id_X, \epsilon_\psi \rangle t$, the previous inequality can be rewritten as

$$P(t)P(\langle id_X, \epsilon_\psi \rangle)(\psi) \le P(\langle id_A, \epsilon_{P(t \times id_Y)(\psi)} \rangle)P(t \times id_Y)(\psi)$$

and therefore

$$P(t)\Sigma_{\pi_X}(\psi) \le \Sigma_{\pi_A} P(t \times id_Y)(\psi)$$

which proves the claim, as the other inequality is standard. $\qquad\square$

Lemmas 4.1 and 4.2 have the following corollary, which is the main theorem of the section.

Proposition 4.3. *Every primary doctrine with the ϵ-operator is existential.*

Proof. Suppose $(\mathcal{C}, P)$ is such a doctrine. After 4.2 it remains to prove that $(\mathcal{C}, P)$ satisfy Frobenius Reciprocity.

Consider a projection $\pi_X \colon X \times Y \longrightarrow X$. For ϕ in $P(X \times Y)$ and β in $P(X)$ abbreviate by ξ the formula $\phi \wedge P(\pi_X)(\beta)$. We have

$$\Sigma_{\pi_X}(\phi \wedge P(\pi_X)(\beta)) = P(\langle id_X, \epsilon_\xi \rangle)(\phi \wedge P(\pi_X)(\beta))$$

Moreover, recalling that $P(\langle id_X, \epsilon_\phi \rangle)P(\pi_X) = P(id_X)$

$$\begin{aligned}
P(\langle id_X, \epsilon_\phi \rangle)(\phi) \wedge \beta &= P(\langle id_X, \epsilon_\phi \rangle)(\phi) \wedge P(id_X)(\beta) \\
&= P(\langle id_X, \epsilon_\phi \rangle)(\phi) \wedge P(\langle id_X, \epsilon_\phi \rangle)P(\pi_X)(\beta) \\
&= P(\langle id_X, \epsilon_\phi \rangle)(\phi \wedge P(\pi_X)(\beta))
\end{aligned}$$

Apply lemma 4.1 to get

$$P(\langle id_X, \epsilon_\phi \rangle)(\phi) \wedge \beta \le P(\langle id_X, \epsilon_\xi \rangle)(\phi \wedge P(\pi_X)(\beta))$$

Hence

$$\Sigma_{\pi_X}\phi \wedge \beta \le \Sigma_{\pi_X}(\phi \wedge P(\pi_X)(\beta))$$

from which the claim follows, as the other inequality is standard. $\qquad\square$

The previous theorem is general enough to cover the class of all doctrines $(\mathcal{C}, P)$ without an empty object in $\mathcal{C}$. We have already seen that the doctrine $(\mathbb{C}^{\mathcal{L}}, LT^{\mathcal{L}})$, built from a first order language $\mathcal{L}$, is such. What happen if the doctrine $(\mathcal{C}, P)$ is such that $\mathcal{C}$ has an empty object? We may aspect that the previous theorem holds provided that the empty object of $\mathcal{C}$ well interacts with the structure P. But which structure? To our knowledge, the structure of primary doctrine is not sufficient to prove that all primary doctrines with the ϵ-operator are existential. To reach this general theorem we work with a class of doctrines which has one more property.

We say that a doctrine $(\mathcal{C}, P)$ has **the false predicate** if for every A in $\mathcal{C}$ the poset $P(A)$ has a bottom element $\bot_A$ and for every $f\colon X \longrightarrow A$ in $\mathcal{C}$ it is

$$P(f)(\bot_A) = \bot_X$$

Example 4.4. The doctrine $(\mathbf{Sets}, \mathbb{P})$ introduced in example 2.3 has the false predicate, since for every set A we have

$$\bot_A = \emptyset \subseteq A$$

and for every function $f\colon X \longrightarrow A$ it is $f^{-1}(\emptyset) = \emptyset$.

Example 4.5. The doctrine $(\mathbb{C}^{\mathcal{L}}, LT^{\mathcal{L}})$ introduced in example 2.4 has the false predicate which is exactly the false predicate $\bot$, which is preserved by substitution.

Suppose $(\mathcal{C}, P)$ is a primary doctrine with the false predicate. An empty object 0 of $\mathcal{C}$ is said **proper** (with respect to P) if

$$P(0) = \{\bot_0\}$$

Example 4.6. The doctrine $(\mathbf{Sets}, \mathbb{P})$ has a proper empty object which is $\emptyset$. In fact the unique subset of the empty set is the empty set itself, i,e,

$$\mathbb{P}(\emptyset) = \{\emptyset\} = \{\bot_\emptyset\}$$

Example 4.7. The doctrine $(\mathbb{C}^{\mathcal{L}}, LT^{\mathcal{L}})$ does not have a proper empty object as the category $\mathbb{C}^{\mathcal{L}}$ does not have an empty object.

For every object X of $\mathcal{C}$ we have a projection $\pi_0\colon X \times 0 \longrightarrow 0$. Since 0 is an empty object $X \times 0$ is isomorphic to 0. We denote by $!_X$ the arrow

$$!_X\colon 0 \simeq X \times 0 \longrightarrow X$$

which is still a projection.

Suppose $(\mathcal{C}, P)$ is a doctrine with a proper empty object 0. Since $P(0)$ is a singleton, the assignment $\bot_0 \mapsto \bot_X$ provides a left adjoint to $P(!_X)$.

To prove the Beck-Chevalley condition, consider an arrow $t\colon A \longrightarrow X$ and the following square

$$
\begin{array}{ccc}
0 & \xrightarrow{\ !_A\ } & A \\
{\scriptstyle id_0}\big\downarrow & & \big\downarrow{\scriptstyle t} \\
0 & \xrightarrow[\ !_X\ ]{} & X
\end{array}
$$

Hence

$$P(t)\Sigma_{!_X}(\bot_0) = P(t)(\bot_X) = \bot_A = \Sigma_{!_A}(\bot_0) = P(id_0)\Sigma_{!_X}(\bot_0)$$

To prove Frobenius reciprocity consider ϕ in $P(X)$

$$\Sigma_{!_X}(\bot_0 \wedge P(!_X)(\phi)) = \Sigma_{!_X}(\bot_0) = \bot_X = \bot_X \wedge \phi = \Sigma_{!_X}(\bot_0) \wedge \phi$$

We have proved the following.

Proposition 4.8. A primary doctrine $(\mathcal{C}, P)$ with the false predicate, the ϵ-operator and a proper empty object in $\mathcal{C}$ is existential.

5 Intuitionistic set theories (naively)

As an application of the theory that we have developed so far, we look at intuitionistic theories of sets with choice. This sentence might appear an oxymoron after the known result of Diaconescu [1], therefore we have to be very careful in formulating the framework into which we aim to work. So far we have written **Sets** to denote the category of sets for an unspecified underlying theory of sets. We want to remain at this naive level and at the same time we want to avoid the Diaconescu's argument to be carried out. So let **Sets**$(\mathcal{T})$ be the category of sets with respect to the theory $\mathcal{T}$, where $\mathcal{T}$ has at least the following set of axioms which we formulate naively

($\mathcal{T}_1$) two sets are equal whenever they have the same elements

($\mathcal{T}_2$) the cartesian product of two sets is a set

($\mathcal{T}_3$) for every set A and every first order property P the collection $B = \{a \in A \mid P\}$ is a set and $x \in B$ if and only if $P(x)$ holds

($\mathcal{T}_4$) every surjection has a section.

We first note that we do not have a powerset axiom. In other words if A is a object of **Sets**$(\mathcal{T})$, i.e. a set for the theory $\mathcal{T}$, the collection of all the inclusions with codomain A, i.e. $\mathbb{P}(A)$, is not a set for the theory $\mathcal{T}$ and therefore is not an object of **Sets**$(\mathcal{T})$. Note that **Sets**$(\mathcal{T})$ might be empty.

Of course we can still collect together all the subsets of a given set A. And we can still denote this collection by $\mathbb{P}(A)$. But we can not say that $\mathbb{P}(A)$ is in **Sets**$(\mathcal{T})$, it lives somewhere else. As we can still order $\mathbb{P}(A)$ by inclusion, we are allow to say that $\mathbb{P}(A)$ still lives in **Pos**. Thus it is immediate to verify that the pair $(\mathbf{Sets}(\mathcal{T}), \mathbb{P})$ is still a doctrine, simply following the arguments in example 2.3.

The theory $\mathcal{T}$ has all the properties that we are going to use in this section along with the advantage that the theorem of Diaconescu is not provable.

Is $(\mathbf{Sets}(\mathcal{T}), \mathbb{P})$ a doctrine with the ϵ-operator? To answer to this question we need to introduce the following fifth axiom.

($\mathcal{T}_5$) for every set A, every subset X of A and every function $f \colon X \longrightarrow B$ where B is not empty, there is a function $k \colon A \longrightarrow B$ which makes the following commute

$$
\begin{array}{ccc}
X & \lhook\joinrel\longrightarrow & A \\
& f \searrow & \downarrow k \\
& & B
\end{array}
$$

Note that using the law of excluded middle, $\mathcal{T}_5$ is derivable and therefore is valid in **Sets**, since we can take as k the following function

$$
k(a) = \begin{cases} f(a) & \text{if } a \in X \\ \\ b & \text{if } a \notin X \end{cases} .
$$

where b is any element of B.

Proposition 5.1. The doctrine $(\mathbf{Sets}(\mathcal{T}), \mathbb{P})$ has the ϵ-operator if and only if $\mathcal{T}_5$ belongs to $\mathcal{T}$.

Suppose that the doctrine $(\mathbf{Sets}(\mathcal{T}), \mathbb{P})$ has the ϵ-operator. Suppose we have $X \subseteq A$ and $f \colon A \longrightarrow B$ where B is not empty. Consider the subset of $A \times B$

$$
\gamma = \{(a, b) \in A \times B \mid f(a) = b \text{ and } a \in X\} \subseteq A \times B
$$

Since $\gamma \in \mathbb{P}(A \times B)$, the ϵ-operator generates a function

$$
\epsilon_\gamma \colon A \longrightarrow B
$$

such that

$$\mathrm{Im}_{\pi_A}\gamma = \langle id_A, \epsilon_\gamma \rangle^{-1}\gamma$$

whence

$$\{a \in A \mid \exists b \in B \; (a,b) \in \gamma \} = \{a \in A \mid (a, \epsilon_\gamma(a)) \in \gamma \}$$

It is easy to verify that X is included into the set on the left hand side. Then for every $x \in X$, the pair $(x, \epsilon_\gamma(x)) \in \gamma$, from which $f(x) = \epsilon_\gamma(x)$.

The converse. Every non empty subset $\phi \subseteq A \times B$ generates an obvious surjection

$$e\colon \phi \longrightarrow \mathrm{Im}_{\pi_A}\phi = \{a \in A \mid \exists b \in B \; (a,b) \in \phi \}$$

which is the function that maps every pair (a,b) in ϕ to the first component a. By the axiom of choice the surjection e has a section

$$s\colon \mathrm{Im}_{\pi_A}\phi \longrightarrow \phi$$

i.e. a function such that $e(s(x)) = x$. This produces the diagram below

$$\mathrm{Im}_{\pi_A}\phi \lhook\joinrel\longrightarrow A$$

By $\mathcal{T}_5$ we can commutatively close the diagram, i.e. there is a function

$$k\colon A \longrightarrow \phi$$

such that for every $a \in \mathrm{Im}_{\pi_A}\phi$ it is $s(a) = k(a)$. Since $\phi \subseteq A \times B$ the section s may be seen as a function with codomain $A \times B$. Define ϵ_ϕ as the composition

$$\epsilon_\phi = \pi_B k\colon A \longrightarrow \phi \subseteq A \times B \longrightarrow B$$

We want to show that ϵ_ϕ is an ϵ-term. We trivially have that $\langle id_A, \epsilon_\phi \rangle^{-1}\phi \subseteq \mathrm{Im}_{\pi_A}\phi$, so it remains to prove the other inclusion. For $a \in \mathrm{Im}_{\pi_A}\phi$ it is $k(a) \in \phi$, thus if we prove that $k(a) = (a, \epsilon_\phi(a))$ we are done.

Since $k(a) \in \phi \subseteq A \times B$, $k(a)$ is of the form $k(a) = (\pi_A(k(a)), \pi_B(k(a)))$, then

$$k(a) = (\pi_A(k(a)), \epsilon_\phi(a))$$

By extensionability $s(a) = k(a)$, hence

$$a = es(a) = ek(a) = e(\pi_A(k(a)), \pi_B(k(a))) = \pi_A(k(a))$$

A substitution yields $k(a) = (a, \epsilon_\phi(a))$ from which the claim.

References

[1] R. Diaconescu. Axiom of choice and complementation. *Proceedings of the American Mathematical Society*, 51(1):176 – 178, 1975.

[2] F. W. Lawvere. Adjointness in foundations. *Dialectica*, 23:281–296, 1969.

[3] Fabio Pasquali. A categorical interpretation of the intuitionistic, typed, first order logic with hilbert's ε-terms. *Logica Universalis*, 10(4):407–418, 2016.

[4] Andrew M. Pitts. Categorical logic. In *Handbook of Logic in Computer Science, Vol. 5*, pages 39–128. Oxford Univ. Press, 2000.

                    Received 31 October 2015

Epsilon Terms in Intuitionistic Sequent Calculus

Giselle Reis
Inria & LIX/École Polytechnique, France

Bruno Woltzenlogel Paleo
Australian National University, Canberra, Australia

Abstract

Skolemization is unsound in intuitionistic logic in the sense that a Skolemization $sk(F)$ of a formula F may be derivable in the intuitionistic sequent calculus **LJ** while F itself is not. This paper defines a transformation T_ε that differs from Skolemization only by its use of ε-terms instead of Skolem terms; and shows that, for a simple locally restricted sequent calculus **LJ***, this transformation is sound: if $T_\varepsilon(F)$ is derivable in **LJ***, then so is F.

1 Introduction

It is well-known that there are formulas whose Skolemizations are derivable in the intuitionistic sequent calculus **LJ** while the formulas themselves are not. Consequently, there exists no immediate method of de-Skolemization, i.e. a method to eliminate Skolem terms from intuitionistic proofs by introducing quantifiers without obtaining just classical proofs. The usual reaction to this fact is to conclude that Skolemization is intrinsically unsound in intuitionistic logic and, consequently, must be either avoided or modified in sophisticated ways [4, 9]. These approaches assume (quite naturally) that provability in **LJ** correctly captures validity in intuitionistic logic even in the presence of Skolem terms.

This paper explores a different approach that regards **LJ** as an unsound calculus for reasoning about formulas containing Skolem terms. From this perspective,

---

The authors would like to thank: Christian Retoré for organizing the workshop on *Hilbert's Epsilon and Tau in Logic, Informatics and Linguistics*; the anonymous reviewers of this workshop for providing useful feedback on an extended abstract of this paper; and Sergei Soloviev, for pointing out an interesting related work of Grigori Mints.

the reason why underivable formulas become derivable in **LJ** after they have been Skolemized is due to **LJ**'s inference rules being too permissive: they fail to recognize the special status of Skolem terms and allow them to be used in ways that should be forbidden. Therefore, the interesting question is not how to modify Skolemization in order to obtain an intuitionistically sound Skolemization-like transformation w.r.t. to **LJ**, but how to modify and restrict **LJ** so that Skolemization is sound w.r.t. the restricted calculus.

The main contribution of this paper is the design of a restricted sequent calculus **LJ*** for which *epsilonization* is sound: if $T_\varepsilon(S)$ (the epsilonization of the sequent S) is derivable in **LJ***, then so is S. In particular, we define a method of de-epsilonization of intuitionistic proofs transforming intuitionistic proofs with ε-terms into ordinary intuitionistic proofs. The transformation T_ε differs from Skolemization mainly in its use of Hilbert's ε-terms instead of Skolem terms. But in contrast to Hilbert's traditional ε-calculus, where all quantifiers are eliminated, T_ε eliminates only strong quantifiers. Skolem terms can be regarded as abbreviations of Hilbert's ε-terms [1]; conversely, ε-terms can be regarded as more informative Skolem terms. **LJ*** restricts the use of ε-terms in the instantiations performed by weak quantifier rules. The restrictions are local and purely syntactic; they use the extra information available in ε-terms but not in Skolem terms.

2 LJ and Epsilonization

We assume the reader is familiarized with the language of first-order logic. The rules of **LJ** are depicted in Figures 1 and 2. $\forall_l$ and $\exists_r$ are called *weak quantifier rules*, while $\forall_r$ and $\exists_l$ are called *strong quantifier rules*. $\forall$-quantifiers of positive polarity and $\exists$-quantifiers of negative polarity are called *strong quantifiers*.

$$\frac{\Gamma_1, A \vdash F \quad \Gamma_2, B \vdash F}{\Gamma_1, \Gamma_2, A \vee B \vdash F} \vee_l \quad \frac{\Gamma \vdash A}{\Gamma \vdash A \vee B} \vee_r^1 \quad \frac{\Gamma \vdash B}{\Gamma \vdash A \vee B} \vee_r^2 \quad \frac{\Gamma \vdash A}{\Gamma, \neg A \vdash} \neg_l \quad \frac{\Gamma, A \vdash}{\Gamma \vdash \neg A} \neg_r$$

$$\frac{\Gamma, A, B \vdash F}{\Gamma, A \wedge B \vdash F} \wedge_l \quad \frac{\Gamma_1 \vdash A \quad \Gamma_2 \vdash B}{\Gamma_1, \Gamma_2 \vdash A \wedge B} \wedge_r \quad \frac{\Gamma_1 \vdash A \quad \Gamma_2, B \vdash F}{\Gamma_1, \Gamma_2, A \rightarrow B \vdash F} \rightarrow_l \quad \frac{\Gamma, A \vdash B}{\Gamma \vdash A \rightarrow B} \rightarrow_r$$

$$\frac{}{A \vdash A} a \; (A \text{ is atomic}) \quad \frac{\Gamma \vdash F}{\Gamma, A \vdash F} w_l \quad \frac{\Gamma \vdash}{\Gamma \vdash A} w_r \quad \frac{\Gamma, A, A \vdash F}{\Gamma, A \vdash F} c_l \quad \frac{\Gamma_1 \vdash A \quad \Gamma_2, A \vdash F}{\Gamma_1, \Gamma_2 \vdash F} cut$$

Figure 1: Propositional and Structural Rules for **LJ**

$$\frac{\Gamma, A[t] \vdash F}{\Gamma, \forall x.A[x] \vdash F} \; \forall_l \qquad \frac{\Gamma \vdash A[\alpha]}{\Gamma \vdash \forall x.A[x]} \; \forall_r \qquad \frac{\Gamma, A[\alpha] \vdash F}{\Gamma, \exists x.A[x] \vdash F} \; \exists_l \qquad \frac{\Gamma \vdash A[t]}{\Gamma \vdash \exists x.A[x]} \; \exists_r$$

where:

- α must satisfy the eigenvariable condition.

Figure 2: Quantifier Rules for **LJ**

Skolemization is a transformation that removes all strong quantifiers from first-order formulas and replaces the variables they quantify by *Skolem terms*. There are various Skolemization methods, which may differ in the proof complexity of the transformed formula [5]. To see that Skolemization does not preserve derivability in the sequent calculus **LJ**, consider the formula $\neg \forall x.P(x) \rightarrow \exists y.\neg P(y)$, in which the $\forall$ quantifier is strong (note that it would be introduced by a $\forall_r$ inference in a sequent calculus proof). While it is clear that $\nvdash_{\mathbf{LJ}} \neg \forall x.P(x) \rightarrow \exists y.\neg P(y)$, the proof below shows that its Skolemization $\neg P(s) \rightarrow \exists y.\neg P(y)$ (where s is a skolem constant) is derivable:

$$\frac{\dfrac{\dfrac{\overline{Ps \vdash Ps}}{Ps, \neg Ps \vdash} \; \neg_l}{\dfrac{\neg Ps \vdash \neg Ps}{\neg Ps \vdash \exists y.\neg Py} \; \exists_r} \; \neg_r}{\vdash \neg Ps \rightarrow \exists y.\neg Py} \; \rightarrow_r$$

In this example, the use of s on the weak quantifier rule could be avoided if we had more information about it. In order to obtain more informative terms, we choose to use ε-terms instead of Skolem terms for replacing the strongly quantified variables of a formula.

ε-terms[1] are formed with two binders: ε and τ. The intended meaning of ε-terms is delimited by the following *epsilon axioms*:

$$\exists x.A[x] \rightarrow A[\varepsilon_x A[x]] \quad \text{and} \quad A[\tau_x A[x]] \rightarrow \forall x.A[x]$$

In classical logic, the following equivalences hold, and hence τ is definable using ε:

$$A[\tau_x A[x]] \leftrightarrow \forall x.A[x] \leftrightarrow \neg \exists x.\neg A[x] \leftrightarrow \neg\neg A[\varepsilon_x \neg A[x]] \leftrightarrow A[\varepsilon_x \neg A[x]]$$

---

[1]We assume that the usual inductively defined *terms* of first-order logic are extended to include ε-terms. Hence, in general, a term may or may not contain ε-binders. ε-terms, on the other hand, are assumed to have ε or τ binders as their outermost symbols. Therefore, every ε-term is a term, but not every term is an ε-term.

In intuitionistic logic, however, the equivalences above do not hold. Therefore, both binders are needed. Epsilonization is analogous to Skolemization, but it uses ε-terms instead of Skolem terms.

Definition 1 (Epsilonization). *An epsilonization $T_\varepsilon(F)$ of a formula F is defined inductively on the structure of F using two functions T_ε^+ and T_ε^-. On the definitions below, $p \in \{+, -\}$ and $\bar{p}$ is $+$ if $p = -$ and $-$ if $p = +$.*

$$
\begin{aligned}
T_\varepsilon(F) &= T_\varepsilon^+(F) \\
T_\varepsilon^p(A) &= A \text{ if } A \text{ is atomic.} \\
T_\varepsilon^p(\neg A) &= \neg T_\varepsilon^{\bar{p}}(A) \\
T_\varepsilon^p(A \wedge B) &= T_\varepsilon^p(A) \wedge T_\varepsilon^p(B) \\
T_\varepsilon^p(A \vee B) &= T_\varepsilon^p(A) \vee T_\varepsilon^p(B) \\
T_\varepsilon^p(A \to B) &= T_\varepsilon^{\bar{p}}(A) \to T_\varepsilon^p(B) \\
T_\varepsilon^+(\exists x.A) &= \exists x.T_\varepsilon^+(A) \\
T_\varepsilon^+(\forall x.A) &= A'\{x \mapsto \tau_x A'\} \text{ for } A' = T_\varepsilon^+(A) \\
T_\varepsilon^-(\forall x.A) &= \forall x.T_\varepsilon^-(A) \\
T_\varepsilon^-(\exists x.A) &= A'\{x \mapsto \varepsilon_x A'\} \text{ for } A' = T_\varepsilon^-(A)
\end{aligned}
$$

Definition 2 (Epsilonization of sequents). *The epsilonization $T_\varepsilon(S)$ of a sequent S of the form $A_1, \ldots, A_n \vdash B_1, \ldots, B_m$ is a sequent of the form $T_\varepsilon^-(A_1), \ldots, T_\varepsilon^-(A_n) \vdash T_\varepsilon^+(B_1), \ldots, T_\varepsilon^+(B_m)$.*

In Skolemization one needs to explicitly keep track of weakly quantified variables in order to add them as arguments of the Skolem function. In epsilonization such book-keeping is not needed. Since the whole formula will be a sub-expression of the ε-term, the weakly quantified variables will occur naturally in the term. In contrast to what is done in Hilbert's ε-calculus [1], the epsilonization procedure defined here does not eliminate the weak quantifiers; therefore ε-terms may contain quantified formulas. Like in the standard ε-calculus, innermost strong quantifiers are removed first. Using this strategy, strong quantifiers will never occur inside an ε-term. Instead, it will contain *nested* ε-terms corresponding to the variables that were bound by those strong quantifiers. An ε-term that is not nested inside another ε-term is a *top-level* ε-term.

Example 1. *Consider the formula $\forall x.\exists y.\exists z.P(x, y, z)$. In a negative context, its epsilonization would be:*

$$\forall x.P(x, \varepsilon_y P(x, y, \varepsilon_z P(x, y, z)), \varepsilon_z P(x, \varepsilon_y P(x, y, \varepsilon_z P(x, y, z)), z))$$

As desired, the weakly quantified variable x naturally occurs inside the ε-terms for y and z. The weak quantifier $\forall x$ remained. The innermost strong quantifier $\exists z$ within the scope of the strong quantifier $\exists y$ resulted in an ε-term for y (i.e. $\varepsilon_y P(x, y, \varepsilon_z P(x, y, z))$) containing a nested ε-term for z (i.e. $\varepsilon_z P(x, y, z)$) as a subterm. The ε-terms $\varepsilon_y P(x, y, \varepsilon_z P(x, y, z))$ and $\varepsilon_z P(x, \varepsilon_y P(x, y, \varepsilon_z P(x, y, z)), z)$ are top-level ε-terms in the formula above. Comparing the epsilonization with a Skolemization of the same formula, such as $\forall x.P(x, sk_y(x), sk_z(x))$, the Skolem terms $sk_y(x)$ and $sk_z(x)$ can be seen as abbreviations for the two top-level ε-terms.

The treatment of strong quantifiers from inside out is compatible to our principal aim: the epsilonization of proofs. Since this procedure (presented in Definition 5) traverses the proof from the axioms to the end-sequent, innermost quantifiers are treated first. The motivation for removing strong quantifier inferences from proofs is due to the CERES method for intuitionistic logic [6, 10], a cut-elimination procedure based on the resolution calculus. To apply this method, the proof must not contain strong quantifier inferences on end-sequent ancestors. This is easily accomplished for classical logic via Skolemization (as we can eventually de-Skolemize the constructed cut-free proof), but it is not straightforward for intuitionistic proofs.

3 LJ*: a restricted LJ

We now define **LJ***, a version of **LJ** with restricted weak quantifier rules, which uses information available in the ε-terms to decide if they can be used on the instantiation of weak quantifiers. In what follows we will use ν to denote any of the ε-binders ε or τ, and $\rightsquigarrow$ as a rewriting relation.

Definition 3. *A term t is* accessible *in a formula F iff:*

- *for any top-level ε-term $\nu_x G$ in t it is the case that $F[\nu_x G \rightsquigarrow x]$ is a sub-formula of G; or*

- *t contains a nested ε-term $\nu_y H$ such that $\nu_y H$ is accessible in F and $t[\nu_y H \rightsquigarrow y]$ is accessible in $F[\nu_y H \rightsquigarrow y]$.*

The recursion in Definition 3 is necessary for coping with arbitrarily nested ε-terms.

Example 2. *Consider the formula F below:*

$$P(w, \varepsilon_y P(w, y, \varepsilon_z P(w, y, z)), \varepsilon_z P(w, \varepsilon_y P(w, y, \varepsilon_z P(w, y, z)), z))$$

Let t_1 be the term $\varepsilon_y P(w, y, \varepsilon_z P(w, y, z))$. The term t_1 is accessible in F, because $F[t_1 \rightsquigarrow y]$ (which is equal to $P(w, y, \varepsilon_z P(w, y, z))$) is a sub-formula of G_1 (where

G_1 is, in accordance with Definition 3, $P(w, y, \varepsilon_z P(w, y, z))$. Let t_2 be the term $\varepsilon_z P(w, \varepsilon_y P(w, y, \varepsilon_z P(w, y, z)), z)$. The term t_2 is accessible in F, because t_1 is accessible in F and $t_2[t_1 \rightsquigarrow y]$ (which is $\varepsilon_z P(w, y, z)$) is accessible in $F[t_1 \rightsquigarrow y]$ (which is $P(w, y, \varepsilon_z P(w, y, z))$), since $F[t_1 \rightsquigarrow y][t_2[t_1 \rightsquigarrow y] \rightsquigarrow z]$ (which is $P(w, y, z)$) is a sub-formula of G_2 (where G_2 is the formula under the scope of the ε-binder in $t_2[t_1 \rightsquigarrow y]$: $P(w, y, z)$).

Definition 4. *A term t is* accessible *in a sequent S iff all top-level ε-terms in t are accessible in some formula occurring in S.*

When thinking about bottom-up proof search, a term is accessible only after the strong quantifier inference introducing its corresponding eigenvariable in a regular **LJ** proof is applied. This means that, at this point, the term (or the eigenvariable) is already available for use in a weak quantifier inference. Take our previous unprovable sequent: $\vdash \neg\forall x.Px \rightarrow \exists x.\neg Px$. As shown before, its Skolemization is provable in **LJ** because the skolem term used for $\forall x.Px$ is available to be used in $\exists x.\neg Px$. The epsilonization of this sequent is: $\vdash \neg P(\tau_x Px) \rightarrow \exists x.\neg Px$. The fact that Px is a sub-formula of $\neg Px$ informs us that the strong quantifier was within the scope of the negation, and therefore a negation inference would have to be applied in order to make the ε-term accessible before it could be used in a weak quantifier inference. Therefore, as desired, the epsilonized sequent is not provable in **LJ***.

Additionally, the ε-terms used in this calculus will contain labels. The purpose of these labels is two-fold.

Firstly, they will restrict the shape of the proofs in **LJ*** in order to make de-epsilonization possible. Without the restriction, the removal of ε-terms and re-introduction of strong quantifiers could generate incorrect **LJ** proofs that violate the eigenvariable condition. Take, for example, the following proof of the epsilonization of $\neg\forall x.\neg Px, \forall z.\forall y.\neg(Pz \wedge Py) \vdash$:

$$
\cfrac{
\cfrac{
\cfrac{
\cfrac{
\cfrac{
\cfrac{
\cfrac{
\cfrac{
\cfrac{P(\tau_x \neg Px) \vdash P(\tau_x \neg Px) \quad P(\tau_x \neg Px) \vdash P(\tau_x \neg Px)}
{P(\tau_x \neg Px), P(\tau_x \neg Px) \vdash P(\tau_x \neg Px) \wedge P(\tau_x \neg Px)} \wedge_r}
{P(\tau_x \neg Px), P(\tau_x \neg Px), \neg(P(\tau_x \neg Px) \wedge P(\tau_x \neg Px)) \vdash} \neg_l}
{P(\tau_x \neg Px), \neg(P(\tau_x \neg Px) \wedge P(\tau_x \neg Px)) \vdash \neg P(\tau_x \neg Px)} \neg_r}
{P(\tau_x \neg Px), \neg\neg P(\tau_x \neg Px), \neg(P(\tau_x \neg Px) \wedge P(\tau_x \neg Px)) \vdash} \neg_l}
{P(\tau_x \neg Px), \neg\neg P(\tau_x \neg Px), \forall y.\neg(P(\tau_x \neg Px) \wedge Py) \vdash} \forall_l}
{P(\tau_x \neg Px), \neg\neg P(\tau_x \neg Px), \forall z.\forall y.\neg(Pz \wedge Py) \vdash} \forall_l}
{\neg\neg P(\tau_x \neg Px), \forall z.\forall y.\neg(Pz \wedge Py) \vdash \neg P(\tau_x \neg Px)} \neg_r}
{\neg\neg P(\tau_x \neg Px), \neg\neg P(\tau_x \neg Px), \forall z.\forall y.\neg(Pz \wedge Py) \vdash} \neg_l}
{\neg\neg P(\tau_x \neg Px), \forall z.\forall y.\neg(Pz \wedge Py) \vdash} c_l
$$

406

When de-epsilonizing, two strong quantifiers need to be introduced in this proof; both of them between $\neg_l$ and $\neg_r$ inferences: one in the second/third level and the other in the sixth/seventh, bottom-up. The proof with the strong quantifiers is:

$$
\cfrac{
\cfrac{
\cfrac{
\cfrac{
\cfrac{
\cfrac{
\cfrac{
\cfrac{
\cfrac{
\cfrac{
\cfrac{\overline{P(\alpha) \vdash P(\alpha)} \quad \overline{P(\beta) \vdash P(\beta)}}{P(\alpha), P(\beta) \vdash P(\alpha) \wedge P(\beta)} \wedge_r
}{P(\alpha), P(\beta), \neg(P(\alpha) \wedge P(\beta)) \vdash} \neg_l
}{P(\alpha), \neg(P(\alpha) \wedge P(\beta)) \vdash \neg P(\beta)} \neg_r
}{P(\alpha), \neg(P(\alpha) \wedge P(\beta)) \vdash \forall x. \neg P(x)} \forall_r *
}{P(\alpha), \neg \forall x. \neg P(x), \neg(P(\alpha) \wedge P(\beta)) \vdash} \neg_l
}{P(\alpha), \neg \forall x. \neg P(x), \forall y. \neg(P(\alpha) \wedge Py) \vdash} \forall_l
}{P(\alpha), \neg \forall x. \neg P(x), \forall z. \forall y. \neg(Pz \wedge Py) \vdash} \forall_l
}{\neg \forall x. \neg P(x), \forall z. \forall y. \neg(Pz \wedge Py) \vdash \neg P(\alpha)} \neg_r
}{\neg \forall x. \neg P(x), \forall z. \forall y. \neg(Pz \wedge Py) \vdash \forall x. \neg P(x)} \forall_r
}{\neg \forall x. \neg P(x), \neg \forall x. \neg P(x), \forall z. \forall y. \neg(Pz \wedge Py) \vdash} \neg_l
}{\neg \forall x. \neg P(x), \forall z. \forall y. \neg(Pz \wedge Py) \vdash} c_l
$$

Note that the top-most $\forall_r *$ inference violates the eigenvariable condition. In fact, as this rule is applied after (above) both weak quantifiers, a violation is unavoidable. The only way of de-epsilonizing the proof into a valid **LJ** proof would be to perform more complex operations, such as re-ordering of inferences. Instead of pursuing a more complicated de-epsilonization procedure, we restrict proof search in **LJ*** by using labels and avoiding the construction of such proofs in the first place. The restriction still preserves completeness.

Secondly, the labels will make epsilonization of **LJ** proofs an injective function. If labels were not used, the two following derivations would map to the same one:

$$
\cfrac{
\cfrac{
\cfrac{\overline{P\alpha, P\beta \vdash}}{P\alpha, \exists x. Px \vdash} \exists_l
}{\exists x. Px, \exists x. Px \vdash} \exists_l
}{\exists x. Px \vdash} c_l
\qquad
\cfrac{
\cfrac{
\cfrac{\overline{P\alpha, P\alpha \vdash}}{P\alpha \vdash} c_l
}{\exists x. Px \vdash} \exists_l
}{}
$$

$$
\Big\downarrow
$$

$$
\cfrac{\overline{P(\varepsilon_x Px), P(\varepsilon_x Px) \vdash}}{P(\varepsilon_x Px) \vdash} c_l
$$

Figure 3 shows the inferences of **LJ*** that are different than those of **LJ**, all others remain the same. The labels in ε-terms can be variables or constants. When epsilonizing a formula according to Definition 1, each ε-term receives a different label variable. When using **LJ*** for proof search, the following conditions must be enforced:

- On the initial rule, the corresponding ε-terms in the antecedent and consequent must have the same labels, and these must be constants.

- On the weak quantifier rules, the term t used for the substitution must be accessible and, *additionally*, its ε-subterms must have constants as labels. If this is not the case, the label variables of the ε-terms in t that occur in accessible positions in the conclusion sequent are substituted by a new (fresh) constant.

- Upon contracting a formula with ε-terms that have a variable label, there are two cases:

  - For accessible ε-terms, the same variable is used in the contracted occurrences in the premise.
  - For inaccessible ε-terms, new variable labels are created to be used in the contracted occurrences in the premise[2].

If the label is a constant, then it was already used by a weak quantifier inference below contraction, which means the term is accessible. In this case, the constant label is simply copied to the contracted occurrences in the premise.

$$\frac{\Gamma, A[t] \vdash F}{\Gamma, \forall x. A[x] \vdash F} \, \forall'_l \qquad \frac{\Gamma \vdash A[t]}{\Gamma \vdash \exists x. A[x]} \, \exists'_r \qquad \frac{}{A[\nu_x^l \, F] \vdash A[\nu_x^l \, F]} \, a$$

where:

- the term t must be accessible in the conclusion sequent (*accessibility condition*).

- accessible occurrences of t or any of its ε-subterms in Γ and F must have a constant as a label (*label condition*).

- l is a constant in a (*initial condition*).

Figure 3: Rules for **LJ***

One might wonder about the (im)possibility to devise a simpler treatment of labels or stronger restrictions on contracted formulas in order to avoid the problems shown before. An immediate thought would be to use always constant labels and force contraction to create two different labels on the premises. But this restriction is too strong and would render the calculus incomplete, if it were adopted. The sequent

---

[2]This means that proofs in **LJ*** might contract formulas with different labels in its ε-terms.

$\exists x.Px \vdash \exists x.(Px \wedge Px)$ is an example. Its epsilonization is $P(\varepsilon_x Px) \vdash \exists x.(Px \wedge Px)$ and a proof in $\mathbf{LJ}^\star$ is shown below:

$$
\cfrac{
\cfrac{
\cfrac{\overline{P(\varepsilon_x Px) \vdash P(\varepsilon_x Px)} \quad \overline{P(\varepsilon_x Px) \vdash P(\varepsilon_x Px)}}{P(\varepsilon_x Px), P(\varepsilon_x Px) \vdash P(\varepsilon_x Px) \wedge P(\varepsilon_x Px)} \wedge_r
}{P(\varepsilon_x Px), P(\varepsilon_x Px) \vdash \exists x.(Px \wedge Px)} \exists_r
}{P(\varepsilon_x Px) \vdash \exists x.(Px \wedge Px)} c_l
$$

If different labels were used when contracting, the sequent would not be provable.

Another simpler potential solution would be to restrict contraction to formulas that only have accessible ε-terms. Unfortunately, this does not work in the general case. Consider the sequent $\neg(\forall x.Px \vee \neg \forall x.Px) \vdash$, whose epsilonization is $\neg(P(\tau_x Px) \vee \neg \forall x.Px) \vdash$. The term $\tau_x Px$ is obviously not accessible, thus should contraction on this formula not be allowed, the sequent would not be provable (whereas the original sequent is intuitionistically valid).

Theorem 1 (Soundness). *For an ε-free formula F, if $\vdash_{\mathbf{LJ}^\star} F$ then $\vdash_{\mathbf{LJ}} F$.*

Proof. Let ψ' be an $\mathbf{LJ}^\star$-proof of F. Then an $\mathbf{LJ}$-proof ψ of F can be constructed simply by replacing $\forall'_l$ and $\exists'_r$ inferences by, respectively, $\forall_l$ and $\exists_r$ inferences. Since F is ε-free, the rules a and c_l are the same as those in $\mathbf{LJ}$. □

$\mathbf{LJ}^\star$ is also sound relative to $\mathbf{LJ}$ for formulas with ε-terms (i.e., if $\vdash_{\mathbf{LJ}^\star} T_\varepsilon(F)$ then $\vdash_{\mathbf{LJ}} T_\varepsilon(F)$). We simply need to ignore the labels when transforming the proof.

Theorem 2 (Completeness). *For an ε-free formula F, if $\vdash_{\mathbf{LJ}} F$ then $\vdash_{\mathbf{LJ}^\star} F$.*

Proof. Since F is ε-free and $\vdash_{\mathbf{LJ}} F$, there is an ε-free $\mathbf{LJ}$-proof ψ of F. An $\mathbf{LJ}^\star$-proof ψ' of F can be constructed simply by replacing all $\forall_l$ and $\exists_r$ inferences by, respectively, $\forall'_l$ and $\exists'_r$ inferences. No accessibility or label violation occurs, because no term in ψ' contains ε-terms. Also, the conditions for the inferences a and c_l are not violated, for the same reason. □

The *epsilonization* of a proof removes the strong quantifier inferences that operate on ancestors of formulas occurring in the end-sequent and replaces the corresponding eigenvariables by ε-terms.

Definition 5 (Epsilonization of proofs). *Let ψ be an $\mathbf{LJ}^\star$ proof of an ε-free sequent S. We define $T_\varepsilon(\psi)$, an $\mathbf{LJ}^\star$ proof of $T_\varepsilon(S)$, inductively on the inference rules.*

Base case: *ψ consists of only one axiom. Then $T_\varepsilon(\psi) = \psi$.*

Step case: *ψ ends with an inference ρ, as in the following cases.*

- ρ is $\forall_r$ or $\exists_l$ applied to an end-sequent ancestor.

  Let $(Qx)F$ be the main formula, ψ' be the proof of ρ's premise and α the eigenvariable used to instantiate the strongly quantified variable x. By induction hypothesis, $T_\varepsilon(\psi')$ is well defined. Then $T_\varepsilon(\psi)$ is $T_\varepsilon(\psi')\{\alpha \mapsto \nu_x^l\ F\}$, where ν is ε if Q is $\exists$ and τ if Q is $\forall$, and l is a fresh constant label.

  Note that strong quantifiers that go to cut-formulas are not replaced.

- ρ is $\forall_l$ or $\exists_r$ applied to an end-sequent ancestor.

  Let $(Qx)Fx$ be the main formula, Ft the auxiliary formula and ψ' the proof of ρ's premise. By induction hypothesis, $T_\varepsilon(\psi')$ is well defined. Then $T_\varepsilon(\psi)$ is $T_\varepsilon(\psi')$ plus the inference $\forall_l$ or $\exists_r$ (depending whether Q is $\forall$ or $\exists$) which introduces the quantifier and replaces t by x in F, including the occurrences of t inside copies of F occurring in ε-terms. The variable x used may not be bound.

- ρ is c_l applied to an end-sequent ancestor, and the formulas contracted contain ε-terms with labels.

  Let ψ' be the proof of ρ's premise. By induction hypothesis, $T_\varepsilon(\psi')$ is well defined. Then $T_\varepsilon(\psi)$ is $T_\varepsilon(\psi')$ plus the contraction, where its main formula will have new variables as labels.

- ρ is another inference. Then $T_\varepsilon(\psi) = \psi$.

Observe that, apart from possibly different labels, contraction will always operate on equal terms, since weak quantifiers also operate on formulas inside ε-terms:

$$
\cfrac{\cfrac{\cfrac{\cfrac{\psi}{P(a,\alpha),P(b,\beta) \vdash}}{\exists y.P(a,y),\exists y.P(b,y) \vdash} \exists_l \times 2}{\forall x.\exists y.P(x,y),\forall x.\exists y.P(x,y) \vdash} \forall_l \times 2}{\forall x.\exists y.P(x,y) \vdash} c_l
\qquad \leadsto \qquad
\cfrac{\cfrac{\cfrac{T_\varepsilon(\psi)}{P(a,\varepsilon_y^{l_1}.P(a,y)),P(b,\varepsilon_y^{l_2}.P(b,y)) \vdash}}{\forall x.P(x,\varepsilon_y^{l_1}.P(x,y)),\forall x.P(x,\varepsilon_y^{l_2}.P(x,y)) \vdash} \forall_l \times 2}{\forall x.P(x,\varepsilon_y^{l}.P(x,y)) \vdash} c_l
$$

Lemma 1. *If an* **LJ****-proof* ψ *has end-sequent* S, *then* $T_\varepsilon(\psi)$ *has end-sequent* $T_\varepsilon(S)$ *(modulo renaming of labels).*

Proof. By induction on the structure of ψ and by Definition 5. $\square$

Lemma 2. *If* ψ *is an* **LJ****-proof of* S, *then no weak quantifier inference in* $T_\varepsilon(\psi)$ *violates the accessibility condition.*

Proof. First of all, note that the order in which the inferences are applied in $T_\varepsilon(\psi)$ is the same as in ψ, with the only difference being that strong quantifier inferences were removed.

Let $Qx.F$ be a strong quantified formula in S, α the eigenvariable used for this strong quantifier in ψ and $Q'x.G$ a weak quantifier in S which is instantiated in ψ with a term containing α. Since ψ is a correct proof, the weak quantifier inference ρ_w on $Q'x.G$ occurs after (above) the strong quantifier inference ρ_s on $Qx.F$.

Now consider the proof $T_\varepsilon(\psi)$. Given Definition 5, at the point where ρ_s was applied, the formula $Qx.F$ will have the shape $F'[\nu_x^l\ F']$, where F' is possibly F without strong quantifiers. Since F' is a sub-formula of F', the ε-term is already accessible. All inferences above this point will either decompose (the outer-most) F' into more sub-formulas or keep it unchanged. In this way, the ε-term $\nu_x^l\ F'$ will remain accessible. As ρ_w is applied after (above) the considered point, the accessibility relation will not be violated. $\square$

Lemma 3. *If ψ is an **LJ*****-proof of S, then no weak quantifier inference in $T_\varepsilon(\psi)$ violates the label condition.*

Proof. By Definition 5, the eigenvariables in a proof are always replaced by ε-terms with constant labels. Since a weak quantifier that uses an eigenvariable α occurs above the strong quantifier that introduced such variable, the label condition will hold in the epsilonized proof. $\square$

Lemma 4. *If ψ is an **LJ*****-proof of S, then no axiom inference in $T_\varepsilon(\psi)$ violates the initial condition.*

Proof. Trivial by Definition 5 and by the fact that there are no inferences operating above axioms. $\square$

Theorem 3. *If $\vdash_{\textbf{LJ}^\star} S$, then $\vdash_{\textbf{LJ}^\star} T_\varepsilon(S)$.*

Proof. Let ψ be an **LJ***-proof of S. Then, by Lemmas 1, 2, 3 and 4 $T_\varepsilon(\psi)$ is a correct **LJ***-proof of $T_\varepsilon(S)$. $\square$

De-epsilonization of proofs, denoted by T_ε^{-1}, replaces ε-terms by eigenvariables and introduces strong quantifier inferences in appropriate places. To detect the appropriate places, the following definition is helpful.

Intuitively, the de-epsilonization procedure will traverse a proof ψ in a top-down manner, re-applying the inference rules from ψ. As this is done, the sequents will contain formulas of the form $A[\nu_x B[x]]$ where $\nu_x B[x]$ is a top-level ε-term, for increasingly more complex A. Thus, $\nu_x B[x]$ is initially accessible in the formula and in

411

the sequent, while A is a subformula of B. Replacing the ε-term by an eigen-variable and introducing a strong quantifier inference for this eigen-variable becomes possible when A becomes exactly equal to B, in which case the ε-term is said to be *ready*. However, to avoid violations of the eigen-variable condition, it is still necessary to postpone the introduction of strong quantifier inferences as much as possible. That is why the de-epsilonization procedure seeks to introduce them just before they become inaccessible. However, introducing them earlier may be necessary if a contraction operates on occurrences of $\nu_x B[x]$ with different labels.

Definition 6 (De-epsilonization of proofs). *Let F be an ε-free formula and ψ an* ***LJ**^** *proof of $T_\varepsilon(F)$. The de-epsilonization $T_\varepsilon^{-1}(\psi)$ is constructed inductively on the inference rules.*

Base case: *ψ consists of only one axiom. Then $T_\varepsilon^{-1}(\psi) = \psi$.*

Step case: *ψ ends with an inference ρ. By the induction hypothesis, the de-epsilonization of ρ's premises: $T_\varepsilon^{-1}(\psi_1)$ and (for the case of binary inferences) $T_\varepsilon^{-1}(\psi_2)$ are well-defined. Then $T_\varepsilon^{-1}(\psi)$ is defined according to the possible cases for ρ:*

- *ρ is a weakening.*

  Then $T_\varepsilon^{-1}(\psi)$ is simply $T_\varepsilon^{-1}(\psi_1)$ followed by the same weakening.

- *ρ is a cut.*

  Then $T_\varepsilon^{-1}(\psi)$ is the proof obtained by applying the same cut on $T_\varepsilon^{-1}(\psi_1)$ and $T_\varepsilon^{-1}(\psi_2)$.

- *ρ is a contraction on a formula F.*

  If F contains no ε-terms, then $T_\varepsilon^{-1}(\psi)$ is defined as $T_\varepsilon^{-1}(\psi_1)$ followed by the contraction. Otherwise, if the contracted formulas contain ε-terms $\nu_x^{l_1} G$ and $\nu_x^{l_2} G$, then $T_\varepsilon^{-1}(\psi)$ depends on the following cases for l_1 and l_2:

  - *The labels l_1 and l_2 are equal, regardless whether they are variables or constants. In this case, $T_\varepsilon^{-1}(\psi)$ is defined as $T_\varepsilon^{-1}(\psi')$ followed by the same contraction.*
  - *The labels l_1 and l_2 are two different constants[3]. In this case, F and G are the same, then $T_\varepsilon^{-1}(\psi)$ is defined as $T_\varepsilon^{-1}(\psi_1)\{\nu_x^{l_1} Fx \mapsto \alpha\}\{\nu_x^{l_2} Fx \mapsto \beta\}$ followed by two strong quantifier inferences ($\forall_r$ if ν is τ and $\exists_l$ if ν is ε) and a contraction on the quantified formulas.*

---

[3]This case occurs for epsilonized proofs, but not in proofs obtained by proof search in **LJ**^*.

The case where l_1 and l_2 are two different variables does not occur for one of two reasons: (1) if ψ was obtained via proof search in $\mathbf{LJ^\star}$, then contraction of formulas with accessible ε-terms copies the variables to the premise, and thus they will be instantiated with the same constant at a later step; or (2) if ψ was obtained via epsilonization of a $\mathbf{LJ}$ proof, the labels will be constants.

- ρ *is a logical inference.*

  If ρ operates on ε-free formulas or all top-level ε-terms in ρ's auxiliary formulas are still accessible in the conclusion, then $T_\varepsilon^{-1}(\psi)$ is defined as ρ applied to the de-epsilonization of its premise(s).

  Otherwise, while there exists a top-level ε-term $\nu_x^l\, F$ that would no longer be accessible in ρ's conclusion, we add the appropriate strong quantifier and apply the replacement $\{\nu_x^l F \mapsto \alpha\}$ to the proof with a fresh variable α as well as the replacements $\{\nu_x^l\, F' \mapsto \alpha\}$ (with the same variable α) for any F' that differs from F only in the presence of nested ε-terms. Finally, when there are no more ε-terms that would become inaccessible, $T_\varepsilon^{-1}(\psi)$ becomes ρ applied to the proof resulting from this iterative quantifier reintroduction procedure.

If after this process the end-sequent still contains ε-terms, then additional strong quantifier inferences are added accordingly.

An example illustrating the need for replacing nested terms and for the while loop in the last case of Definition 6 is available in Section 4.3.

We can now prove soundness of the epsilonization method.

Lemma 5. *If ψ is an $\mathbf{LJ^\star}$-proof of an end-sequent $T_\varepsilon(S)$, then the end-sequent of $T_\varepsilon^{-1}(\psi)$ is S.*

Proof. In Definition 6, all ε-terms from ψ are replaced by eigenvariables and strong quantifier rules are applied, so that eventually formulas of the form $A[\varepsilon_x A[x]]$ (or $A[\tau_x A[x]]$) in $T_\varepsilon(S)$ are replaced by $\exists x.A[x]$ (or, respectively, $\forall x.A[x]$), innermost subformulas first. Notice that, at the time of the introduction of the strong quantifier, the outer formulas and those bound by the ε-term are indeed the same, since possibly nested ε-terms correspond to innermost quantifiers which will have been already introduced above in the proof. $\square$

Lemma 6. *If ψ is an $\mathbf{LJ^\star}$-proof of an end-sequent $T_\varepsilon(S)$, then there is an $\mathbf{LJ^\star}$-proof ψ' of S obtainable from $T_\varepsilon^{-1}(\psi)$ by reductive cut-elimination.*

Proof. The key point is to show that any violation of the eigenvariable condition in $T_\varepsilon^{-1}(\psi)$ can be removed by reductive cut-elimination. Assume that there is a strong quantifier inference ρ in $T_\varepsilon^{-1}(\psi)$ that violates the eigenvariable condition. This means that $T_\varepsilon^{-1}(\psi)$ has one of the following forms near ρ:

$$\frac{\vdots}{\dfrac{\Gamma, A[\alpha] \vdash B[\alpha]}{\Gamma, \exists x.A[x] \vdash B[\alpha]}} \ \rho : \exists_l$$

$$\vdots$$

$$\frac{\dfrac{\vdots}{\Gamma, B[\alpha], A[\alpha] \vdash C}}{\Gamma, B[\alpha], \exists x.A[x] \vdash C} \ \rho : \exists_l$$

$$\vdots$$

$$\frac{\dfrac{\vdots}{\Gamma, B[\alpha] \vdash A[\alpha]}}{\Gamma, B[\alpha] \vdash \forall x.A[x]} \ \rho : \forall_r$$

$$\vdots$$

For each of the cases above, there are four potential subcases. We show below that three of them cannot occur, because they would lead to contradictions, whereas the fourth can be fixed by reductive cut-elimination:

- $B[\alpha]$ propagates down to the end-sequent: α would then occur in the end-sequent of $T_\varepsilon^{-1}(\psi)$, but this would contradict Lemma 5.

- $B[\alpha]$ propagates down to a strong quantifier inference ρ' which has eigenvariable α: this case cannot occur, because ψ would then violate the label condition, thus contradicting the assumption that ψ is a correct **LJ***-proof.

- $B[\alpha]$ propagates down to a weak quantifier inference ρ' with an auxiliary formula $D[t[\alpha]]$: then ρ' would have auxiliary formula $D[t[\varepsilon_x B'[x]]]$ in ψ. If B were a proper super-formula of B', the term $t[\varepsilon_x B'[x]]$ would not be accessible and ρ' would be violating the accessibility condition. If B were equal to B', then ρ' would be occurring below ρ, which contradicts the fact that, in Definition 6, strong quantifier inferences such as ρ are introduced as low as possible. Indeed, notice that as the weak quantifier inference occurs in the proof with

414

ep-terms, it will be applied during de-epsilonization in the same place, while the strong quantifier is only added when absolutely necessary (i.e., the term is no longer accessible or at the end-sequent).

- $B[\alpha]$ propagates down to a cut: in this case, the eigenvariable violation can be removed by shifting the cut upward, using Gentzen's reductive cut-elimination method.

$\square$

Theorem 4 (Soundness of Epsilonization). *If* $\vdash_{\textbf{LJ}^\star} T_\varepsilon(S)$, *then* $\vdash_{\textbf{LJ}} S$.

Proof. Let ψ be an $\textbf{LJ}^\star$-proof of $T_\varepsilon(S)$. Then, by Lemmas 5 and 6, $T_\varepsilon^{-1}(\psi)$ is a correct $\textbf{LJ}^\star$-proof of S. By Theorem 1, $\vdash_{\textbf{LJ}} S$. $\square$

4 Examples

This section presents a set of examples that help understand the epsilonization and de-epsilonization of proofs. Each example demonstrates the need for some aspect of the definitions.

4.1 Labels, Contractions and Inaccessible ε-terms

This section illustrates the need for different labels when contracting formulas with inaccessible ε-terms. We start with an end-sequent already considered before:

$$\neg\forall x.\neg Px, \forall z.\forall y.\neg(Pz \wedge Py) \vdash$$

whose epsilonization is

$$\neg\neg P(\tau_x\neg Px), \forall z.\forall y.\neg(Pz \wedge Py) \vdash$$

We have seen that, had labels not been used, the later sequent would admit a proof whose de-epsilonization would generate a proof with eigenvariable violations. Taking the labels into account, the proof found by proof search in $\textbf{LJ}^\star$ is the following:

$$\cfrac{\cfrac{\cfrac{\cfrac{\cfrac{\cfrac{\cfrac{P(\tau_x^{l_1}.\neg Px) \vdash P(\tau_x^{l_1}.\neg Px) \quad P(\tau_x^{l_2}.\neg Px) \vdash P(\tau_x^{l_2}.\neg Px)}{P(\tau_x^{l_2}.\neg Px), P(\tau_x^{l_1}.\neg Px) \vdash P(\tau_x^{l_1}.\neg Px) \land P(\tau_x^{l_2}.\neg Px)} \land_r}{P(\tau_x^{l_2}.\neg Px), P(\tau_x^{l_1}.\neg Px), \neg(P(\tau_x^{l_1}.\neg Px) \land P(\tau_x^{l_2}.\neg Px)) \vdash} \neg_l}{P(\tau_x^{l_2}.\neg Px), P(\tau_x^{l_1}.\neg Px), \forall y.\neg(P(\tau_x^{l_1}.\neg Px) \land Py) \vdash} \forall_l'}{P(\tau_x^{l_1}.\neg Px), \forall y.\neg(P(\tau_x^{l_1}.\neg Px) \land Py) \vdash \neg P(\tau_x^{l_2}.\neg Px)} \neg_r}{P(\tau_x^{l_1}.\neg Px), \neg\neg P(\tau_x^{l_2}.\neg Px), \forall y.\neg(P(\tau_x^{l_1}.\neg Px) \land Py) \vdash} \neg_l}{P(\tau_x^{l_1}.\neg Px), \neg\neg P(\tau_x^{l_2}.\neg Px), \forall z.\forall y.\neg(Pz \land Py) \vdash} \forall_l'}{\cfrac{\cfrac{\neg\neg P(\tau_x^{l_2}.\neg Px), \forall z.\forall y.\neg(Pz \land Py) \vdash \neg P(\tau_x^{l_1}.\neg Px)}{\neg\neg P(\tau_x^{l_1}.\neg Px), \neg\neg P(\tau_x^{l_2}.\neg Px), \forall z.\forall y.\neg(Pz \land Py) \vdash} \neg_l}{\neg\neg P(\tau_x^{l}.\neg Px), \forall z.\forall y.\neg(Pz \land Py) \vdash} c_l}$$

(the last two lines connected by $\neg_r$ and c_l)

Note how the ε-term labelled with l_2 is not available for the weak quantifier $\forall y$. Observe also how the two labels of the contracted formulas need to be different. Had they been the same, we would be able to obtain the same proof as before, which de-epsilonizes to an incorrect proof.

The de-epsilonization procedure constructs, in a top-down manner, the same proof up to this point:

$$\cfrac{\cfrac{\cfrac{\cfrac{\cfrac{P(\tau_x^{l_1}.\neg Px) \vdash P(\tau_x^{l_1}.\neg Px) \quad P(\tau_x^{l_2}.\neg Px) \vdash P(\tau_x^{l_2}.\neg Px)}{P(\tau_x^{l_2}.\neg Px), P(\tau_x^{l_1}.\neg Px) \vdash P(\tau_x^{l_1}.\neg Px) \land P(\tau_x^{l_2}.\neg Px)} \land_r}{P(\tau_x^{l_2}.\neg Px), P(\tau_x^{l_1}.\neg Px), \neg(P(\tau_x^{l_1}.\neg Px) \land P(\tau_x^{l_2}.\neg Px)) \vdash} \neg_l}{P(\tau_x^{l_2}.\neg Px), P(\tau_x^{l_1}.\neg Px), \forall y.\neg(P(\tau_x^{l_1}.\neg Px) \land Py) \vdash} \forall_l'}{P(\tau_x^{l_1}.\neg Px), \forall y.\neg(P(\tau_x^{l_1}.\neg Px) \land Py) \vdash \neg P(\tau_x^{l_2}.\neg Px)} \neg_r$$

If the next inference, $\neg_l$, were applied, the ε-term $\tau_x^{l_2}.\neg Px$ would no longer be accessible. Therefore, it is time to introduce a strong quantifier. Since the ε-term is bound by τ, the de-epsilonization procedure introduces a $\forall_r$ inference and replaces $\tau_x^{l_2}.\neg Px$ by a new fresh variable α.

$$\cfrac{\cfrac{\cfrac{\cfrac{\cfrac{P(\tau_x^{l_1}.\neg Px) \vdash P(\tau_x^{l_1}.\neg Px) \quad P\alpha \vdash P\alpha}{P\alpha, P(\tau_x^{l_1}.\neg Px) \vdash P(\tau_x^{l_1}.\neg Px) \land P\alpha} \land_r}{P\alpha, P(\tau_x^{l_1}.\neg Px), \neg(P(\tau_x^{l_1}.\neg Px) \land P\alpha) \vdash} \neg_l}{P\alpha, P(\tau_x^{l_1}.\neg Px), \forall y.\neg(P(\tau_x^{l_1}.\neg Px) \land Py) \vdash} \forall_l'}{P(\tau_x^{l_1}.\neg Px), \forall y.\neg(P(\tau_x^{l_1}.\neg Px) \land Py) \vdash \neg P\alpha} \neg_r}{P(\tau_x^{l_1}.\neg Px), \forall y.\neg(P(\tau_x^{l_1}.\neg Px) \land Py) \vdash \forall x.\neg Px} \forall_r$$

416

The re-construction of the proof is continued until the next point where a strong quantifier is needed, for the same reason as before. The same procedure is followed, now replacing $\tau_x^{l_1}.\neg Px$ by a new variable β. The final result is the following valid $\mathbf{LJ}^\star$ (and also $\mathbf{LJ}$) proof:

$$
\cfrac{
\cfrac{
\cfrac{
\cfrac{
\cfrac{
\cfrac{
\cfrac{
\cfrac{
\cfrac{
\cfrac{
\cfrac{\overline{P\beta \vdash P\beta} \quad \overline{P\alpha \vdash P\alpha}}{P\alpha, P\beta \vdash P\beta \wedge P\alpha}\ {\wedge_r}
}{P\alpha, P\beta, \neg(P\beta \wedge P\alpha) \vdash}\ {\neg_l}
}{P\alpha, P\beta, \forall y.\neg(P\beta \wedge Py) \vdash}\ {\forall'_l}
}{P\beta, \forall y.\neg(P\beta \wedge Py) \vdash \neg P\alpha}\ {\neg_r}
}{P\beta, \forall y.\neg(P\beta \wedge Py) \vdash \forall x.\neg Px}\ {\forall_r}
}{P\beta, \neg\forall x.\neg Px, \forall y.\neg(P\beta \wedge Py) \vdash}\ {\neg_l}
}{P\beta, \neg\forall x.\neg Px, \forall z.\forall y.\neg(Pz \wedge Py) \vdash}\ {\forall'_l}
}{\neg\forall x.\neg Px, \forall z.\forall y.\neg(Pz \wedge Py) \vdash \neg P\beta}\ {\neg_r}
}{\neg\forall x.\neg Px, \forall z.\forall y.\neg(Pz \wedge Py) \vdash \forall x.\neg Px}\ {\forall_r}
}{\neg\forall x.\neg Px, \neg\forall x.\neg Px, \forall z.\forall y.\neg(Pz \wedge Py) \vdash}\ {\neg_l}
}{\neg\forall x.\neg Px, \forall z.\forall y.\neg(Pz \wedge Py) \vdash}\ {c_l}
$$

4.2   Contraction with Distinct Labels

The example in this section illustrates the need for allowing contraction on formulas with different constant labels. Consider the following $\mathbf{LJ}^\star$ proof of an ε-free end-sequent:

$$
\cfrac{
\cfrac{
\cfrac{
\cfrac{
\cfrac{\overline{P\alpha \vdash P\alpha}}{P\alpha \vdash \exists x.Px}\ {\exists_r} \quad \cfrac{\overline{P\beta \vdash P\beta}}{P\beta \vdash \exists x.Px}\ {\exists_r}
}{P\alpha, P\beta \vdash \exists x.Px \wedge \exists x.Px}\ {\wedge_r}
}{P\alpha, \exists x.Px \vdash \exists x.Px \wedge \exists x.Px}\ {\exists_l}
}{\exists x.Px, \exists x.Px \vdash \exists x.Px \wedge \exists x.Px}\ {\exists_l}
}{\exists x.Px \vdash \exists x.Px \wedge \exists x.Px}\ {c_l}
$$

Following the epsilonization procedure, the proof obtained is:

$$\dfrac{\dfrac{P(\varepsilon_x^a.Px) \vdash P(\varepsilon_x^a.Px)}{P(\varepsilon_x^a.Px) \vdash \exists x.Px}\,\exists_r \quad \dfrac{P(\varepsilon_x^b.Px) \vdash P(\varepsilon_x^b.Px)}{P(\varepsilon_x^b.Px) \vdash \exists x.Px}\,\exists_r}{\dfrac{P(\varepsilon_x^a.Px), P(\varepsilon_x^b.Px) \vdash \exists x.Px \wedge \exists x.Px}{P(\varepsilon_x^l.Px) \vdash \exists x.Px \wedge \exists x.Px}\,c_l}\,\wedge_r$$

Note how contraction must allow the two ε-terms to have different constant labels. The label in the conclusion can be arbitrary. Such flexibility makes it possible to map the epsilonized proof to the exact intention of the original proof, which was to use two different eigenvariables for the strong quantifier. Interestingly, this situation only occurs if a proof is epsilonized. Had we searched for a proof of the same end-sequent in **LJ***, only one "eigenvariable" would have been used.

4.3   Nested ε-terms

When sequents contain blocks of strong quantifiers or strong quantifiers inside the scope of other strong quantifiers, epsilonization results in sequents with nested ε-terms. In this section, we look at an example of this kind.

Let F be $\forall x.(\exists y.\exists z.P(x,y,z) \rightarrow \exists w.\exists v.\exists q.P(q,v,w))$. Then $T_\varepsilon(F)$ is:

$$P(t_x, t_y, t_z) \rightarrow \exists w.\exists v.\exists q.P(q,v,w)$$

where:

- $t_z = \delta_z(\delta_x, \delta_y(\delta_x))$
- $t_y = \delta_y(\delta_x)$
- $t_x = \delta_x$
- $\delta_x = \gamma_x[\delta_y(x), \delta_z(x, \delta_y(x))]$
- $\delta_y(x) = \gamma_y[\delta_z(x,y)](x)$
- $\delta_z(x,y) = \gamma_z(x,y) = \varepsilon_z P(x,y,z)$
- $\gamma_y[t](x) = \varepsilon_y P(x,y,t)$
- $\gamma_x[t_1, t_2] = \tau_x(P(x, t_1, t_2) \rightarrow \exists w.\exists v.\exists q.P(q,v,w))$

Let ψ be the following **LJ***-proof of $T_\varepsilon(F)$:

$$\dfrac{\dfrac{\dfrac{\dfrac{P(t_x, t_y, t_z) \vdash P(t_x, t_y, t_z)}{P(t_x, t_y, t_z) \vdash \exists q.P(q, t_y, t_z)}\,\exists_r}{P(t_x, t_y, t_z) \vdash \exists v.\exists q.P(q, v, t_z)}\,\exists_r}{P(t_x, t_y, t_z) \vdash \exists w.\exists v.\exists q.P(q, v, w)}\,\exists_r}{\vdash P(t_x, t_y, t_z) \rightarrow \exists w.\exists v.\exists q.P(q, v, w)}\,\rightarrow_r$$

418

During the top-down construction of $T_\varepsilon^{-1}(\psi)$, initially the three $\exists_r$ inferences are simply reapplied:

$$
\cfrac{
\cfrac{
\cfrac{
\cfrac{P(t_x, t_y, t_z) \vdash P(t_x, t_y, t_z)}{P(t_x, t_y, t_z) \vdash \exists q.P(q, t_y, t_z)} \; \exists_r
}{P(t_x, t_y, t_z) \vdash \exists v.\exists q.P(q, v, t_z)} \; \exists_r
}{P(t_x, t_y, t_z) \vdash \exists w.\exists v.\exists q.P(q, v, w)} \; \exists_r
$$

At this point, t_z is ready, but applying $\to_R$ would make it inaccessible. Therefore, it is time to introduce the strong quantifier for z:

$$
\cfrac{
\cfrac{
\cfrac{
\cfrac{P(t_x'[\alpha_z], t_y'[\alpha_z], \alpha_z) \vdash P(t_x'[\alpha_z], t_y'[\alpha_z], \alpha_z)}{P(t_x'[\alpha_z], t_y'[\alpha_z], \alpha_z) \vdash \exists q.P(q, t_y'[\alpha_z], \alpha_z)} \; \exists_r
}{P(t_x'[\alpha_z], t_y'[\alpha_z], \alpha_z) \vdash \exists v.\exists q.P(q, v, \alpha_z)} \; \exists_r
}{P(t_x'[\alpha_z], t_y'[\alpha_z], \alpha_z) \vdash \exists w.\exists v.\exists q.P(q, v, w)} \; \exists_r
}{\exists z.P(t_x'[z], t_y'[z], z) \vdash \exists w.\exists v.\exists q.P(q, v, w)} \; \exists_l
$$

where $t_y'[z] = t_y\{\delta_z(.,.) \mapsto z\}$ and $t_x'[z] = t_x\{\delta_z(.,.) \mapsto z\}$. Note that the replacement of terms of the general form $\delta_z(.,.)$ by z de-epsilonizes occurrences of $\varepsilon_z P(.,.,z)$ nested inside t_x and t_y. This illustrates the need for substituting not only top-level ε-terms, but also nested ε-terms in the last case of Definition 6.

Now $t_y'[z]$ becomes ready and it would not be accessible anymore after application of $\to_R$. Therefore, it is time to introduce the strong quantifier inference for y:

$$
\cfrac{
\cfrac{
\cfrac{
\cfrac{
\cfrac{P(t_x''[\alpha_y], \alpha_y, \alpha_z) \vdash P(t_x''[\alpha_y], \alpha_y, \alpha_z)}{P(t_x''[\alpha_y], \alpha_y, \alpha_z) \vdash \exists q.P(q, \alpha_y, \alpha_z)} \; \exists_r
}{P(t_x''[\alpha_y], \alpha_y, \alpha_z) \vdash \exists v.\exists q.P(q, v, \alpha_z)} \; \exists_r
}{P(t_x''[\alpha_y], \alpha_y, \alpha_z) \vdash \exists w.\exists v.\exists q.P(q, v, w)} \; \exists_r
}{\exists z.P(t_x''[\alpha_y], \alpha_y, z) \vdash \exists w.\exists v.\exists q.P(q, v, w)} \; \exists_l
}{\exists y.\exists z.P(t_x''[y], y, z) \vdash \exists w.\exists v.\exists q.P(q, v, w)} \; \exists_l
$$

$t_x''[y]$ is not ready yet, and the $\to_r$ inference rule can be applied:

$$
\cfrac{
\cfrac{
\cfrac{
\cfrac{
\cfrac{
\cfrac{
\overline{P(t''_x[\alpha_y], \alpha_y, \alpha_z) \vdash P(t''_x[\alpha_y], \alpha_y, \alpha_z)}
}{P(t''_x[\alpha_y], \alpha_y, \alpha_z) \vdash \exists q.P(q, \alpha_y, \alpha_z)} \exists_r
}{P(t''_x[\alpha_y], \alpha_y, \alpha_z) \vdash \exists v.\exists q.P(q, v, \alpha_z)} \exists_r
}{P(t''_x[\alpha_y], \alpha_y, \alpha_z) \vdash \exists w.\exists v.\exists q.P(q, v, w)} \exists_r
}{\exists z.P(t''_x[\alpha_y], \alpha_y, z) \vdash \exists w.\exists v.\exists q.P(q, v, w)} \exists_l
}{\exists y.\exists z.P(t''_x[y], y, z) \vdash \exists w.\exists v.\exists q.P(q, v, w)} \exists_l
}{\vdash \exists y.\exists z.P(t''_x[y], y, z) \to \exists w.\exists v.\exists q.P(q, v, w)} \to_r
$$

The fact that we had to introduce two strong quantifier inferences, de-epsilonizing two different ε-terms, before being able to reapply a logical inference rule illustrates the need for a while loop in the last case of Definition 6.

Finally, a $\forall_r$ inference rule has to be applied, because t_x is now ready in the end sequent and there are no further inferences from ψ to be reapplied:

$$
\cfrac{
\cfrac{
\cfrac{
\cfrac{
\cfrac{
\cfrac{
\cfrac{
\overline{P(\alpha_x, \alpha_y, \alpha_z) \vdash P(\alpha_x, \alpha_y, \alpha_z)}
}{P(\alpha_x, \alpha_y, \alpha_z) \vdash \exists q.P(q, \alpha_y, \alpha_z)} \exists_r
}{P(\alpha_x, \alpha_y, \alpha_z) \vdash \exists v.\exists q.P(q, v, \alpha_z)} \exists_r
}{P(\alpha_x, \alpha_y, \alpha_z) \vdash \exists w.\exists v.\exists q.P(q, v, w)} \exists_r
}{\exists z.P(\alpha_x, \alpha_y, z) \vdash \exists w.\exists v.\exists q.P(q, v, w)} \exists_l
}{\exists y.\exists z.P(\alpha_x, y, z) \vdash \exists w.\exists v.\exists q.P(q, v, w)} \exists_l
}{\vdash \exists y.\exists z.P(\alpha_x, y, z) \to \exists w.\exists v.\exists q.P(q, v, w)} \to_r
}{\vdash \forall x.(\exists y.\exists z.P(x, y, z) \to \exists w.\exists v.\exists q.P(q, v, w))} \forall_r
$$

5 Related Work

Other methods for a Skolemization-like procedure for intuitionistic logic have been investigated. This has been the topic of a series of papers by Baaz and Iemhoff that study the use of an existence predicate, introduced by Scott [11], for Skolemization. They start by defining *eSkolemization* [2], a process for removing strong existential quantifiers in intuitionistic logic. In the same paper there is a semantical proof of completeness of eSkolemization and later on they provide a proof-theoretic proof [4]. In [3] the authors extend the method for strong universal quantifiers, but the solution is more ad-hoc, as it requires the addition of a pre-order to the logic and introduces weak quantifiers. Roughly, the eSkolemization method replaces strong occurrences of $\forall \bar{x}.A\bar{x}$ by $E(f(\bar{x})) \to A(f(\bar{x}))$ and strong occurrences of $\exists \bar{x}.A\bar{x}$ by

$E(f(\bar{x})) \wedge A(f(\bar{x}))$, where f is a new function symbol. In contrast to our approach, which only adds to the language the ε and τ operators, eSkolemization requires extending the language with infinitely many symbols, including a predicate. Moreover, the treatment of existential and universal quantifiers is not uniform whereas in our method those treatments are naturally dual. The calculus LJE, presented in [4], contains different rules for the quantifiers which add the existence predicate to the premises. Therefore, it does not have the sub-formula property. Also, the rules for $\forall_l$ and $\exists_r$ are binary, adding yet another complexity for proof search. It is worth noting that **LJ*** presents none of these issues.

The approach that comes closer to what is presented here is that of Mints [7, 8]. Although the precise relation is not easy to pinpoint and describe, it is straightforward to note important differences. Firstly, whereas Mints is concerned with the extension of **LJ** by an epsilonization rule in the calculus (which acts only at whole formulas), we consider epsilonization as a pre-processing step, acting deeply on all strongly quantified subformulas in the end-sequent. In Mints' calculus, the epsilonization rule is essentially a strong quantifier rule that instantiates the variable by an ε-term instead of an eigen-variable. In contrast, **LJ***-proofs of epsilonized end-sequents contain no inferences that act as strong quantifier inferences in disguise. It was a significant challenge, and one of the main distinguishing contributions of this paper, to discover that ε-terms are informative enough to tell where strong quantifier inferences need to be introduced when de-epsilonizing. Mints also describes a condition for the correctness of proofs, requiring that all sequents are *intelligent*[4]. The definition of *intelligence* is related to the definition of *accessibility* presented here. However, the definition of *intelligence* is not local: to decide whether a sequent S is intelligent in a proof ψ, it may be necessary to look at every sequent S' occurring below S in ψ. This is undesirable in the context of bottom-up proof search, because the whole derivation may have to be traversed and checked in order to decide if an inference is allowed. The definition of *accessibility*, on the other hand, is local: to decide if a weak quantifier inference is allowed, only its conclusion sequent needs to be checked. Furthermore, while Mints [7, 8] restricts all inference rules (by requiring that all sequents be *intelligent*), in the **LJ*** calculus presented here, only the weak quantifier rules need to be restricted. Therefore, the restrictions described here are weaker. Another difference is that Mints [7, 8] considers only the ε binder, whereas here τ is also taken into account.

Decades later, Mints [9] proposed a new calculus where he dropped the global intelligibility condition and adopted binary weak quantifier rules (thus following

---

[4]Mints used the adjective осмысленный in the Russian original [7]. This was translated as *intelligent* in [8]. In [12], Soloviev uses the better translation *meaningful*.

the trend of [4]) whose left premises require proving that the instantiating term is *defined*. While the notion of *defined* is arguably more local than the notion of *intelligent*, it requires proof search and is semantically inspired. Moreover, the definition of *defined* is incomplete because it is defined only for top-level ε-terms. It is not clear what should be done, for example, when the instantiating term is not a top-level ε-term but contains an ε-term as a sub-term. Furthermore, in Mints' new calculus, epsilonization is still treated as an inference rule, not as a pre-processing step.

6 Conclusion

We have shown that, whereas Skolemization is unsound for **LJ** (as is well-known), the new epsilonization transformation defined here is sound for the restricted calculus **LJ*** proposed. Although the definitions and proofs are technically complex, the underlying idea is conceptually very simple. The unsoundness of Skolemization for **LJ** is essentially due to violations of the eigenvariable condition, which happen implicitly and unnoticed, because Skolemization replaces eigenvariables by Skolem terms. In the case of epsilonization, on the other hand, ε-terms are informative enough to allow us to know where strong quantifier inferences introducing their corresponding eigenvariables would be located if the sequent had not been epsilonized. This information allows us to restrict the weak quantifier rules in **LJ*** that use ε-terms, so that they only occur above those implicit strong quantifier inferences' locations. Consequently, as desired, de-epsilonizing **LJ*** proofs never results in violations of the eigenvariable condition.

The approach presented here distinguishes itself from related work primarily by being the only purely syntactic, deterministic (not requiring additional proof search) and local restriction of the intuitionistic sequent calculus where a Skolemization-like pre-processing transformation is sound.

References

[1] J. Avigad and R. Zach. The Epsilon Calculus. In E. N. Zalta, editor, *The Stanford Encyclopedia of Philosophy*. Winter 2013 edition, 2013.

[2] M. Baaz and R. Iemhoff. The Skolemization of existential quantifiers in intuitionistic logic. *Annals of Pure and Applied Logic*, 142(1–3):269 – 295, 2006.

[3] M. Baaz and R. Iemhoff. On Skolemization in Constructive Theories. *The Journal of Symbolic Logic*, 73(3):pp. 969–998, 2008.

[4] M. Baaz and R. Iemhoff. Eskolemization in Intuitionistic Logic. *J. Log. Comput.*, 21(4):625–638, 2011.

[5] M. Baaz and A. Leitsch. On Skolemization and Proof Complexity. *Fundam. Inform.*, 20(4):353–379, 1994.

[6] A. Leitsch, G. Reis, and B. Woltzenlogel Paleo. Towards CERes in intuitionistic logic. In P. Cégielski and A. Durand, editors, *CSL*, volume 16 of *LIPIcs*, pages 485–499. Schloss Dagstuhl - Leibniz-Zentrum fuer Informatik, 2012.

[7] G. Mints. Heyting Predicate Calculus with Epsilon Symbol. *Zapiski Nauchnykh Seminarov Leningradskogo Otdeleniya Matematicheskogo Instituta im. V. A. Steklova AN SSSR*, 40:110–118, 1974.

[8] G. Mints. Heyting Predicate Calculus with Epsilon Symbol. *Journal of Soviet Mathematics*, 8(3):317–323, 1977.

[9] G. Mints. Intuitionistic Existential Instantiation and Epsilon Symbol. *CoRR*, abs/1208.0861, 2012.

[10] G. Reis. *Cut-elimination by resolution in intuitionistic logic*. PhD thesis, Vienna University of Technology, July 2014.

[11] D. Scott. Identity and existence in intuitionistic logic. In M. Fourman, C. Mulvey, and D. Scott, editors, *Applications of Sheaves*, volume 753 of *Lecture Notes in Mathematics*, pages 660–696. Springer Berlin Heidelberg, 1979.

[12] S. Soloviev. Studies of Hilbert's ε-operator in the USSR. *Journal of Logic and its Applications*, 2016.

 Received 23 October 2015

Studies of Hilbert's ε-operator in the USSR

Sergei Soloviev

IRIT, Université de Toulouse, Toulouse, France; associated researcher at ITMO University, St.-Petersburg, Russia
soloviev@irit.fr

Abstract

The aim of this paper is to give a short survey of the studies that concern the notion of Hilbert's ε-operator and its applications by the researchers in the USSR and continuation of their works abroad in post-soviet time.

Keywords: Hilbert's epsilon-operator, history of logic, research in the USSR.

The aim of this paper is to give a short survey of the studies that concern the notion of Hilbert's ε-operator and its applications by the researchers in the USSR. These works mostly belong to the domain of mathematical logics, but interactions with philosophical logics and mathematical linguistics cannot be ignored, all three domains being quite active.

The paper is based on a systematic historical investigation and my personal recollections, that helped to organize the search for information. I knew personally Grigori Mints, Albert Dragalin and Vladimir Smirnov mentioned below. Mints was the adviser of my graduate work, and I met Dragalin and Smirnov at various conferences. It is possible that my investigation is not completely exhaustive, but I recall whose names were mentioned then in connection with the ε-operator, and this knowledge was confirmed when I did look for references. I am reasonably sure that there were no other authors who did significant study of the ε-operator in the USSR.

My personal recollections were used also to reconstruct to some extent the atmosphere of Soviet times, at least as far as the relationship with the international science is concerned.

Three main aspects will be addressed.

The author would like to thank Christian Retoré for organizing the workshop on *Hilbert's Epsilon and Tau in Logic, Informatics and Linguistics* and Giselle Reis, Bruno Woltzenlogel Paleo, and Hans Leiss for fruitful discussions. Thanks also are due to the anonimous referees. This work was partially supported by the Goverment of the Russian Federation Grant 074-U01.

- How the interest for ε-operator had arisen, in particular, the external sources of logical research in the USSR.

- Principal works by soviet researchers who studied ε-operator and continuation of their works in post-soviet time. There are three main names: A. G. Dragalin (1941-1998), G. E. Mints (1939-2014) and V. A. Smirnov (1931-1996).

- Influence of these works on the research worldwide in their own time, and how they influence contemporary research.

Two volumes of German edition of "Grundlagen der Mathematik" by Hilbert and Bernays [5] were known to leading researchers in the USSR. Russian translation was done in 1968/70 and published in 1979/82 by "Nauka" (transl. from German by N. M. Nagorny). The full English translation still does not exist.

In general, the translation of scientific literature into Russian at this period was extremely active, and this fact mostly answers the question about sources. One may mention the publication of Kleene's "Introduction to Metamathematics" in 1957 (translated by Essenin-Volpin), the translation of A. Robinson's "Introduction to Model Theory and to the Metamathematics of Algebra" (1967), the translation of selected foundational papers in proof theory published as "Mathematical theory of logical deduction" (edited by A. V. Idelson and G.E. Mints) in 1967. In 1973 Kleene's "Mathematical Logic" (translated by Yu. Gastev and edited by G. Mints) was published. In 1981 G. Kreisel's selected papers in proof theory translated by Gastev and Mints [15] were published. For scientific literature at this time the interval between publication of an original and its Russian translation was often 5-7 years. The number of copies was usually at least several thousands, *e.g.*, 7800 for [15]. Partly (but only partly) this may be explained by the fact that the USSR did not sign most of the international copyright agreements [1].

What else was accessible? Just as an illustration, one may mention that the library of Steklov Mathematical Institute included almost all "Mathematische Annalen" until June 1941 (with Gentzen's foundational paper of 1936, and Ackermann's paper of 1940), and again since 1949. It got also "Dissertationes Mathematicae" (published by Polish Academy), all issues 1953 - 1989 (including such papers as [13] by Kreisel and Takeuti)...

Personal contacts also should not be underestimated. For example, many prominent logicians attended the International Congress of Mathematicians in Moscow (1966), among them Tarski, Church, Kleene, Curry, Schütte, Feferman, Cohen.

---

[1] A curious fact, mentioned in [8], is that some Finnish universities, for example, the University of Turku, purchased in 1960s and 1970s Russian translations of American or West-European research due to financial reasons, since they were much cheaper than the original editions.

An one-hour talk was given by Schütte, 30 minutes talks by Cohen, Ershov (from Novosibirsk), joint talk by Shanin (the head of the logic group at Leningrad Branch of Steklov Mathematical Institute), Tseitin, and Zaslavski. Junior logicians from Leningrad (Maslov, Matiyasevich, Mints, Orevkov, Slissenko) participated in ICM with 15 minutes talks and had many opportunities to discuss logical problems with western colleagues. Tarski mentions [4] his dinner at "Praha" restaurant with Mal'cev (Maltsev), Markov, Shanin, Ershov, Kleene, Curry, Chang, Feferman. After the ICM Tarski visited Leningrad and gave a talk. Earlier, in 1965, both Moscow and Leningrad were visited by John McCarthy. G. Mints, whose works on the ε-symbol we shall consider below, was greatly influenced by G. Kreisel. Mints considered Kreisel (who did several important works concerning the ε-symbol himself) as one of his teachers. Kreisel visited the USSR in 1976, but actively communicated with Mints before (cf. [14]).

I cannot go too much into detail in this short paper, but would like to recommend the book [19] for a more general picture of the East-West scientific and cultural exchanges during Cold War times.

Another illustration of active interaction of soviet researchers with worldwide research community is a series of biannual Finnish-Soviet Logic Conferences that started in 1976. "Somewhere around 1975 J. Hintikka and V.A. Smirnov have agreed to hold Finnish-Soviet Conference on logic" [9] [2].

The proceedings of the first Soviet-Finnish Logic Conference included 6 papers by soviet participants. Among other contributions let us mention the papers by S. Feferman, J. Hintikka, J. Ketonen, G. Kreisel, D. Prawitz, R. Statman, D. Van Dalen. As Karpenko writes [9], the second Finnish-Soviet Logic Conference was held in Moscow, at the Institute of Philosophy, in 1979. The first Finnish-Soviet-Polish Logic Conference at Polanica-Zdrój was held in 1981. The series of Finnish-Soviet Logic Conferences continued (Helsinki, 1983; Telavi, Georgia, 1985; Helsinki, 1987; Moscow, 1989). It continued even after the fall of the USSR (until 1997, and restarted in 2012) [3].

Among early works on the ε-operator one may cite V. A. Smirnov [38] (from Institute of Philosophy), and G. E. Mints [20] (from Leningrad Branch of Mathematical Institute). In Smirnov's paper [38] (published in French) Hilbert, Bernays, Quine,

---

[2] Finland played a special role in this interaction. It was a liberal democracy, but it was not a NATO member and had a special relationship with the USSR because of the conditions of the Agreement of Friendship, Cooperation, and Mutual Assistance (1948). In times of the Cold War it was considered as a kind of "neutral ground". For example, in 1975 it was the venue of the Conference on Security and Co-operation in Europe.

[3] I guess that the interruption was connected with the changes in the finance of scientific research.

Russell, Sloupiétzski[4] were mentioned, but there is no bibliography. Mints [20] cited several publications, in particular [3], [16], [17], [18], [36]. From this brief outline it is clear that a large variety of sources was accessible, at least to researchers working in academic institutions. It is interesting to notice the role of Japanese and Polish sources, in addition to German, English and American.

We may agree that the most important contributions of the USSR researchers concerning the ε-symbol belong to mathematical logics. This is true also for continuation of this line of research after the fall of the USSR. As B. H. Slater writes: "A good deal of technical work has been done, as a result of such proofs, to create epsilon extensions of Intuitionistic Logic which are conservative." [37] It is more difficult to agree with Slater that this work is now mostly of academic interest. The research on ε in the post-soviet time included the studies of ε-substitutions for analysis [25] and other theories [1], [29] (see the discussion in Stanford Encyclopedia of Philosophy [2]).

As said above, the first paper on the ε-operator by a soviet researcher was the paper [38] published in "Revue Internationale de Philosophie" in 1971. In 1974 two more mathematical papers were published: the paper [3] by Albert Dragalin and [20] by Grigori Mints.

Here is a brief outline of the main points of Mints' paper (which takes into account the two others):

- It is known, that adding the ε-axiom $A[t] \to A[\varepsilon x A]$ to Heyting's (intuitionistic) predicate calculus HA gives a non-conservative extension - for example, the formula $\exists x(\neg Px \to \neg Pb \wedge \neg Pa)$ becomes derivable[5].

- We know (says Mints) two conservative ε-extensions of HA. In one of them, due to V. A. Smirnov, a rather limiting constraint is imposed on the notion of proof. We shall use another formulation, due to A. G. Dragalin; in this formulation functions defined by ε-expressions, are seen as partially defined.

- A. G. Dragalin...uses model-theoretical methods; we shall use proof-theoretical methods, that may be extended to stronger systems, and obtain a supplementary theorem about cut-elimination for proofs of arbitrary formulas, not only ε-free.

- These results admit natural extension to Intuitionistic Predicate Calculus with decidable equality (*i.e.*, with supplementary axiom $\forall x \forall y(x = y \vee \neg x = y)$ and

---

[4]In fact it has to be Jerzy Słupecki, who published a book with Ludwik Borkowski on logic and set theory; its russian edition appeared in 1965.

[5]Here HA may be a misprint, since below Mints speaks about Heyting's arithmetic, and calls Heyting's predicate calculus HPC.

to Heyting's arithmetic with *free* functional variables and a choice principle in the form

$$\frac{\Gamma \to \forall x \exists y A \quad \forall x A_y[f(x)], \Gamma \to C}{\Gamma \to C}$$

where f does not occur in Γ, C, A. An extension to Heyting's arithmetic with *bound* variables of higher types and with a respective choice principle requires new ideas.

- The system studied in the paper is denoted HPC^ε. It is obtained from Heyting Predicate Calculus HPC with two sorts of variables (free and bound), with functional symbols but without equality by addition of the following term formation rule: for a formula A, a free and x bound variable $\varepsilon x A_a[x]$ is a term. The system HPC^ε has the same postulates as Hentzen's LJ (except modified $\exists \to$), but the definition of proof is different.

- An occurrence V of some sequent in a tree-form figure (of deduction) is called meaningful[6] if for every quasiterm $\varepsilon x A$ in V, there is in the antecedent of V or in some sequent lying below V a member

$$(1) \quad \forall \alpha_1 ... \forall \alpha_n \exists x A$$

where $\alpha_1, ..., \alpha_n$ is the full list of free variables of quasiformula $\exists x A$. The (1) is denoted $!\varepsilon x A$.

- The figure (of deduction) that is built from axioms $A \to A$ using deduction rules is called meaningful if all sequents in it are meaningful.

- There are three theorems: the cut elimination theorem for HPC^ε, the conservativity theorem for HPC^ε w.r.t. HPC and the theorem that the following rule

$$\frac{\forall x A_y[f(x)], \Gamma \to C}{\forall x \exists y A, \Gamma \to C}$$

is admissible in HPC^ε and HPC for function symbols f that do not occur in A, Γ, C.

This work by Mints keeps its relevance for modern research. A surprising fact is that later he might underestimate its relevance himself. Bruno Woltzenlogel Paleo met Mints in 2012 and discussed his current work about epsilonization (in collaboraion with Giselle Reis). As he wrote to the author, concerning [20]: "the work you pointed to us is much more relevant than the work Grigori Mints recommended. He must have forgotten."[7] Mints recommended [34].

---

[6]"Osmyslennyi" in Russian.
[7]E-mail to the author, 6 Oct. 2015.

Next paper by V. A. Smirnov [39], presented initially as a talk at the First Soviet-Finnish Conference, was published in 1979. In this paper Smirnov recalls the history of his intuitionistic system of natural deduction with the ε-symbol (a conservative extension of the system without ε). Dragalin's technique is used to obtain a similar result for the system of natural deduction with identity. At the same conference was presented also the talk by Yu. Gladkich "Singular Terms, Existence and Truth: Some Remarks on a First Order Logic of Existence" published later in the same proceedings as [39].

In 1982 and 1989 Mints published two papers [21], [22].

- **Zbl 0523.03043** [21]: The paper contains a simplified proof of Ackermann's theorem about the consistency of Peano Arithmetic. The proof is based on Hilbert's idea to apply so called ε-substitutions to systems of arithmetical formulas (...). A careful analysis of the behaviour of the above substitutions and (...) give a proof of the convergence of a suitable system of ε-substitutions. Hence Ackermann's theorem follows. (E. Adamowicz)

- **Zbl 0677.03040** [22]: The Hilbert epsilon-substitution method is extended to some formalizations of the theory of hereditarily finite sets. Applying the methodology developed in an earlier paper (...) the convergence (...) for the theory of hereditarily finite sets is established which generalizes the Ackermann theorem of the convergence of the epsilon-substitution method for first-order arithmetic. (B.R. Boricić)

After the end of the USSR, G. Mints (with some co-authors of the next generation) was, no doubt the most active (among the researchers formed in the USSR) to pursue the studies of ε. V. A. Smirnov also continued to work in this direction [40], [41], [42][8].

The papers of this period are easily accessible, so there is less need for a detailed presentation.

G. Mints did work on versions of ε-substitution method in various systems, problems of completeness, termination and cut elimination [24], [25], [26], [27], [28], [29], [30], [31], [32], [33], [34], [35]. Among his collaborators were W. Buchholz, D. Sarenac, S. Tupailo.

Some of these papers concerned other aspects of ε-methods, such as their relationship to Kripke semantics. Here is the abstract of [34]:

---

[8]It must be noticed that in the USSR and post-soviet Russia mathematical logic interacted very little with philosophical logic and philosophy of language. The V. A. Smirnov's works had considerable influence on philosophical community.

- A natural deduction system for intuitionistic predicate logic with the existential instantiation rule presented here uses Hilbert's ε-symbol. It is conservative over intuitionistic predicate logic. We provide a completeness proof for a suitable Kripke semantics, sketch an approach to a normalization proof, survey related work and state some open problems. Our system extends intuitionistic systems with the ε-symbol due to A. Dragalin and Sh. Maehara.

In the paper with Sarenac [31] Mints studied context-dependent descriptions and so called "salience hierarchy". Here is the abstract:

- Epsilon terms indexed by contexts were used by K. von Heusinger[9] to represent definite and indefinite noun phrases as well as some other constructs of natural language. We provide a language and a complete first order system allowing to formalize basic aspects of this representation. The main axiom says that for any finite collection $S_1, ..., S_k$ of distinct definable sets and elements $a_1, ..., a_k$ of these sets there exists a choice function assigning a_i to S_i for all $i \leq k$. We prove soundness and completeness theorems for this system S_{ε_i}.

The method of the proof proposed in [31] is based on a modification of the completeness proof for first-order predicate calculus given by Henkin [7]. Recently a critical analysis of [31] was presented in a talk by Hans Leiss[10]. His criticism concerns two important points: one technical (there is a gap in the proof), and one conceptual (whether the first-order theory proposed by Mints and Sarenac is an adequate representation of the second-order theory of von Heusinger).

It is clear, however, that independently of the results of this controversy [31] turns out to be a very stimulating work.

In his talk at Montpellier conference mentioned above Hans Leiss proposed what he himself described as a non-conclusive counterexample, and tried to "fill the gap" in the Mints-Sarenac proof (to modify the construction of term model).

An attempt to reflect a second-order theory in a first-order system also looks promising even if the details need to be fixed. One may hope that this theme will see new interesting developments.

It reminds the story of another proof proposed by G. Mints that was included in [43]. One may read there (p. 281): "Our treatment follows Mints [1992d], with a correction in proposition 8.4.12. (The correction was formulated after exchanges between Mints, Solovjov and the authors.)" It is meant here a "short proof" of so

---

[9]in his paper [6]

[10]At the same conference where the initial version of this work was presented, *Epsilon 2015. Hilbert's epsilon and tau in logic, informatics and linguistics*, Montpellier, LIRMM, 10-12 June 2015.

called Coherence Theorem for Cartesian Closed Categories. First proof that was considerably longer was published by the author of this article and A. Babaev in 1979. G. Mints suggested a much shorter proof in 1992.

Conclusion. If I wanted to make a "lesson" from the history of studies of the ε-symbol outlined above, I would stress the role of continuity and interaction. Indeed, the papers by Smirnov, Mints and Dragalin could hardly appear in 1970s without continuity of the development of logic inside and outside the USSR, and strong collaboration between researchers on an international level.

The role of these factors remains vital even after the end of the USSR, as it is clearly seen from the examples above.

References

[1] Arai, T. Epsilon substitution method for theories of jump hierarchies. Archive for Mathematical Logic, 2, 2002, 123-153.

[2] Avigad, J., Zach, R. The Epsilon Calculus. (First published May 3, 2002; substantive revision Nov 27, 2013.) Stanford Encyclopedia of Philosophy. http://plato.stanford.edu/entries/epsilon-calculus/

[3] Dragalin, A. Intuitionistic Logic and Hilbert's ε-symbol, (Russian) Istoriia i Metodologiia Estestvennykh Nauk, Moscow, 1974, 78-84. Republished in: Albert Grigorevich Dragalin, Konstruktivnaia Teoriia Dokazatelstv i Nestandartnyi Analiz, s. 255-263, Moscow, Editorial Publ.,2003.

[4] Fefereman, A., Feferman, S. Alfred Tarski. Life and Logic. Cambridge University Press, 2004.

[5] Hilbert, D., Bernays, P. Grundlagen der Mathematik. I, Die Grundlehren der mathematischen Wissenschaften 40, 1934. II, Die Grundlehren der mathematischen Wissenschaften 50, 1939. Berlin, New York: Springer-Verlag.

[6] Von Heusinger, K. Reference and Salience. In F. Hamm, J. Kolb, A. von Stechow (eds.). Proc. Workshop on Recent Developments in the Theory of Natural Language Semantics, 149-172. SfS, Tübingen 1995.

[7] Henkin, L. The completeness of the first-order functional calculus. J. Symbolic Logic, 14 (1949), 159-166.

[8] Kaataja, S. Expert Groups Closing the Divide: Estonian-Finnish Computing Cooperation since the 1960s. In [19], pp.101-121.

[9] Karpenko, A.S. Preface. (The History of Finnish-Soviet Logic Colloquium.) Logical Investigations (Logicheskie Issledovaniya). 20. Moscow, July 2013.

[10] Kreisel, G. On the interpretation of non-finitist proofs, part I. *J. of Symbolic Logic*, 16, 1951, 241-267.

[11] Kreisel, G. On the interpretation of non-finitist proofs, part II. *J. of Symbolic Logic*, 17, 1952, 43-58.

[12] Kreisel, G. Mathematical logic. *Lectures in Modern Mathematics III*(T.L. Saaty ed.), Wiley and Sons, New york, 1965, 95-195.

[13] Kreisel, G., Takeuti, G. Formally self-referential propositions for cut-free analysis and related systems. *Dissertationes mathematicae*, 118, Warsaw, 1974, 50 pp.

[14] Kreisel, G., Mints, G. E., and Simpson, S. G. The use of abstract language in elementary metamathematics: some pedagogic examples. Springer Lecture Notes, 453 (1975), 38-131.

[15] . Kreisel, G. Issledovaniya po theorii dokazatelstv. Moscow, "Mir", 1981.

[16] Leisenring, A. C. Mathematical logic and Hilbert's ε-symbol. London, MacDonald, 1969.

[17] Leivant, D. Existential instantiation in a system of natural deduction for intuitionistic arithmetics. Technical Report ZW 13/73, Stichtung Mathematische Centrum, Amsterdam, 1973

[18] Maehara, Sh. A General Theory of Completeness Proofs. Ann. Jap. Assoc. Phil. Sci., no. 3, 1970, p. 242-256.

[19] Mikkonen, S. and Koivunen, P. (eds.). Beyond the Divide. Entangled Histories of Cold War Europe. Berghahn, N.-Y., Oxford, 2015, viii+324 pp.

[20] Mints, G. Heyting Predicate Calculus with Epsilon Symbol (Russian). Zapiski Nauchnykh Seminarov Leningradskogo Otdeleniya Matematicheskogo Instituta im. V. A. Steklova AN SSSR, Vol. 40, 1974, pp. 110-118. English Translation in [23] p. 97-104

[21] Mints, G. A simplified consistency proof for arithmetic. (Russian, English summary), Izv. Akad. Nauk. Ehst. SSR, Fiz., Mat. 31, 376-381, 1982.

[22] Mints, G. Epsilon substitution method for the theory of hereditarily finite sets (Russian, English Summary). Izv. Akad. Nauk Ehst. SSR, Fiz., Math. 38, N 2, 154-164, 1989.

[23] Mints, G. Selected Papers in Proof Theory, Bibliopolis/North-Holland, 1992.

[24] Mints, G. Hilbert's Substitution Method and Gentzen-type Systems. In Proc. of 9th International Congress of Logic, Methododology and Philososophy of Science, Uppsala, Sweden, 1994.

[25] Mints, G., Tupailo, S., Buchholz, W. Epsilon substitution method for elementary analysis. Archive for Mathematical Logic, 35, 1996, p. 103-130.

[26] Mints, G. Strong termination for the epsilon substitution method. J. of Symb. Logic, 61:1193-1205, 1996.

[27] Mints, G. Thoralf Skolem and the epsilon substitution method for predicate logic. Nordic J. of Philos. Logic, 1:133-146, 1996.

[28] Mints, G. Strong termination for the epsilon substitution method for predicate logic. J. of Symb. Logic, 1998.

[29] Mints, G., Tupailo, S. Epsilon-substitution method for the ramified language and Δ_1^1-comprehension rule. In A. Cantini et al. (eds.), Logic and Foundations of Mathematics (Florence, 1995), Dordrecht: Kluwer, 1999, p. 107-130.

[30] Mints, G. A termination proof for epsilon substitution using partial derivations. Theor. Comput. Sci. 1(303): 187-213 (2003)

[31] Mints, G., Sarenac, D. Completeness of indexed epsilon-calculus. Arch. Math. Log. 42(7): 617-625 (2003)

[32] Mints, G. Cut Elimination for a Simple Formulation of PAepsilon. Electr. Notes Theor. Comput. Sci. 143: 159-169 (2006)

[33] Mints, G. Cut elimination for a simple formulation of epsilon calculus. Ann. Pure Appl. Logic 152(1-3): 148-160 (2008)

[34] Mints, G. Intuitionistic Existential Instantiation and Epsilon Symbol. arXiv:1208.0861v1 [math.LO] 3 Aug 2012.

[35] Mints, G. Epsilon substitution for first- and second-order predicate logic. Ann. Pure Appl. Logic 164(6): 733-739 (2013)

[36] Shirai, K. Intuitionistic Predicate Calculus with ε-symbol. Annals of the Japan Association for Philosophy, no.4, 1971, p.49-67.

[37] Slater, B. H. Hilbert's Epsilon Calculus and its Successors. In: Dov Gabbay (ed.), The Handbook of the History of Logic. Elsevier. 385-448 (2009)

[38] Smirnov, V. A. Elimination des termes ε dans la logique intuitioniste. "Revue internationale de Philosophie", 98, 1971, 512-519.

[39] Smirnov, V. A. Theory of Quantification and ε-calculi. Proceedings of the Fourth Scandinavian Logic Symposium and the First Soviet-Finnish Conference, Jyväskulë, Finland, June 29 - July 6, 1976. Essays on mathematical and philosophical logic. Synthese Library, vol.122, D. Reidel P.C.1979.

[40] Smirnov, V. A. (with Anisov, A. M., Bystrov, P. I.). Logika i Komputer. Moscow, 1990 (Russian). In this book Smirnov contributed a paper on proof search.

[41] Smirnov, V. A. Cut elimination in ε-calculus // Book of abstracts, N. 1. XIX World Congress of Philosophy. Section 4. Moscow, 1993.

[42] Smirnov, V. A.. Proof search in intuitionistic natural deduction with ε-symbol and existence predicate. (Russian) // Logical investigations, 3, Moscow, 1995, 163-173.

[43] Troelstra, A. S., Schwichtenberg, H. Basic Proof Theory (second edition). *Cambridge Tracts in Theoretical Computer Science*, 2000.

Received 23 October 2015

A Simplified and Improved Free-Variable Framework for Hilbert's epsilon as an Operator of Indefinite Committed Choice

Claus-Peter Wirth

Dept. of Math., ETH Zurich, Rämistr. 101, 8092 Zürich, Switzerland
wirth@logic.at

Abstract

Free variables occur frequently in mathematics and computer science with *ad hoc* and altering semantics. We present here the most recent version of our free-variable framework for two-valued logics with properly improved functionality, but only two kinds of free variables left (instead of three): implicitly universally and implicitly existentially quantified ones, now simply called "free atoms" and "free variables", respectively. The quantificational expressiveness and the problem-solving facilities of our framework exceed standard first-order logic and even higher-order modal logics, and directly support Fermat's *descente infinie*. With the improved version of our framework, we can now model also Henkin quantification, neither using any binders (such as quantifiers or epsilons) nor raising (Skolemization). Based only on the traditional ε-formula of Hilbert–Bernays, we present our flexible and elegant semantics for Hilbert's ε as a choice operator with the following features: We avoid overspecification (such as right-uniqueness), but admit indefinite choice, committed choice, and classical logics. Moreover, our semantics for the ε supports reductive proof search optimally.

Keywords: Logical Foundations, Theories of Truth and Validity, Formalized Mathematics, Human-Oriented Interactive Theorem Proving, Automated Theorem Proving, Choice, Hilbert's epsilon-Operator, Henkin Quantification, Fermat's Descente Infinie.

1 Overview

Driven by a weakness in representing Henkin quantification (cf. [111, § 6.4.1]) and inspired by nominal terms (cf. e.g. [101]), in this paper we significantly improve our semantic free-variable framework for two-valued logics:

This is an invited paper.

1. We have replaced the two-layered construction of free δ^+-variables on top of free γ-variables over free δ^--variables of [106; 108; 111] with a one-layered construction of *free variables* over *free atoms*:

   - Free variables without choice-condition now play the former rôle of the γ-variables.
   - Free variables with choice-condition play the former rôle of the δ^+-variables.
   - Free atoms now play the former rôle of the δ^--variables.

2. As a consequence, the proofs of the lemmas and theorems have shortened by more than a factor of 2. Therefore, we now can present all the proofs in this paper and make it self-contained in this aspect; whereas in [108; 111], we had to point to [106] for most of the proofs.

3. The difference between free variables and atoms and their names are now more standard and more clear than those of the different free variables before; cf. § 2.1.

4. Compared to [106], besides shortening the proofs, we have made the meta-level presuppositions more explicit in this paper; cf. § 5.8.

5. Last but not least, we can now treat Henkin quantification in a direct way; cf. § 5.11.

Taking all these points together, the version of our free-variable framework presented in this paper is the version we recommend for further reference, development, and application: it is indeed much easier to handle than its predecessors.

And so we found it appropriate to present most of the material from [108; 111] in this paper in the improved form; we have omitted only the discussions on the tailoring of operators similar to our ε, and on the analysis of natural-language semantics. The material on mathematical induction in the style of Fermat's *descente infinie* in our framework of [106] is to be reorganized accordingly in a later publication.

This paper is organized as follows. There are three introductory sections: to our free variables and atoms (§ 2), to their relation to our reductive inference rules (§ 3), and to Hilbert's ε (§ 4). Afterward we explain and formalize our novel approach to the semantics of our free variables and atoms and the ε (§ 5), and summarize and discuss it (§ 6). We conclude in § 7. In an appendix, the reader can find a discussion of the literature on extended semantics given to Hilbert's ε-operator in the 2nd half of the 20th century, and on Leisenring's axiom (E2) (§ A). The proofs of all lemmas and theorems can be found in § B.

2 Introduction to Free Variables and Atoms

2.1 Outline

Free variables or free atoms occur frequently in practice of mathematics and computer science. The logical function of these free symbols varies locally; it is typically determined *ad hoc* by the context. And the intended semantics is given only *implicitly* and varies from context to context. In this paper, however, we will make the semantics of our free variables and atoms *explicit* by using disjoint sets of symbols for different semantic functions; namely we will use the following sets of symbols:

$\mathbb{V}$ (the set of *free variables*),
$\mathbb{A}$ (the set of *free atoms*),
$\mathbb{B}$ (the set of *bound*[1] *atoms*).

An *atom* typically stands for an arbitrary object in a proof attempt or in a discourse. Nothing else is known on any atom. Atoms are invariant under renaming. And we will never want to know anything about a possible atom but whether it is an atom, and, if yes, whether it is identical to another atom or not. In our context here, for reasons of efficiency, we would also like to know whether an atom is a free or a bound one. The name "atom" for such an object has a tradition in set theories with atoms. (In German, besides "Atom", an atom is also called an "Urelement", but that alternative name puts some emphasis on the origin of creation, in which we are not interested here.)

A *variable*, however, in the sense we will use the word in this paper, is a place-holder in a proof attempt or in a discourse, which gathers and stores information and which may be replaced with a definition or a description during the discourse or proof attempt. The name "free variable" for such a place-holder has a tradition in free-variable semantic tableaus; cf. [20; 21].

Both variables and atoms may be instantiated with terms. Only variables, however, may refer to other free variables and atoms, or may depend on them; and only variables have the following properties w.r.t. instantiation:

1. If a variable is instantiated, then this affects *all* of its occurrences in the entire state of the proof attempt (i.e. it is *rigid* in the terminology of semantic tableaus). Thus, if the instantiation is executed eagerly, the variable must be replaced *globally* in all terms of the entire state of the proof attempt with the same term; afterwards the variable can be eliminated from the resulting proof

---

[1]"Bound" atoms (or variables) should actually be called "bindable" instead of "bound", because we will always have to treat some unbound occurrences of "bound" atoms. When the name of the notion of bound variables was coined, however, neither "bindable" nor the German "bindbar" were considered to be proper words of their respective languages, cf. [62, § 4].

forest completely — without any further effect on the chance to complete it into a successful proof.

2. The instantiation may be relevant for the consequences of a proof because the global replacement may strengthen the input proposition (or query) by providing a witnessing term for an existential property stated in the proposition (or by providing an answer to the query).

By contrast to these properties of variables, atoms cannot refer to any other symbols, nor depend on them in any form. Moreover, free atoms have the following properties w.r.t. instantiation:

1. A free atom may be

   - globally renamed, or else
   - locally and possibly repeatedly instantiated with arbitrary different terms in the application of lemmas or induction hypotheses (provided that the instantiation is admissible in the sense of Theorem 5.27(7)).

   We cannot eliminate a free atom safely, however. Indeed, neither global renaming nor local instantiation can achieve that completely.

2. The question with which terms an atom was actually instantiated can never influence the consequences of a proof (whereas it may be relevant for book-keeping or for a replay mechanism).

2.2 Notation

The classification as a (free) $\underline{v}$ariable, (free) $\underline{a}$tom, or $\underline{b}$ound atom will be indicated by adjoining a "$\mathbb{V}$", an "$\mathbb{A}$", or a "$\mathbb{B}$", respectively, as a label to the upper right of the meta-variable for the symbol. If a meta-variable stands for a symbol of the union of some of these sets, we will indicate this by listing all possible sets; e.g. "$x^{\mathbb{V}\mathbb{A}}$" is a meta-variable for a symbol that may be either a free variable or a free atom.

Meta-variables with disjoint labels always denote different symbols; e.g. "$x^{\mathbb{V}}$" and "$x^{\mathbb{A}}$" will always denote different symbols, whereas "$x^{\mathbb{V}\mathbb{A}}$" may denote the same symbol as "$x^{\mathbb{A}}$". In formal discussions, also "$x^{\mathbb{A}}$" and "$y^{\mathbb{A}}$" may denote the same symbol. In concrete examples, however, we will implicitly assume that different meta-variables denote different symbols.

2.3 Semantics of Free Variables and Atoms

2.3.1 Semantics of Free Atoms

As already noted by Russell in 1919 [89, p.155], free symbols of a formula often have an obviously universal intention in mathematical practice, such as the free symbols m, p, and q in the formula

$$(m)^{(p+q)} = (m)^{(p)} * (m)^{(q)}.$$

Moreover, the formula itself is not meant to denote a propositional function, but actually stands for the explicitly universally quantified, closed formula

$$\forall m^{\mathbb{B}}, p^{\mathbb{B}}, q^{\mathbb{B}}. \left((m^{\mathbb{B}})^{(p^{\mathbb{B}}+q^{\mathbb{B}})} = (m^{\mathbb{B}})^{(p^{\mathbb{B}})} * (m^{\mathbb{B}})^{(q^{\mathbb{B}})} \right).$$

In this paper, however, we indicate by

$$(m^{\mathbb{A}})^{(p^{\mathbb{A}}+q^{\mathbb{A}})} = (m^{\mathbb{A}})^{(p^{\mathbb{A}})} * (m^{\mathbb{A}})^{(q^{\mathbb{A}})},$$

a proper formula with *free atoms*, which — independent of its context — is equivalent to the explicitly universally quantified formula, but which also admits the reference to the free atoms, which is required for mathematical induction in the style of Fermat's *descente infinie*, and which may also be beneficial for solving reference problems in the analysis of natural language. So the third version combines the practical advantages of the first version with the semantic clarity of the second version.

2.3.2 Semantics of Free Variables

Changing from universal to existential intention, it is somehow clear that the linear system of the formula

$$\begin{pmatrix} 2 & 3 \\ 5 & 7 \end{pmatrix} \begin{pmatrix} x \\ y \end{pmatrix} = \begin{pmatrix} 11 \\ 13 \end{pmatrix}$$

asks us to find the set of solutions for x and y, say $(x, y) \in \{(-38, 29)\}$. The mere existence of such solutions is expressed by the explicitly existentially quantified, closed formula

$$\exists x^{\mathbb{B}}, y^{\mathbb{B}}. \left(\begin{pmatrix} 2 & 3 \\ 5 & 7 \end{pmatrix} \begin{pmatrix} x^{\mathbb{B}} \\ y^{\mathbb{B}} \end{pmatrix} = \begin{pmatrix} 11 \\ 13 \end{pmatrix} \right).$$

In this paper, however, we indicate by

$$\begin{pmatrix} 2 & 3 \\ 5 & 7 \end{pmatrix} \begin{pmatrix} x^{\mathbb{V}} \\ y^{\mathbb{V}} \end{pmatrix} = \begin{pmatrix} 11 \\ 13 \end{pmatrix}$$

a proper formula with free variables, which — independent of its context — is equivalent to the explicitly existentially quantified formula, but which admits also the reference to the free variables, which is required for retrieving solutions for $x^{\mathbb{V}}$ and $y^{\mathbb{V}}$ as instantiations for $x^{\mathbb{V}}$ and $y^{\mathbb{V}}$ chosen in a formal proof. So the third version again combines the practical advantages of the first with the semantic clarity of the second.

3 Reductive Inference Rules

We will now present the essential reductive inference rules for our free-variable framework. Regarding form and notation, please note the following:

We choose a sequent-calculus representation to enhance the readability of the rules and the explicitness of eliminability of formulas. As we restrict ourselves to two-valued logics, we just take the right-hand side of standard sequents. This means that our *sequents* are just disjunctive lists of formulas.

We assume that all binders have minimal scope; e.g.

$$\forall x^{\mathrm{B}}, y^{\mathrm{B}}.\ A \wedge B$$

reads

$$(\forall x^{\mathrm{B}}.\ \forall y^{\mathrm{B}}.\ A) \wedge B.$$

Our reductive inference rules will be written "reductively" in the sense that passing the line means reduction. Note that in the good old days when trees grew upward, Gerhard Gentzen (1909–1945) would have inverted the inference rules such that passing the line means consequence. In our case, passing the line means reduction, and trees grow downward.

Raymond M. Smullyan (1919–2017) has classified reductive inference rules into α-, β-, γ-, and δ-rules, and invented a uniform notation for them [98].

In the following rules, let A always be a formula and Γ and Π be sequents.

3.1 α- and β-Rules

α-**rules** are the non-branching propositional rules, such as

$$\frac{\Gamma \quad \neg\neg A \quad \Pi}{\Gamma \quad A \quad \Pi} \qquad\qquad \frac{\Gamma \quad A \Rightarrow B \quad \Pi}{\Gamma \quad \neg A \quad B \quad \Pi}$$

β-**rules** are the branching propositional rules, which reduce a sequent to several sequents, such as

$$\frac{\Gamma \quad \neg(A \Rightarrow B) \quad \Pi}{\begin{array}{ccc} \Gamma & A & \Pi \\ \Gamma & \neg B & \Pi \end{array}}$$

3.2 γ-Rules

Suppose we want to prove an existential proposition $\exists y^{\mathrm{B}}.\ A$. Here "$y^{\mathrm{B}}$" is a bound variable according to standard terminology, but as it is an atom according to our classification of § 2.1, we will speak of a "bound atom" instead. Then the γ-rules of old-fashioned inference systems (such as Gentzen's [30] or Smullyan's [98]) enforce the choice of a witnessing term t as a substitution for the bound atom *immediately* when eliminating the quantifier.

γ-rules: Let t be any term:

$$\frac{\quad\quad\quad \Gamma \quad \exists y^{\mathsf{B}}.\, A \quad \Pi}{A\{y^{\mathsf{B}}\mapsto t\} \quad \Gamma \quad \exists y^{\mathsf{B}}.\, A \quad \Pi} \qquad\qquad \frac{\quad\quad\quad \Gamma \quad \neg\forall y^{\mathsf{B}}.\, A \quad \Pi}{\neg A\{y^{\mathsf{B}}\mapsto t\} \quad \Gamma \quad \neg\forall y^{\mathsf{B}}.\, A \quad \Pi}$$

More modern inference systems (such as the ones in Fitting's [21]) enable us to delay the crucial choice of the term t until the state of the proof attempt may provide more information to make a successful decision. This delay is achieved by introducing a special kind of variable.

This special kind of variable is called "dummy" in Prawitz' [84] and Kanger's [66], "free variable" in Fitting's [20; 21] and in Footnote 11 of Prawitz' [84], "meta variable" in the field of planning and constraint solving, and "free γ-variable" in Wirth's [106; 107; 108; 109; 111; 112] and Wirth &al.'s [114; 115].

In this paper, we call these variables simply "free variables" and write them like "y^{V}". When these additional variables are available, we can reduce $\exists y^{\mathsf{B}}.\, A$ first to $A\{y^{\mathsf{B}} \mapsto y^{\mathsf{V}}\}$ and then sometime later in the proof we may globally replace y^{V} with an appropriate term.

The addition of these free variables changes the notion of a term, but not the notation of the γ-rules, whereas it will become visible in the δ-rules.

3.3 δ^--Rules

A δ-rule may introduce either a free atom (δ^--*rule*) or an ε-constrained free variable (δ^+-*rule*, cf. §3.4).

δ^--rules: Let x^{A} be a fresh free atom:

$$\frac{\Gamma \quad \forall x^{\mathsf{B}}.\, A \quad \Pi}{A\{x^{\mathsf{B}}\mapsto x^{\mathsf{A}}\} \quad \Gamma \quad \Pi} \quad \mathbb{V}(\Gamma\ \forall x^{\mathsf{B}}.\, A\ \Pi)\times\{x^{\mathsf{A}}\}$$

$$\frac{\Gamma \quad \neg\exists x^{\mathsf{B}}.\, A \quad \Pi}{\neg A\{x^{\mathsf{B}}\mapsto x^{\mathsf{A}}\} \quad \Gamma \quad \Pi} \quad \mathbb{V}(\Gamma\ \neg\exists x^{\mathsf{B}}.\, A\ \Pi)\times\{x^{\mathsf{A}}\}$$

Note that $\mathbb{V}(\Gamma\ \forall x^{\mathsf{B}}.\, A\ \Pi)$ stands for the set of all symbols from $\mathbb{V}$ (in this case: the free variables) that occur in the sequent $\Gamma\ \forall x^{\mathsf{B}}.\, A\ \Pi$.

Let us recall that a free atom typically stands for an arbitrary object in a discourse of which nothing else is known. The free atom x^A introduced by the δ^--rules is sometimes also called "parameter", "eigenvariable", or "free δ-variable". In Hilbert-calculi, however, this free atom is called a "free *variable*", because the non-reductive (i.e. generative) deduction in Hilbert-calculi admits its unrestricted instantiation by the substitution rule, cf. p. 63 of Hilbert–Bernays' [57] or p. 62 of Hilbert–Bernays' [59; 62]. The equivalents of the δ^--rules in Hilbert–Bernays' predicate calculus are Schemata (α) and (β) on p. 103f. of [57] or on p. 102f. of [59; 62].

The occurrence of the free atom x^A of the δ^--rules must be disallowed in the terms that may be used to replace those free variables which have already been in use when x^A was introduced by application of the δ^--rule, i.e. the free variables of the upper sequent to which the δ^--rule was applied. The reason for this restriction of instantiation of free variables is that the dependencies (or scoping) of the quantifiers must be somehow reflected in the dependencies of the free variables on the free atoms. In our framework, these dependencies are to be captured in binary relations on the free variables and the free atoms, called *variable-conditions*.

Indeed, it is sometimes unsound to instantiate a free variable x^V with a term containing a free atom y^A that was introduced later than x^V:

Example 3.1 (Soundness of δ^--rule)
The formula
$$\exists y^\text{B}.\ \forall x^\text{B}.\ (y^\text{B} = x^\text{B})$$
is not universally valid. We can start a reductive proof attempt as follows:

γ-step: $\forall x^\text{B}.\ (y^\text{V} = x^\text{B}),\quad \exists y^\text{B}.\ \forall x^\text{B}.\ (y^\text{B} = x^\text{B})$

δ^--step: $(y^\text{V} = x^\text{A}),\quad \exists y^\text{B}.\ \forall x^\text{B}.\ (y^\text{B} = x^\text{B})$

Now, if the free variable y^V could be replaced with the free atom x^A, then we would get the tautology $(x^\text{A} = x^\text{A})$, i.e. we would have proved an invalid formula. To prevent this, as indicated to the lower right of the bar of the first of the δ^--rules, the δ^--step has to record
$$\mathbb{V}(\forall x^\text{B}.\ (y^\text{V} = x^\text{B}),\ \exists y^\text{B}.\ \forall x^\text{B}.\ (y^\text{B} = x^\text{B})) \times \{x^\text{A}\}\quad =\quad \{(y^\text{V}, x^\text{A})\}$$
in a variable-condition, where (y^V, x^A) means that y^V is somehow "necessarily older" than x^A, so that we may never instantiate the free variable y^V with a term containing the free atom x^A.

Starting with an empty variable-condition, we extend the variable-condition during proof attempts by δ-steps and by global instantiations of free variables. Roughly speaking, this kind of global instantiation of these *rigid* free variables is *consistent* if the resulting variable-condition (seen as a directed graph) has no cycle after adding, for each free variable y^V instantiated with a term t and for each free variable or atom x^VA occurring in t, the pair $(x^\text{VA}, y^\text{V})$. This consistency, however, would be violated by the cycle between y^V and x^A if we instantiated y^V with x^A.

3.4 δ^+-Rules

There are basically two different versions of the δ-rules: standard δ^--rules (also simply called "δ-rules") and δ^+-rules (also called *"liberalized δ-rules"*). They differ in the kind of symbol they introduce and — crucially — in the way they enlarge the variable-condition, depicted to the lower right of the bar:

δ^+-**rules:** Let x^{V} be a fresh free variable:

$$\frac{\Gamma \quad \forall x^{\mathrm{B}}.\, A \quad \Pi}{A\{x^{\mathrm{B}}\!\mapsto\! x^{\mathrm{V}}\} \quad \Gamma \quad \Pi} \qquad \begin{array}{c} (\ x^{\mathrm{V}}, \ \varepsilon x^{\mathrm{B}}.\, \neg A\ ) \\ \mathbb{VA}(\forall x^{\mathrm{B}}.\, A) \times \{x^{\mathrm{V}}\} \end{array}$$

$$\frac{\Gamma \quad \neg \exists x^{\mathrm{B}}.\, A \quad \Pi}{\neg A\{x^{\mathrm{B}}\!\mapsto\! x^{\mathrm{V}}\} \quad \Gamma \quad \Pi} \qquad \begin{array}{c} (\ x^{\mathrm{V}}, \ \varepsilon x^{\mathrm{B}}.\, A\ ) \\ \mathbb{VA}(\neg \exists x^{\mathrm{B}}.\, A) \times \{x^{\mathrm{V}}\} \end{array}$$

While in the (first) δ^--rule, $\mathbb{V}(\Gamma \ \forall x^{\mathrm{B}}.\, A \ \Pi)$ denotes the set of the free variables occurring in the *entire* upper sequent, in the (first) δ^+-rule, $\mathbb{VA}(\forall x^{\mathrm{B}}.\, A)$ denotes the set of all free variables and all free atoms, but only the ones occurring in particular in the *principal*[2] *formula* $\forall x^{\mathrm{B}}.\, A$.

Therefore, the variable-conditions generated by the δ^+-rules are typically smaller than the ones generated by the δ^--rules. Smaller variable-conditions permit additional proofs. Indeed, the δ^+-rules enable additional proofs on the same level of γ-*multiplicity* (i.e. the maximal number of repeated γ-steps applied to the identical principal formula); cf. e.g. [106, Example 2.8, p. 21]. For certain classes of theorems, these proofs are exponentially and even non-elementarily shorter than the shortest proofs which apply only δ^--rules; for a short survey cf. [106, § 2.1.5]. Moreover, the δ^+-rules provide additional proofs that are not only shorter but also more natural and easier to find, both automatically and for human beings; see the discussion on design goals for inference systems in [106, § 1.2.1], and the formal proof of the limit theorem for + in [107; 110]. All in all, the name "liberalized" for the δ^+-rules is indeed justified: They provide more freedom to the prover.[3]

Moreover, note that the pairs indicated to the upper right of the bar of the δ^+-rules are to augment another global binary relation besides the variable-condi-

---

[2]The notions of a *principal formula* (in German: Hauptformel) and a *side formula* (Seitenformel) were introduced in [30] and refined in [90]. Very roughly speaking, the principal formula of an inference rule is the formula that is reduced by that rule, and the side formulas are the resulting pieces replacing the the principle formula. In our reductive inference rules here, the principal formulas are the formulas above the lines except the ones in Γ, Π (which are called *parametric formulas*, in German: Nebenformeln), and the side formulas are the formulas below the lines except the ones in Γ, Π.

[3]Cf. § C.1.

tion, namely a function called the *choice-condition*.

Roughly speaking, the addition of an element $(x^{\mathrm{V}},\ \varepsilon x^{\mathrm{B}}.\ \neg A)$ to the current choice-condition — as required by the first of the δ^+-rules — is to be interpreted as the addition of the equational constraint $x^{\mathrm{V}} = \varepsilon x^{\mathrm{B}}.\ \neg A$. To preserve the soundness of the δ^+-step under subsequent global instantiation of the free variable x^{V}, this constraint must be observed in such instantiations. What this actually means will be explained in § 4.12.

All of the three following systems are sound and complete for first-order logic: The one that has (besides the straightforward propositional rules (α-, β-rules) and the γ-rules) only the δ^--rules, the one that has only the δ^+-rules, and the one that has *both* the δ^-- and δ^+-rules.

For a replay of Example 3.1 using the δ^+-rule instead of the δ^--rule, see Example 4.12 in § 4.12.

3.5 Skolemization

Note that there is a popular alternative to variable-conditions, namely Skolemization, where the δ^-- and δ^+-rules introduce functions (i.e. the logical order of the replacements for the bound atoms is incremented) which are given the free variables of $\mathbb{V}(\Gamma\ \forall x^{\mathrm{B}}.\ A\ \Pi)$ and $\mathbb{V}(\forall x^{\mathrm{B}}.\ A)$ as initial arguments, respectively. Then, the occur-check of unification implements the restrictions on the instantiation of free variables, which are required for soundness. In some inference systems, however, Skolemization is unsound (e.g. for higher-order systems such as the one in [68] or the system in [106] for *descente infinie*) or inappropriate (e.g. in the matrix systems of [102]).

We prefer inference systems that include variable-conditions to inference systems that offer only Skolemization. Indeed, this inclusion provides a more general and often simpler approach, which never results in a necessary reduction in efficiency. Moreover, note that variable-conditions cannot add unnecessary complications here:

- If, in some application, variable-conditions are superfluous, then we can work with empty variable-conditions as if there would be no variable-conditions at all.

- We will need the variable-conditions anyway for our choice-conditions, which again are needed to formalize our approach to Hilbert's ε-operator.

4 Introduction to Hilbert's ε

4.1 Motivation

Hilbert's ε-symbol is an operator or binder that forms terms, just like Peano's ι-symbol. Roughly speaking, the term $\varepsilon x^\mathrm{B}.\, A$, formed from a bound atom (or "bound variable") x^B and a formula A, denotes *just some* object that is *chosen* such that — if possible — A (seen as a predicate on x^B) holds for this object.

For Ackermann, Bernays, and Hilbert, the ε was an intermediate tool in proof theory, to be eliminated in the end. Instead of giving a model-theoretic semantics for the ε, they just specified those axioms which were essential in their proof transformations. These axioms did not provide a complete definition, but left the ε underspecified.

Descriptive terms such as $\varepsilon x^\mathrm{B}.\, A$ and $\iota x^\mathrm{B}.\, A$ are of universal interest and applicability. Our more elegant and flexible treatment turns out to be useful in many areas where logic is designed or applied as a tool for description and reasoning.

4.2 Requirements Specification

For the usefulness of such descriptive terms we consider the following requirements to be the most important ones.

Requirement I (Indication of Commitment):
> The syntax must clearly express where exactly a *commitment* to a choice of a particular object is required, and where, to the contrary, different objects corresponding with the description may be chosen for different occurrences of the same descriptive term.

Requirement II (Reasoning):
> It must be possible to replace a descriptive term with a term that corresponds with its description. The correctness of such a replacement must be expressible and should be verifiable in the original calculus.

Requirement III (Semantics):
> The semantics should be simple, straightforward, natural, formal, and model-based. Overspecification should be carefully avoided. Furthermore, the semantics should be modular and abstract in the sense that it adds the operator to a variety of logics, independent of the details of a concrete logic.

Our more elegant and flexible, indefinite treatment of the ε-operator is compatible with Hilbert's original one and satisfies these requirements. As it involves novel semantic techniques, it may also serve as the paradigm for the design of similar operators.

4.3 Overview

In §A of the appendix, the reader can find an update of our review form [108; 111] of the literature on extended semantics given to Hilbert's ε-operator in the 2nd half of the 20th century. In the current §4, we will now introduce to the ι and the ε (§§ 4.4 and 4.5), to the ε's proof-theoretic origin (§ 4.6), and to our more general semantic objective (§ 4.7) with its emphasis on *indefinite* and *committed choice* (§ 4.8).

4.4 From the ι to the ε

As the ε-operator was developed as an improvement over the still very popular ι-operator, a careful discussion of the ι in this section is required for a deeper understanding of the ε.

4.4.1 The Symbols for the ι-Operator

The probably first descriptive ι-operator occurs in Frege's [23, Vol. I], written as a boldface backslash. As a *boldface* version of the backslash is not easily available in standard typesetting, we will use a simple backslash (\) in § 4.4.4.

A slightly different ι-operator occurs in Peano's [79], written as "$\bar{\iota}$", i.e. as an overlined ι. In its German translation [81], we also find an alternative symbol with the same denotation, namely an upside-down ι-symbol. Both symbols are meant to indicate the inverse of Peano's ι-function, which constructs the set of its single argument.

Nowadays, however, "$\{y\}$" is written for Peano's "ιy", and thus — as a simplifying convention to avoid problems in typesetting and automatic indexing — a simple ι should be used to designate the descriptive ι-operator, without overlining or inversion.

4.4.2 The Essential Idea of the ι-Operator

Let us define the quantifier of *unique existence* by
$$\exists! x^{\mathbb{B}}.\, A \quad := \quad \exists y^{\mathbb{B}}.\, \forall x^{\mathbb{B}}.\, ((y^{\mathbb{B}}{=}x^{\mathbb{B}}) \Leftrightarrow A),$$
for some fresh $y^{\mathbb{B}}$. All the slightly differing specifications of the ι-operator agree in the following point: If there is a unique $x^{\mathbb{B}}$ such that the formula A (seen as a predicate on $x^{\mathbb{B}}$) holds, then the ι-term $\iota x^{\mathbb{B}}.\, A$ denotes this unique object:
$$\exists! x^{\mathbb{B}}.\, A \quad \Rightarrow \quad A\{x^{\mathbb{B}} \mapsto \iota x^{\mathbb{B}}.\, A\} \qquad (\iota_0)$$
or in different notation $\quad (\exists! x^{\mathbb{B}}.\, (A(x^{\mathbb{B}}))) \quad \Rightarrow \quad A(\iota x^{\mathbb{B}}.\, (A(x^{\mathbb{B}})))$.

Example 4.1 (ι-operator)
For an informal introduction to the ι-operator, consider Father to be a predicate for which Father(Heinrich III, Heinrich IV) holds, i.e. "Heinrich III is father of Heinrich IV". Now, "*the* father of Heinrich IV" is designated by $\iota x^{\mathtt{B}}$. Father($x^{\mathtt{B}}$, Heinrich IV), and because this is nobody but Heinrich III, i.e. Heinrich III = $\iota x^{\mathtt{B}}$. Father($x^{\mathtt{B}}$, Heinrich IV), we know that Father($\iota x^{\mathtt{B}}$. Father($x^{\mathtt{B}}$, Heinrich IV), Heinrich IV). Similarly,

$$\text{Father}(\iota x^{\mathtt{B}}.\ \text{Father}(x^{\mathtt{B}}, \text{Adam}), \text{Adam}), \tag{4.1.1}$$

and thus $\exists y^{\mathtt{B}}$. Father($y^{\mathtt{B}}$, Adam), but, oops! Adam and Eve do not have any fathers. If you do not agree, you probably appreciate the following problem that occurs when somebody has God as an additional father.

$$\text{Father}(\text{Holy Ghost}, \text{Jesus}) \quad \wedge \quad \text{Father}(\text{Joseph}, \text{Jesus}). \tag{4.1.2}$$

Then the Holy Ghost is *the* father of Jesus and Joseph is *the* father of Jesus:

$$\text{Holy Ghost} = \iota x^{\mathtt{B}}.\ \text{Father}(x^{\mathtt{B}}, \text{Jesus}) \quad \wedge \quad \text{Joseph} = \iota x^{\mathtt{B}}.\ \text{Father}(x^{\mathtt{B}}, \text{Jesus}) \tag{4.1.3}$$

This implies something *the* Pope may not accept, namely Holy Ghost = Joseph, and he anathematized Heinrich IV in the year 1076:

$$\text{Anathematized}(\iota x^{\mathtt{B}}.\ \text{Pope}(x^{\mathtt{B}}), \text{Heinrich IV}, 1076). \tag{4.1.4}$$

4.4.3 Elementary Semantics Without Straightforward Overspecification

Semantics without a straightforward form of overspecification can be given to the ι-terms in the following three elementary ways:

Russell's non-referring ι-operator of 1905 in [88]:

In *Principia Mathematica* [103] by Bertrand Russell (1872–1970) and Alfred North Whitehead (1861–1947), an ι-term is given a meaning only in form of quantifications over contexts $C[\cdots]$ of the occurrences of the ι-term: $C[\iota x^{\mathtt{B}}.\ A]$ is defined as a short form for

$$\exists y^{\mathtt{B}}.\ \left(\forall x^{\mathtt{B}}.\ \left((y^{\mathtt{B}}{=}x^{\mathtt{B}}) \Leftrightarrow A\right) \wedge C[y^{\mathtt{B}}]\right).$$

This definition is peculiar because the *definiens* is not of the expected form $C[t]$ (for some term t), and because an ι-term on its own — i.e. without a context $C[\cdots]$ — cannot *directly refer* to an object that it may be intended to denote.

This was first presented in 1905 as a linguistic theory of descriptions in Russell's [88] — but without using any symbol for the ι.

Russell's *On Denoting* [88] became so popular that the term "non-referring" had to be introduced to make aware of the fact that Russell's ι-terms are not denoting (in spite of the title), and that Russell's theory of descriptions ignores the fundamental reference aspect of descriptive terms, cf. Strawson's *On Referring* [100].

Hilbert–Bernays' presuppositional ι-operator of 1934 in [57]:

To overcome the complex difficulties of Russell's non-referring semantics, in § 8 of the first volume of the two-volume monograph *Foundations of Mathematics* (*Grundlagen der Mathematik*, 1st edn. 1934 [57], 2nd edn. 1968 [59]) by David Hilbert (1862–1943) and Paul Bernays (1888–1977), a completed proof of $\exists! x^{\mathsf{B}}$. A is required to precede each formation of a term ιx^{B}. A, which otherwise is not considered a well-formed term at all.

This way of defining the ι is nowadays called *"presuppositional"*. This word occurs in relation to Hilbert–Bernays' ι in [94] of 2007 and in [96, §§ 1, 6, and 8f.] of 2009, but it does not occur in Russell's [100], and we do not know where it occurs first with this meaning.

Peano's partially specified ι-operator of 1896 in [79]:

Since Hilbert–Bernays' presuppositional treatment makes the ι quite impractical and the formal syntax of logic undecidable in general, in § 1 of the second volume of Hilbert–Bernays' *Foundations of Mathematics* (1st edn. 1939 [58], 2nd edn. 1970 [60]), Hilbert's ε, however, is already given a more flexible treatment: The simple idea is to leave the ε-terms uninterpreted. This will be described below. In this paper, we will present this more flexible treatment also for the ι.

After all, this treatment is the original one of Peano's ι, found already in 1896 in the article *Studii di Logica Matematica* [79].[4]

---

[4] In [79] of 1896f., Guiseppe Peano (1858–1932) wrote $\bar{\iota}$ instead of the ι of Example 4.1, and $\bar{\iota}\{\,x\mid A\,\}$ instead of $\iota x.\, A$. (Note that we have changed the class notation to modern standard here. Peano actually wrote $\overline{x\in}A$ instead of $\{\,x\mid A\,\}$ in [79].)
The bar above the ι (just as the alternative inversion the symbol) were to indicated that $\bar{\iota}$ was implicitly defined as the inverse operator of the operator ι defined by $\iota y := \{y\}$, which occurred already occurs in Peano's [78] of 1890 and still in Quine's [85] of 1981.
The definition of $\bar{\iota}$ reads literally [79, Definition 22]:
$$a \in K \ . \ \exists a : x, y \in a \ . \ \supset_{x,y} \ . \ x = y : \supset : x = \bar{\iota}a \ . = . \ a = \iota x$$
This straightforwardly translates into more modern notation as follows:
$$\text{For any class } a: \quad a \neq \emptyset \wedge \forall x, y. \ (x, y \in a \ \Rightarrow \ x = y) \ \Rightarrow \ \forall x. \ (x = \bar{\iota}a \ \Leftrightarrow \ a = \iota x)$$
Giving up the flavor of an explicit definition of $\ "x = \bar{\iota}a\ "$, this can be simplified to the following equivalent form: $\qquad \text{For any class } a: \quad \exists! x. \ x \in a \ \Rightarrow \ \bar{\iota}a \in a \qquad (\bar{\iota}_0)$
Besides notational difference, this is (ι_0) of our § 4.4.2.

It cannot surprise that it was Peano — interested in written languages for specification and communication, but hardly in calculi — who came up with the only practical specification of ι-terms (unlike Russell and Hilbert–Bernays).

Moreover, by the partiality of his specification, Peano avoided also the other pitfall, namely overspecification, and all its unintended consequences (unlike Frege and Quine, cf. § 4.4.4). As the symbol "ι" was invented by Peano as well (cf. § 4.4.1), we have good reason to speak of "Peano's ι", at least as much as we have reason to speak of "Hilbert's ε".

It must not be overlooked that Peano's ι — in spite of its partiality — always denotes: It is not a partial operator, it is just partially specified.

At least in non-modal classical logics, it is a well justified standard that *each term denotes*. More precisely — in each model or structure $\mathcal{S}$ under consideration — each occurrence of a proper term must denote an object in the universe of $\mathcal{S}$. Following that standard, to be able to write down $\iota x^{\mathbb{B}}. A$ without further consideration, we have to treat $\iota x^{\mathbb{B}}. A$ as an uninterpreted term about which we only know axiom (ι_0) from § 4.4.2.

With (ι_0) as the only axiom for the ι, the term $\iota x^{\mathbb{B}}. A$ has to satisfy A (seen as a predicate on $x^{\mathbb{B}}$) only if there exists a unique object such that A holds for it. The price, however, we have to pay for the avoidance of non-referringness, presuppositionality, and overspecification is that — roughly speaking — the term $\iota x^{\mathbb{B}}. A$ is of no use unless the unique existence $\exists! x^{\mathbb{B}}. A$ can be derived.

Finally, let us come back to Example 4.1 of § 4.4.2. The problems presented there do not actually appear if (ι_0) is the only axiom for the ι, because (4.1.1) and (4.1.3) are not valid. Indeed, the description of (4.1.1) lacks existence and the descriptions of (4.1.3) and (4.1.4) lack uniqueness.

4.4.4 Overspecified ι-Operators

From Frege to Quine, we find a multitude of ι-operators with definitions that overspecify the ι in different ways for the sake of *complete definedness* and *syntactic eliminability*. As we already stated in Requirement III (Semantics) of § 4.2, overspecification should be carefully avoided. Indeed, any overspecification leads to puzzling, arbitrary consequences, which may cause harm to the successful application of descriptive operators in practice.

Frege's arbitrarily overspecified ι-operator of 1893 in [23]:

The first occurrence of a descriptive ι-operator in the literature seems to be in 1893, namely in § 11 of the first volume of the two-volume monograph *Grundgesetze der Arithmetik — Begriffsschriftlich abgeleitet* [23] by Gottlob Frege (1848–1925):

For A seen as a function from objects to truth values, $\backslash A$ (in our notation ιA or $\iota x^{\mathbb{B}}.\,A$) is defined to be the object Δ if A is extensionally equal to the function that checks for equality to Δ, i.e. if $A = \lambda x^{\mathbb{B}}.\,(\Delta = x^{\mathbb{B}})$.

In the case that there is no such Δ, Frege overspecified his ι-operator pretty arbitrarily by defining $\backslash A$ to be A, which is not even an object, but a function. (Note that Frege actually wrote an ε (having nothing to do with the ε-operator) instead of our $x^{\mathbb{B}}$, and a *spiritus lenis* over it instead of a modern λ-operator before and a dot after it. Moreover, he wrote a ξ for the A.)

Set theory's overspecified ι-operator:

In set theories without urelements of 1981, such as in Quine's [85] of 1981, the ι-operator can be defined by something like
$$\iota x^{\mathbb{B}}.\,A \quad := \quad \{\ z^{\mathbb{B}}\ |\ \exists y^{\mathbb{B}}.\ (\ \forall x^{\mathbb{B}}.\ ((y^{\mathbb{B}}{=}x^{\mathbb{B}}) \Leftrightarrow A)\ \wedge\ z^{\mathbb{B}} \in y^{\mathbb{B}}\ )\ \},$$
for fresh $y^{\mathbb{B}}$ and $z^{\mathbb{B}}$.

This is again an overspecification resulting in $\iota x^{\mathbb{B}}.\,A = \emptyset$ if there is no such $y^{\mathbb{B}}$ (which otherwise is always unique).

4.4.5 A Completely Defined, but Not Overspecified ι-Operator

The complete definitions of the ι in § 4.4.4 take place in *possibly inconsistent* logical frameworks, namely Frege's Begriffsschrift and Quine's set theory.

That neither overspecification nor possible inconsistency are necessary for complete definitions of the ι is witnessed by the following complete, but non-elementary definition of the ι, which is also referring and non-presuppositional.

The ε-calculus' ι-operator:

In the ε-calculus, which is a conservative extension of first-order predicate calculus, first elaborated in 1939 in the second volume of Hilbert–Bernays' *Foundations of Mathematics* [58], we can define the ι simply by
$$\iota x^{\mathbb{B}}.\,A \quad := \quad \varepsilon y^{\mathbb{B}}.\ \forall x^{\mathbb{B}}.\ ((y^{\mathbb{B}} = x^{\mathbb{B}}) \Leftrightarrow A)$$
(for a fresh $y^{\mathbb{B}}$), i.e. as the unique $x^{\mathbb{B}}$ such that A holds (provided there is such an $x^{\mathbb{B}}$).

This definition is non-elementary, however, because it introduces ε-terms, which cannot be eliminated in first-order logic in general.

Note that this definition is — to the best of our knowledge — the most useful and elegant way to introduce the ι, although it is somehow *ex eventu*, because the development of the ε was started two dozen years after the first publications on Frege's and Peano's ι-operators.

4.5 The ε as an Improvement over the ι

Compared to the ι, the ε is more useful because — instead of (ι_0) — it comes with the stronger axiom

$$\exists x^{\mathrm{B}}.\, A \;\Rightarrow\; A\{x^{\mathrm{B}} \mapsto \varepsilon x^{\mathrm{B}}.\, A\} \tag{ε_0}$$

More specifically, as the formula $\exists x^{\mathrm{B}}.\, A$ (which has to be true to guarantee an interpretation of the ε-term $\varepsilon x^{\mathrm{B}}.\, A$ that is meaningful in the sense that it satisfies its formula A) is weaker than the corresponding formula $\exists! x^{\mathrm{B}}.\, A$ (for the respective ι-term), the area of useful application is wider for the ε- than for the ι-operator. Indeed, we have already seen in §4.4.5 that the ι can be defined in terms of the ε, but not vice versa.

Moreover, in case of $\exists! x^{\mathrm{B}}.\, A$, the ε-operator picks the same element as the ι-operator:

$$\exists! x^{\mathrm{B}}.\, A \;\Rightarrow\; (\;\varepsilon x^{\mathrm{B}}.\, A \;=\; \iota x^{\mathrm{B}}.\, A\;).$$

Thus, unless eliminability is relevant, we should replace all useful occurrences of the ι with the ε: As a consequence, among other advantages, the arising proof obligations become weaker and both human and automated generation and generalization of proofs become more efficient.

4.6 On the ε's Proof-Theoretic Origin

4.6.1 The ε-Formula and the Historical Sources of the ε

The main historical source on the ε is the second volume of the *Foundations of Mathematics* [57; 58; 59; 60], the fundamental work which summarizes the foundational and proof-theoretic contributions of David Hilbert and his mathematical-logic group.

The preferred specification for Hilbert's ε in proof-theoretic investigations is not the axiom (ε_0), but actually the following formula:

$$A\{x^{\mathrm{B}} \mapsto x^{\mathrm{A}}\} \;\Rightarrow\; A\{x^{\mathrm{B}} \mapsto \varepsilon x^{\mathrm{B}}.\, A\} \tag{ε-formula}$$

The ε-formula is equivalent to (ε_0), but it gets along without any quantifier.

The name "ε-formula" originates in [58, p. 13], where the ε-operator is simply called "Hilbert's ε-symbol".

For historical correctness, note that the notation in the original is closer to

$$A(x^{\mathrm{A}}) \Rightarrow A(\varepsilon x^{\mathrm{B}}.\, A(x^{\mathrm{B}})),$$

where the A is a concrete singular predicate atom (called "formula variable" in the original) and comes with several extra rules for its instantiation, cf. [58, p. 13f.].

The exact notation actually is
$$A(a) \Rightarrow A(\varepsilon_x\, A(x)),$$
and the deductive equivalence is straightforward to the exact notation of (ε_0), i.e. to
$$(Ex)\, A(x) \Rightarrow A(\varepsilon_x\, A(x)),$$
cf. [58, pp. 13–15].

In our notation, however, (ε_0) and the ε-formula are axiom *schemata* where the A is a meta-variable for a formula (which, contrary to the predicate atom, may contain occurrences of x^A). Nevertheless, their deductive equivalence is given for versions of (ε_0) and the ε-formula where the A is replaced with $A\{x^\mathrm{A}\!\mapsto\! y^\mathrm{A}\}$ for some fresh (free) atom y^A, from which both (ε_0) and the ε-formula can be obtained by instantiation.

The ε-formula already occurs, however under different names, in the pioneering papers on the ε, i.e. in Ackermann's [1] of 1925 as "transfinite axiom 1", in Hilbert's [47] of 1926 as "axiom of choice" (in the operator form $A(a) \Rightarrow A(\varepsilon A)$, where the ε is called "transfinite logical choice function"), and in Hilbert's [48] of 1928 as "logical ε-axiom" (again in operator form, where the ε is called "logical ε-function").

4.6.2 The Original Explanation of the ε

As the basic methodology of Hilbert's program is to treat all symbols as meaningless, no semantics is required besides the one given by the single axiom (ε_0). To further the understanding, however, we read on p.12 of [58; 60]:

> $\varepsilon x^\mathrm{B}.\ A\ \ldots$ "ist ein Ding des Individuenbereichs, und zwar ist dieses Ding gemäß der inhaltlichen Übersetzung der Formel (ε_0) *ein solches, auf das jenes Prädikat A zutrifft, vorausgesetzt, daß es überhaupt auf ein Ding des Individuenbereichs zutrifft.*"

> $\varepsilon x^\mathrm{B}.\ A\ \ldots$ "is a thing of the domain of individuals for which — according to the contentual translation of the formula (ε_0) — *the predicate A holds, provided that A holds for any thing of the domain of individuals at all.*"

Example 4.2 (ε instead of ι) *(continuing Example 4.1 of § 4.4.2)*
Just as for the ι, for the ε we have $\mathsf{Heinrich\,III} = \varepsilon x^\mathrm{B}.\ \mathsf{Father}(x^\mathrm{B}, \mathsf{Heinrich\,IV})$ and
$$\mathsf{Father}(\varepsilon x^\mathrm{B}.\ \mathsf{Father}(x^\mathrm{B}, \mathsf{Heinrich\,IV}), \mathsf{Heinrich\,IV}).$$
But, from the contrapositive of (ε_0) and $\neg\mathsf{Father}(\varepsilon x^\mathrm{B}.\ \mathsf{Father}(x^\mathrm{B}, \mathsf{Adam}), \mathsf{Adam})$, we now conclude that $\neg\exists y^\mathrm{B}.\ \mathsf{Father}(y^\mathrm{B}, \mathsf{Adam})$.

4.6.3 Defining the Quantifiers via the ε

Hilbert and Bernays did not need any semantics or precise intention for the ε-symbol because it was introduced merely as a formal syntactic device to facilitate proof-theoretic investigations, motivated by the possibility to get rid of the existential and universal quantifiers via two direct consequences of axiom (ε_0):

$$\exists x^{\mathsf{B}}.\, A \quad \Leftrightarrow \quad A\{x^{\mathsf{B}} \mapsto \varepsilon x^{\mathsf{B}}.\, A\} \qquad\qquad (\varepsilon_1)$$

$$\forall x^{\mathsf{B}}.\, A \quad \Leftrightarrow \quad A\{x^{\mathsf{B}} \mapsto \varepsilon x^{\mathsf{B}}.\, \neg A\} \qquad\qquad (\varepsilon_2)$$

These equivalences can be seen as definitions of the quantifiers because innermost rewriting with (ε_1), (ε_2) yields a normal form after as many steps as there are quantifiers in the input formula. Moreover, also arbitrary rewriting is confluent and terminating, cf. [113].

It should be noted, however, that rewriting with (ε_1), (ε_2) must not be taken for granted *under* modal operators, at least not under the assumption that ε-terms are to remain *rigid,* i.e. independent in their interpretation from their modal contexts. For this assumption there are very good reasons, nicely explained e.g. in [94; 96].

Example 4.3   Consider the first-order modal logic formula
$$\Box \exists x^{\mathsf{B}}.\, A.$$
Moreover, to simplify matters, let us assume that we have constant domains, i.e. that all modal contexts have the same domain of individuals.

Under this condition and for a formula of this structure, it is suggested in [94, p.153] to apply (ε_1) to the considered formula, resulting in
$$\Box A\{x^{\mathsf{B}} \mapsto \varepsilon x^{\mathsf{B}}.\, A\},$$
from which we can doubtlessly conclude
$$\exists x^{\mathsf{B}}.\, \Box A,$$
e.g. by Formula (a) in § 4.6.4.

Let us interpret the $\Box$ as "believes" and A as "x^{B} is the number of rice corns in my car", and let our constant domain be the one of the standard model of the natural numbers. Note that I do not believe of any concrete and definite number that it numbers the rice corns in my car just because I believe that their number is finite.

This interpretation shows that our rewriting with (ε_1) *under* the operator $\Box$ is incorrect for modal logic in general, at least for rigid ε-terms.

On the other hand, rewriting with (ε_1), (ε_2) *above* modal operators is uncritical: $\exists x^{\mathsf{B}}.\, \Box A$ is indeed equivalent to $\Box A\{x^{\mathsf{B}} \mapsto \varepsilon x^{\mathsf{B}}.\, \Box A\}$.

4.6.4 The ε-Theorems

When we remove all quantifiers in a derivation of the Hilbert-style predicate calculus of the *Foundations of Mathematics* along (ε_1) and (ε_2), the following transformations occur:

- Tautologies are turned into tautologies.

- The axioms

$$A\{x^{\mathbb{B}} \mapsto x^{\mathbb{A}}\} \quad \Rightarrow \quad \exists x^{\mathbb{B}}.\ A \qquad\qquad (\textit{Formula (a)})$$

  and

$$\forall x^{\mathbb{B}}.\ A \quad \Rightarrow \quad A\{x^{\mathbb{B}} \mapsto x^{\mathbb{A}}\} \qquad\qquad (\textit{Formula (b)})$$

  (cf. p. 100f. of [57] or p. 99f. of [59; 62]), are turned into the ε-formula (cf. § 4.6.1) and, roughly speaking, its contrapositive, respectively.

  Indeed, for the case of Formula (b), we can replace first all A with $\neg A$, and after applying (ε_2), replace $\neg\neg A$ with A, and thus obtain the contrapositive of the ε-formula.

- The inference steps are turned into inference steps: the inference schema [of modus ponens] into the inference schema; the substitution rule for free atoms as well as quantifier introduction (Schemata (α) and (β) on p. 103f. of [57] or on p. 102f. of [59; 62]) into the substitution rule including ε-terms.

- Finally, the ε-formula is taken as a new axiom scheme instead of (ε_0) because it has the advantage of being free of quantifiers.

The argumentation of the previous paragraphs is actually part of the proof transformation that constructively proves the first of Hilbert–Bernays' two theorems on ε-elimination in first-order logic, the so-called *1st ε-Theorem*. In its sharpened form, this theorem can be stated as follows. Note that the original speaks of "bound variables" instead of "bound atoms" and of "formula variables" instead of "predicate atoms", because what we call (free) "variables" is not part of the formula languages of Hilbert–Bernays.

Theorem 4.4 (Sharpened 1ˢᵗ ε-Theorem, [58, p.79f.])
From a derivation of $\exists x_1^{\mathrm{B}}. \ldots \exists x_r^{\mathrm{B}}. A$ (containing no bound atoms besides the ones bound by the prefix $\exists x_1^{\mathrm{B}}. \ldots \exists x_r^{\mathrm{B}}.$) from the formulas $P_1, \ldots, P_k$ (containing neither predicate atoms nor bound atoms) in the predicate calculus (incl. the ε-formula and $=$-substitutability as axiom schemes, plus $=$-reflexivity), we can construct a (finite) disjunction of the form $\bigvee_{i=0}^{s} A\{x_1^{\mathrm{B}}, \ldots, x_r^{\mathrm{B}} \mapsto t_{i,1}, \ldots, t_{i,r}\}$ and a derivation of it

- *in which bound atoms do not occur at all*

- *from $P_1, \ldots, P_k$ and $=$-axioms (containing neither predicate atoms nor bound atoms)*

- *in the quantifier-free predicate calculus (i.e. tautologies plus the inference schema [of modus ponens] and the substitution rule).*

Note that r, s range over natural numbers including 0, and that A, $t_{i,j}$, and P_i are ε-free because otherwise they would have to include (additional) bound atoms.

Moreover, the *2ⁿᵈ ε-Theorem* (in [58; 60]) states that the ε (just as the ι, cf. [57; 59]) is a conservative extension of the predicate calculus in the sense that each formal proof of an ε-free formula can be transformed into a formal proof that does not use the ε at all.

For logics different from classical axiomatic first-order predicate logic, however, it is not a conservative extension when we add the ε either with (ε_0), with (ε_1), or with the ε-formula to other first-order logics — may they be weaker such as *intuitionist* first-order logic, or stronger such as first-order set theories with axiom schemes over arbitrary terms *including the ε*; cf. [108, § 3.1.3]. Moreover, even in classical first-order logic there is no translation from the formulas containing the ε to formulas not containing it.

4.7   Our Objective

While the historiographical and technical research on the ε-theorems is still going on and the methods of ε-elimination and ε-substitution did not die with Hilbert's program, this is not our subject here. We are less interested in Hilbert's formal program and the consistency of mathematics than in the powerful use of logic in creative processes. And, instead of the tedious syntactic proof transformations, which easily lose their usefulness and elegance within their technical complexity and which — more importantly — can only refer to an already existing logic, we look for *model-theoretic* means for finding new logics and new applications. And the question that still has to be answered in this field is:

What would be a proper semantics for Hilbert's ε?

4.8 Indefinite and Committed Choice

Just as the ι-symbol is usually taken to be the referential interpretation of the *definite* articles in natural languages, it is our opinion that the ε-symbol should be that of the *indefinite* determiners (articles and pronouns) such as "a(n)" or "some".

Example 4.5 (ε instead of ι again)   *(continuing Example 4.1)*
It may well be the case that
$$\text{Holy Ghost} \; = \; \varepsilon x^{\text{B}}. \; \text{Father}(x^{\text{B}}, \text{Jesus}) \quad \wedge \quad \text{Joseph} \; = \; \varepsilon x^{\text{B}}. \; \text{Father}(x^{\text{B}}, \text{Jesus})$$
i.e. that "The Holy Ghost is $\underline{a}$ father of Jesus and Joseph is $\underline{a}$ father of Jesus."
But this does not bring us into trouble with the Pope because we do not know whether all fathers of Jesus are equal. This will become clearer when we reconsider this in Example 4.14.

Closely connected to indefinite choice (also called "indeterminism" or "don't care nondeterminism") is the notion of *committed choice*. For example, when we have a new telephone, we typically *don't care* which number we get, but once a number has been chosen for our telephone, we will insist on a *commitment to this choice*, so that our phone number is not changed between two incoming calls.

Example 4.6 (Committed choice)
Suppose we want to prove $\exists x^{\text{B}}. \; (x^{\text{B}} \neq x^{\text{B}})$
According to (ε_1) from § 4.6 this reduces to $\varepsilon x^{\text{B}}. \; (x^{\text{B}} \neq x^{\text{B}}) \neq \varepsilon x^{\text{B}}. \; (x^{\text{B}} \neq x^{\text{B}})$
Since there is no solution to $x^{\text{B}} \neq x^{\text{B}}$ we can replace
$\varepsilon x^{\text{B}}. \; (x^{\text{B}} \neq x^{\text{B}})$ with anything. Thus, the above reduces to $0 \neq \varepsilon x^{\text{B}}. \; (x^{\text{B}} \neq x^{\text{B}})$
and then, by exactly the same argumentation, to $0 \neq 1$
which is true in the natural numbers.
Thus, we have proved our original formula $\exists x^{\text{B}}. \; (x^{\text{B}} \neq x^{\text{B}})$, which, however, is false. What went wrong? Of course, we have to commit to our choice for all occurrences of the ε-term introduced when eliminating the existential quantifier: If we choose 0 on the left-hand side, we have to commit to the choice of 0 on the right-hand side as well.

4.9 Quantifier Elimination and Subordinate ε-terms

Before we can introduce to our treatment of the ε, we also have to get more acquainted with the ε in general.

The elimination of $\forall$- and $\exists$-quantifiers with the help of ε-terms (cf. § 4.6) may be more difficult than expected when some ε-terms become "subordinate" to others.

Definition 4.7 (Subordinate) An ε-term $\varepsilon v^{\text{B}}.\, B$ (or, more generally, a binder on v^{B} together with its scope B) is *superordinate* to an (occurrence of an) ε-term $\varepsilon x^{\text{B}}.\, A$ if

1. $\varepsilon x^{\text{B}}.\, A$ is a subterm of B and

2. an occurrence of the bound atom v^{B} in $\varepsilon x^{\text{B}}.\, A$ is free in B
   (i.e. the binder on v^{B} binds an occurrence of v^{B} in $\varepsilon x^{\text{B}}.\, A$).

An (occurrence of an) ε-term $\varepsilon x^{\text{B}}.\, A$ is *subordinate* to an ε-term $\varepsilon v^{\text{B}}.\, B$ (or, more generally, to a binder on v^{B} together with its scope B) if $\varepsilon v^{\text{B}}.\, B$ is superordinate to $\varepsilon x^{\text{B}}.\, A$.

On p. 24 of [58; 60], these subordinate ε-terms, which are responsible for the difficulty to prove the ε-theorems constructively, are called "untergeordnete ε-Ausdrücke". Note that — contrary to Hilbert–Bernays — we do not use a special name for ε-terms with free occurrences of bound atoms here — such as "ε-Ausdrücke" ("ε-expressions" or "quasi ε-terms") instead of "ε-Terme" ("ε-terms") — but simply call them "ε-terms" as well.

Example 4.8 (Quantifier Elimination and Subordinate ε-Terms)
Let us repeat the formulas (ε_1) and (ε_2) from § 4.6 here:
$$\exists x^{\text{B}}.\, A \quad\Leftrightarrow\quad A\{x^{\text{B}} \mapsto \varepsilon x^{\text{B}}.\, A\} \tag{ε_1}$$
$$\forall x^{\text{B}}.\, A \quad\Leftrightarrow\quad A\{x^{\text{B}} \mapsto \varepsilon x^{\text{B}}.\, \neg A\} \tag{ε_2}$$
Let us consider the formula
$$\exists w^{\text{B}}.\, \forall x^{\text{B}}.\, \exists y^{\text{B}}.\, \forall z^{\text{B}}.\, \mathsf{P}(w^{\text{B}}, x^{\text{B}}, y^{\text{B}}, z^{\text{B}})$$
and apply (ε_1) and (ε_2) to remove the four quantifiers completely.

We introduce the following abbreviations, where w^{B}, x^{B}, y^{B}, w_a^{B}, x_a^{B}, y_a^{B}, z_a^{B} are bound atoms and w_a, x_a, y_a, z_a are meta-level symbols for functions from terms to terms:

$$
\begin{aligned}
z_a(w^{\text{B}})(x^{\text{B}})(y^{\text{B}}) &= \varepsilon z_a^{\text{B}}.\, \neg\mathsf{P}(w^{\text{B}}, x^{\text{B}}, y^{\text{B}}, z_a^{\text{B}})\\
y_a(w^{\text{B}})(x^{\text{B}}) &= \varepsilon y_a^{\text{B}}.\, \mathsf{P}(w^{\text{B}}, x^{\text{B}}, y_a^{\text{B}}, z_a(w^{\text{B}})(x^{\text{B}})(y_a^{\text{B}}))\\
x_a(w^{\text{B}}) &= \varepsilon x_a^{\text{B}}.\, \neg\mathsf{P}(w^{\text{B}}, x_a^{\text{B}}, y_a(w^{\text{B}})(x_a^{\text{B}}), z_a(w^{\text{B}})(x_a^{\text{B}})(y_a(w^{\text{B}})(x_a^{\text{B}})))\\
w_a &= \varepsilon w_a^{\text{B}}.\, \mathsf{P}(w_a^{\text{B}}, x_a(w_a^{\text{B}}), y_a(w_a^{\text{B}})(x_a(w_a^{\text{B}})), z_a(w_a^{\text{B}})(x_a(w_a^{\text{B}}))(y_a(w_a^{\text{B}})(x_a(w_a^{\text{B}}))))
\end{aligned}
$$

Innermost rewriting with (ε_1) and (ε_2) results in a unique normal form after at most as many steps as there are quantifiers. Thus, we eliminate inside-out, i.e. we start with the elimination of $\forall z^{\text{B}}$. The transformation is:

$$
\begin{aligned}
&\exists w^{\text{B}}.\, \forall x^{\text{B}}.\, \exists y^{\text{B}}.\, \forall z^{\text{B}}.\, \mathsf{P}(w^{\text{B}}, x^{\text{B}}, y^{\text{B}}, z^{\text{B}}),\\
&\exists w^{\text{B}}.\, \forall x^{\text{B}}.\, \exists y^{\text{B}}.\quad \mathsf{P}(w^{\text{B}}, x^{\text{B}}, y^{\text{B}}, z_a(w^{\text{B}})(x^{\text{B}})(y^{\text{B}})),\\
&\exists w^{\text{B}}.\, \forall x^{\text{B}}.\quad \mathsf{P}(w^{\text{B}}, x^{\text{B}}, y_a(w^{\text{B}})(x^{\text{B}}), z_a(w^{\text{B}})(x^{\text{B}})(y_a(w^{\text{B}})(x^{\text{B}}))),\\
&\exists w^{\text{B}}.\quad \mathsf{P}(w^{\text{B}}, x_a(w^{\text{B}}), y_a(w^{\text{B}})(x_a(w^{\text{B}})), z_a(w^{\text{B}})(x_a(w^{\text{B}}))(y_a(w^{\text{B}})(x_a(w^{\text{B}})))),\\
&\quad \mathsf{P}(w_a, x_a(w_a), y_a(w_a)(x_a(w_a)), z_a(w_a)(x_a(w_a))(y_a(w_a)(x_a(w_a)))).
\end{aligned}
$$

Note that the resulting formula is quite deep and has more than one thousand occurrences of the ε-binder. Indeed, in general, n nested quantifiers result in an ε-nesting depth of $2^n - 1$.

To understand this, let us have a closer look a the resulting formula. Let us write it as

$$\mathsf{P}(w_a, x_b, y_d, z_h) \tag{4.8.1}$$

then (after renaming some bound atoms) we have

$$
\begin{aligned}
z_h &= \varepsilon z_h^{\mathrm{B}}.\ \neg\mathsf{P}(w_a, x_b, y_d, z_h^{\mathrm{B}}), &(4.8.2)\\
y_d &= \varepsilon y_d^{\mathrm{B}}.\ \mathsf{P}(w_a, x_b, y_d^{\mathrm{B}}, z_g(y_d^{\mathrm{B}})) &(4.8.3)\\
&\quad \text{with } z_g(y_d^{\mathrm{B}}) = \varepsilon z_g^{\mathrm{B}}.\ \neg\mathsf{P}(w_a, x_b, y_d^{\mathrm{B}}, z_g^{\mathrm{B}}), &(4.8.4)\\
x_b &= \varepsilon x_b^{\mathrm{B}}.\ \neg\mathsf{P}(w_a, x_b^{\mathrm{B}}, y_c(x_b^{\mathrm{B}}), z_f(x_b^{\mathrm{B}})) &(4.8.5)\\
&\quad \text{with } z_f(x_b^{\mathrm{B}}) = \varepsilon z_f^{\mathrm{B}}.\ \neg\mathsf{P}(w_a, x_b^{\mathrm{B}}, y_c(x_b^{\mathrm{B}}), z_f^{\mathrm{B}}) &(4.8.6)\\
&\quad \text{and } y_c(x_b^{\mathrm{B}}) = \varepsilon y_c^{\mathrm{B}}.\ \mathsf{P}(w_a, x_b^{\mathrm{B}}, y_c^{\mathrm{B}}, z_e(x_b^{\mathrm{B}})(y_c^{\mathrm{B}})) &(4.8.7)\\
&\qquad \text{with } z_e(x_b^{\mathrm{B}})(y_c^{\mathrm{B}}) = \varepsilon z_e^{\mathrm{B}}.\ \neg\mathsf{P}(w_a, x_b^{\mathrm{B}}, y_c^{\mathrm{B}}, z_e^{\mathrm{B}}), &(4.8.8)\\
w_a &= \varepsilon w_a^{\mathrm{B}}.\ \mathsf{P}(w_a^{\mathrm{B}}, x_a(w_a^{\mathrm{B}}), y_b(w_a^{\mathrm{B}}), z_d(w_a^{\mathrm{B}})) &(4.8.9)\\
&\quad \text{with } z_d(w_a^{\mathrm{B}}) = \varepsilon z_d^{\mathrm{B}}.\ \neg\mathsf{P}(w_a^{\mathrm{B}}, x_a(w_a^{\mathrm{B}}), y_b(w_a^{\mathrm{B}}), z_d^{\mathrm{B}}) &(4.8.10)\\
&\quad \text{and } y_b(w_a^{\mathrm{B}}) = \varepsilon y_b^{\mathrm{B}}.\ \mathsf{P}(w_a^{\mathrm{B}}, x_a(w_a^{\mathrm{B}}), y_b^{\mathrm{B}}, z_c(w_a^{\mathrm{B}})(y_b^{\mathrm{B}})) &(4.8.11)\\
&\qquad \text{with } z_c(w_a^{\mathrm{B}})(y_b^{\mathrm{B}}) = \varepsilon z_c^{\mathrm{B}}.\ \neg\mathsf{P}(w_a^{\mathrm{B}}, x_a(w_a^{\mathrm{B}}), y_b^{\mathrm{B}}, z_c^{\mathrm{B}}), &(4.8.12)\\
&\qquad x_a(w_a^{\mathrm{B}}) = \varepsilon x_a^{\mathrm{B}}.\ \neg\mathsf{P}(w_a^{\mathrm{B}}, x_a^{\mathrm{B}}, y_a(w_a^{\mathrm{B}})(x_a^{\mathrm{B}}), z_b(w_a^{\mathrm{B}})(x_a^{\mathrm{B}})) &(4.8.13)\\
&\qquad\quad \text{with } z_b(w_a^{\mathrm{B}})(x_a^{\mathrm{B}}) = \varepsilon z_b^{\mathrm{B}}.\ \neg\mathsf{P}(w_a^{\mathrm{B}}, x_a^{\mathrm{B}}, y_a(w_a^{\mathrm{B}})(x_a^{\mathrm{B}}), z_b^{\mathrm{B}}) &(4.8.14)\\
&\qquad\quad \text{and } y_a(w_a^{\mathrm{B}})(x_a^{\mathrm{B}}) = \varepsilon y_a^{\mathrm{B}}.\ \mathsf{P}(w_a^{\mathrm{B}}, x_a^{\mathrm{B}}, y_a^{\mathrm{B}}, z_a(w_a^{\mathrm{B}})(x_a^{\mathrm{B}})(y_a^{\mathrm{B}})) &(4.8.15)\\
&\qquad\qquad \text{with } z_a(w_a^{\mathrm{B}})(x_a^{\mathrm{B}})(y_a^{\mathrm{B}}) = &(4.8.16)\\
&\qquad\qquad\qquad \varepsilon z_a^{\mathrm{B}}.\ \neg\mathsf{P}(w_a^{\mathrm{B}}, x_a^{\mathrm{B}}, y_a^{\mathrm{B}}, z_a^{\mathrm{B}}).
\end{aligned}
$$

First of all, note that the bound atoms with free occurrences in the indented ε-terms (i.e., in the order of their appearance, the bound atoms y_d^{B}, x_b^{B}, y_c^{B}, w_a^{B}, y_b^{B}, x_a^{B}, y_a^{B}) are actually bound by the next ε to the left, to which the respective ε-terms thus become subordinate. For example, the ε-term $z_g(y_d^{\mathrm{B}})$ is subordinate to the ε-term y_d binding y_d^{B}.

Moreover, the ε-terms defined by the above equations are exactly those that require a commitment to their choice. This means that each of z_a, z_b, z_c, z_d, z_e, z_f, z_g, z_h, each of y_a, y_b, y_c, y_d, and each of x_a, x_b may be chosen differently without affecting soundness of the equivalence transformation. Note that the variables are strictly nested into each other; so we must choose in the order of

$$z_a,\ y_a,\ z_b,\ x_a,\ z_c,\ y_b,\ z_d,\ w_a,\ z_e,\ y_c,\ z_f,\ x_b,\ z_g,\ y_d,\ z_h.$$

Furthermore, in case of all ε-terms except w_a, x_b, y_d, z_h, we actually have to choose a function instead of a simple object.

In Hilbert–Bernays' view, however, there are neither functions nor objects at all, but only terms (and expressions with free occurrences of bound atoms):

In the standard notation the term $x_a(w_a^{\mathrm{B}})$ reads

$$\varepsilon x_a^{\scriptscriptstyle B}.\ \neg\mathsf{P}\left(\begin{array}{l} w_a^{\scriptscriptstyle B}, \\ x_a^{\scriptscriptstyle B}, \\ \varepsilon y_a^{\scriptscriptstyle B}.\ \mathsf{P}\big(\ w_a^{\scriptscriptstyle B},\quad x_a^{\scriptscriptstyle B},\quad y_a^{\scriptscriptstyle B},\quad \varepsilon z_a^{\scriptscriptstyle B}.\ \neg\mathsf{P}(w_a^{\scriptscriptstyle B},x_a^{\scriptscriptstyle B},y_a^{\scriptscriptstyle B},z_a^{\scriptscriptstyle B})\ \big), \\ \varepsilon z_b^{\scriptscriptstyle B}.\ \neg\mathsf{P}\big(\ w_a^{\scriptscriptstyle B},\quad x_a^{\scriptscriptstyle B},\quad \varepsilon y_a^{\scriptscriptstyle B}.\ \mathsf{P}\big(\ w_a^{\scriptscriptstyle B},\quad x_a^{\scriptscriptstyle B},\quad y_a^{\scriptscriptstyle B},\quad \varepsilon z_a^{\scriptscriptstyle B}.\ \neg\mathsf{P}(w_a^{\scriptscriptstyle B},x_a^{\scriptscriptstyle B},y_a^{\scriptscriptstyle B},z_a^{\scriptscriptstyle B})\ \big)\ \big),\quad z_b^{\scriptscriptstyle B}\big) \end{array}\right).$$

Moreover, $y_b(w_a^{\scriptscriptstyle B})$ reads

$$\varepsilon y_b^{\scriptscriptstyle B}.\ \neg\mathsf{P}\left(\begin{array}{l} w_a^{\scriptscriptstyle B}, \\ \varepsilon x_a^{\scriptscriptstyle B}.\ \neg\mathsf{P}\left(\begin{array}{l} w_a^{\scriptscriptstyle B}, \\ x_a^{\scriptscriptstyle B}, \\ \varepsilon y_a^{\scriptscriptstyle B}.\ \mathsf{P}\big(\ w_a^{\scriptscriptstyle B},\ x_a^{\scriptscriptstyle B},\ y_a^{\scriptscriptstyle B},\ \varepsilon z_a^{\scriptscriptstyle B}.\ \neg\mathsf{P}(w_a^{\scriptscriptstyle B},x_a^{\scriptscriptstyle B},y_a^{\scriptscriptstyle B},z_a^{\scriptscriptstyle B})\ \big), \\ \varepsilon z_b^{\scriptscriptstyle B}.\ \neg\mathsf{P}\big(\ w_a^{\scriptscriptstyle B},\ x_a^{\scriptscriptstyle B},\ \varepsilon y_a^{\scriptscriptstyle B}.\ \mathsf{P}\big(\ w_a^{\scriptscriptstyle B},\ x_a^{\scriptscriptstyle B},\ y_a^{\scriptscriptstyle B},\ \varepsilon z_a^{\scriptscriptstyle B}.\ \neg\mathsf{P}(w_a^{\scriptscriptstyle B},x_a^{\scriptscriptstyle B},y_a^{\scriptscriptstyle B},z_a^{\scriptscriptstyle B})\ \big)\ \big),\ z_b^{\scriptscriptstyle B}\big) \end{array}\right), \\ y_b^{\scriptscriptstyle B}, \\ \varepsilon z_c^{\scriptscriptstyle B}.\ \neg\mathsf{P}\left(\begin{array}{l} \varepsilon x_a^{\scriptscriptstyle B}.\ \neg\mathsf{P}\left(\begin{array}{l} w_a^{\scriptscriptstyle B}, \\ x_a^{\scriptscriptstyle B}, \\ \varepsilon y_a^{\scriptscriptstyle B}.\ \mathsf{P}\big(\ w_a^{\scriptscriptstyle B},\ x_a^{\scriptscriptstyle B},\ y_a^{\scriptscriptstyle B},\ \varepsilon z_a^{\scriptscriptstyle B}.\ \neg\mathsf{P}(w_a^{\scriptscriptstyle B},x_a^{\scriptscriptstyle B},y_a^{\scriptscriptstyle B},z_a^{\scriptscriptstyle B})\ \big), \\ \varepsilon z_b^{\scriptscriptstyle B}.\ \neg\mathsf{P}\big(\ w_a^{\scriptscriptstyle B},\ x_a^{\scriptscriptstyle B},\ \varepsilon y_a^{\scriptscriptstyle B}.\ \mathsf{P}\big(\ w_a^{\scriptscriptstyle B},\ x_a^{\scriptscriptstyle B},\ y_a^{\scriptscriptstyle B},\ \varepsilon z_a^{\scriptscriptstyle B}.\ \neg\mathsf{P}(w_a^{\scriptscriptstyle B},x_a^{\scriptscriptstyle B},y_a^{\scriptscriptstyle B},z_a^{\scriptscriptstyle B})\ \big)\ \big),\ z_b^{\scriptscriptstyle B}\big) \end{array}\right), \\ y_b^{\scriptscriptstyle B}, \\ z_c^{\scriptscriptstyle B} \end{array}\right).$$

Condensed data on the above terms read as follows:

	ε-nesting depth	number of ε-binders	Ackermann rank	Ackermann degree
$z_a(w_a^{\scriptscriptstyle B})(x_a^{\scriptscriptstyle B})(y_a^{\scriptscriptstyle B})$	1	1	1	undefined
$y_a(w_a^{\scriptscriptstyle B})(x_a^{\scriptscriptstyle B})$	2	2	2	undefined
$z_b(w_a^{\scriptscriptstyle B})(x_a^{\scriptscriptstyle B})$	3	3	1	undefined
$x_a(w_a^{\scriptscriptstyle B})$	4	6	3	undefined
$z_c(w_a^{\scriptscriptstyle B})(y_b^{\scriptscriptstyle B})$	5	7	1	undefined
$y_b(w_a^{\scriptscriptstyle B})$	6	14	2	undefined
$z_d(w_a^{\scriptscriptstyle B})$	7	21	1	undefined
w_a	8	42	4	1
$z_e(y_c^{\scriptscriptstyle B})(w_a^{\scriptscriptstyle B})$	9	43	1	undefined
$y_c(x_b^{\scriptscriptstyle B})$	10	86	2	undefined
$z_f(x_b^{\scriptscriptstyle B})$	11	129	1	undefined
x_b	12	258	3	2
$z_g(y_d^{\scriptscriptstyle B})$	13	301	1	undefined
y_d	14	602	2	3
z_h	15	903	1	4
$\mathsf{P}(w_a,x_b,y_d,z_h)$	15	1805	undefined	undefined

For

$$\forall w^{\scriptscriptstyle B}.\ \forall x^{\scriptscriptstyle B}.\ \forall y^{\scriptscriptstyle B}.\ \forall z^{\scriptscriptstyle B}.\ \mathsf{P}(w^{\scriptscriptstyle B},x^{\scriptscriptstyle B},y^{\scriptscriptstyle B},z^{\scriptscriptstyle B})$$

instead of $\exists w^{\scriptscriptstyle B}.\ \forall x^{\scriptscriptstyle B}.\ \exists y^{\scriptscriptstyle B}.\ \forall z^{\scriptscriptstyle B}.\ \mathsf{P}(w^{\scriptscriptstyle B},x^{\scriptscriptstyle B},y^{\scriptscriptstyle B},z^{\scriptscriptstyle B})$, we get the same exponential growth of nesting depth as in the example above, when we completely eliminate the quantifiers using (ε_2). The only difference is that we get additional occurrences of '$\neg$'. If we have quantifiers of the same kind, however, we had better choose them in parallel; e.g., for $\forall w^{\scriptscriptstyle B}.\ \forall x^{\scriptscriptstyle B}.\ \forall y^{\scriptscriptstyle B}.\ \forall z^{\scriptscriptstyle B}.\ \mathsf{P}(w^{\scriptscriptstyle B},x^{\scriptscriptstyle B},y^{\scriptscriptstyle B},z^{\scriptscriptstyle B})$, we choose

$$v_a\ :=\ \varepsilon v^{\scriptscriptstyle B}.\ \neg\mathsf{P}\big(1^{\mathrm{st}}(v^{\scriptscriptstyle B}),2^{\mathrm{nd}}(v^{\scriptscriptstyle B}),3^{\mathrm{rd}}(v^{\scriptscriptstyle B}),4^{\mathrm{th}}(v^{\scriptscriptstyle B})\big),$$

and then take $\mathsf{P}\big(1^{\mathrm{st}}(v_a),2^{\mathrm{nd}}(v_a),3^{\mathrm{rd}}(v_a),4^{\mathrm{th}}(v_a)\big)$ as result of the elimination.

Roughly speaking, in today's automated theorem proving, cf. e.g. [21], the exponential explosion of term depth of the example above is avoided by an outside-in removal of δ-quantifiers *without removing the quantifiers below ε-binders* and by a replacement of γ-quantified variables with free variables without choice-conditions. For the formula of Example 4.8, this yields $P(w^{\vee}, x_e, y^{\vee}, z_e)$ with $x_e = \varepsilon x_e^{\text{B}}. \neg \exists y^{\text{B}}.$ $\forall z^{\text{B}}. P(w^{\vee}, x_e^{\text{B}}, y^{\text{B}}, z^{\text{B}})$ and $z_e = \varepsilon z_e^{\text{B}}. \neg P(w^{\vee}, x_e, y^{\vee}, z_e^{\text{B}})$. Thus, in general, the nesting of binders for the complete elimination of a prenex of n quantifiers does not become deeper than $\frac{1}{4}(n+1)^2$.

Moreover, if we are only interested in reduction and not in equivalence transformation of a formula, we can abstract Skolem terms from the ε-terms and just reduce to the formula $P(w^{\vee}, x^{\text{A}}(w^{\vee}), y^{\vee}, z^{\text{A}}(w^{\vee})(y^{\vee}))$. In non-Skolemizing inference systems with variable-conditions we get $P(w^{\vee}, x^{\text{A}}, y^{\vee}, z^{\text{A}})$ instead, with $\{(w^{\vee}, x^{\text{A}}), (w^{\vee}, z^{\text{A}}), (y^{\vee}, z^{\text{A}})\}$ as an extension to the variable-condition. Note that with Skolemization or variable-conditions we have no growth of nesting depth at all, and the same will be the case for our approach to ε-terms.

4.10 Do not be afraid of Indefiniteness!

From the discussion in § 4.8, one could get the impression that an indefinite logical treatment of the ε is not easy to find. Indeed, on the first sight, there is the problem that some standard axiom schemes cannot be taken for granted, such as substitutability

$$s = t \quad \Rightarrow \quad f(s) = f(t)$$

and reflexivity

$$t = t$$

Note that substitutability is similar to the *extensionality axiom*

$$\forall x^{\text{B}}. (A_0 \Leftrightarrow A_1) \quad \Rightarrow \quad \varepsilon x^{\text{B}}. A_0 = \varepsilon x^{\text{B}}. A_1 \qquad \text{(E2)}$$

(cf. § A.1.1) when we take logical equivalence as equality. Moreover, note that

$$\varepsilon x^{\text{B}}. \text{ true} = \varepsilon x^{\text{B}}. \text{ true} \qquad \text{(REFLEX)}$$

is an instance of reflexivity.

Thus, it seems that — in case of an indefinite ε — the replacement of a subterm with an equal term is problematic, and so is the equality of syntactically equal terms.

It may be interesting to see that — in computer programs — we are quite used to committed choice and to an indefinite behavior of choosing, and that the violation of substitutability and even reflexivity is no problem there:

460

Example 4.9 (Violation of Substitutability and Reflexivity in Programs)
In the implementation of the specification of the web-based hypertext system of [73], we needed a function that chooses an element from a set implemented as a list. Its ML code is:

```
fun choose s = case s of Set (i :: _) => i | _ => raise Empty;
```

And, of course, it simply returns the first element of the list. For another set that is equal — but where the list may have another order — the result may be different. Thus, the behavior of the function `choose` is indefinite for a given set, but any time it is called for an implemented set, it chooses a particular element and *commits to this choice,* i.e.: when called again, it returns the same value. In this case we have `choose s = choose s`, but `s = t` does not imply `choose s = choose t`. In an implementation where some parallel reordering of lists may take place, even `choose s = choose s` may be wrong.

From this example we may learn that the question of `choose s = choose s` may be indefinite until the choice steps have actually been performed. *This is exactly how we will treat our ε.* The steps that are performed in logic are related to proving: Reductive inference steps that make proof trees grow toward the leaves, and choice steps that instantiate variables and atoms for various purposes.

Thus, on the one hand, when we want to prove

$$\varepsilon x^{\mathtt{B}}. \mathsf{true} \; = \; \varepsilon x^{\mathtt{B}}. \mathsf{true}$$

we can choose 0 for both occurrences of $\varepsilon x^{\mathtt{B}}.\mathsf{true}$, get $0\!=\!0$, and the proof is successful.

On the other hand, when we want to prove

$$\varepsilon x^{\mathtt{B}}. \mathsf{true} \; \neq \; \varepsilon x^{\mathtt{B}}. \mathsf{true}$$

we can choose 0 for one occurrence and 1 for the other, get $0\!\neq\!1$, and the proof is successful again.

This procedure may seem wondrous again, but is very similar to something quite common for free variables with empty choice-conditions:

On the one hand, when we want to prove

$$x^{\mathtt{V}} = y^{\mathtt{V}}$$

we can choose 0 to replace both $x^{\mathtt{V}}$ and $y^{\mathtt{V}}$, get $0\!=\!0$, and the proof is successful.

On the other hand, when we want to prove

$$x^{\mathtt{V}} \neq y^{\mathtt{V}}$$

we can choose 0 to replace $x^{\mathtt{V}}$ and 1 to replace $y^{\mathtt{V}}$, get $0\!\neq\!1$, and the proof is successful again.

4.11 Replacing ε-terms with Free Variables

There is an important difference between the inequations $\varepsilon x^{\text{B}}.\,\text{true} \neq \varepsilon x^{\text{B}}.\,\text{true}$ and $x^{\text{V}} \neq y^{\text{V}}$ at the end of § 4.10: The latter does not violate the reflexivity axiom! And we are going to cure the violation of the former immediately with the help of our free variables, but now with non-empty choice-conditions. Instead of $\varepsilon x^{\text{B}}.\,\text{true} \neq \varepsilon x^{\text{B}}.\,\text{true}$ we write $x^{\text{V}} \neq y^{\text{V}}$ and remember what these free variables stand for by storing this into a function C, called a *choice-condition*:

$$C(x^{\text{V}}) \ :=\ \varepsilon x^{\text{B}}.\,\text{true},$$

$$C(y^{\text{V}}) \ :=\ \varepsilon x^{\text{B}}.\,\text{true}.$$

For a first step, suppose that our ε-terms are not subordinate to any outside binder (cf. Definition 4.7). Then, we can replace an ε-term $\varepsilon z^{\text{B}}.\,A$ with a fresh free variable z^{V} and extend the partial function C by

$$C(z^{\text{V}}) \ :=\ \varepsilon z^{\text{B}}.\,A.$$

By this procedure we can eliminate all ε-terms without loosing any syntactic information.

As a first consequence of this elimination, the substitutability and reflexivity axioms are immediately regained, and the problems discussed in § 4.10 disappear.

A second reason for replacing the ε-terms with free variables is that the latter can solve the question whether a committed choice is required: We can express

committed choice by repeatedly using the same free variable, and

choice without commitment by using several variables with the same choice-condition.

Indeed, this also solves our problems with committed choice of Example 4.6 of § 4.8: Now, again using (ε_1), $\exists x^{\text{B}}.\,(x^{\text{B}} \neq x^{\text{B}})$ reduces to $x^{\text{V}} \neq x^{\text{V}}$ with

$$C(x^{\text{V}}) \ :=\ \varepsilon x^{\text{B}}.\,(x^{\text{B}} \neq x^{\text{B}})$$

and the proof attempt immediately fails because of the now regained reflexivity axiom.

As the second step, we still have to explain what to do with subordinate ε-terms. If the ε-term $\varepsilon v_l^{\text{B}}.\,A$ contains free occurrences of exactly the distinct bound atoms $v_0^{\text{B}}, \ldots, v_{l-1}^{\text{B}}$, then we have to replace this ε-term with the application term $z^{\text{V}}(v_0^{\text{B}})\cdots(v_{l-1}^{\text{B}})$ of the same type as v_l^{B} (for a fresh free variable z^{V}) and to extend the choice-condition C by

$$C(z^{\text{V}}) \ :=\ \lambda v_0^{\text{B}}.\ \ldots \lambda v_{l-1}^{\text{B}}.\ \varepsilon v_l^{\text{B}}.\,A.$$

Example 4.10 (Higher-Order Choice-Condition)

(continuing Example 4.8 of § 4.9)

In our framework, the complete elimination of ε-terms in (4.8.1) of Example 4.8 results in

$$\mathsf{P}(w_a^{\mathrm{V}}, x_b^{\mathrm{V}}, y_d^{\mathrm{V}}, z_h^{\mathrm{V}}) \tag{cf. (4.8.1)!}$$

with the following higher-order choice-condition:

$$C(z_h^{\mathrm{V}}) \; := \; \varepsilon z_h^{\mathrm{B}}.\ \neg\mathsf{P}(w_a^{\mathrm{V}}, x_b^{\mathrm{V}}, y_d^{\mathrm{V}}, z_h^{\mathrm{B}}) \tag{cf. (4.8.2)!}$$

$$C(y_d^{\mathrm{V}}) \; := \; \varepsilon y_d^{\mathrm{B}}.\ \mathsf{P}(w_a^{\mathrm{V}}, x_b^{\mathrm{V}}, y_d^{\mathrm{B}}, z_c^{\mathrm{V}}(y_d^{\mathrm{B}})) \tag{cf. (4.8.3)!}$$

$$C(z_g^{\mathrm{V}}) \; := \; \lambda y_d^{\mathrm{B}}.\ \varepsilon z_g^{\mathrm{B}}.\ \neg\mathsf{P}(w_a^{\mathrm{V}}, x_b^{\mathrm{V}}, y_d^{\mathrm{B}}, z_g^{\mathrm{B}}) \tag{cf. (4.8.4)!}$$

$$C(x_b^{\mathrm{V}}) \; := \; \varepsilon x_b^{\mathrm{B}}.\ \neg\mathsf{P}(w_a^{\mathrm{V}}, x_b^{\mathrm{B}}, y_c^{\mathrm{V}}(x_b^{\mathrm{B}}), z_f^{\mathrm{V}}(x_b^{\mathrm{B}})) \tag{cf. (4.8.5)!}$$

$$C(z_f^{\mathrm{V}}) \; := \; \lambda x_b^{\mathrm{B}}.\ \varepsilon z_f^{\mathrm{B}}.\ \neg\mathsf{P}(w_a^{\mathrm{V}}, x_b^{\mathrm{B}}, y_c^{\mathrm{V}}(x_b^{\mathrm{B}}), z_f^{\mathrm{B}}) \tag{cf. (4.8.6)!}$$

$$C(y_c^{\mathrm{V}}) \; := \; \lambda x_b^{\mathrm{B}}.\ \varepsilon y_c^{\mathrm{B}}.\ \mathsf{P}(w_a^{\mathrm{V}}, x_b^{\mathrm{B}}, y_c^{\mathrm{B}}, z_e^{\mathrm{V}}(x_b^{\mathrm{B}})(y_c^{\mathrm{B}})) \tag{cf. (4.8.7)!}$$

$$C(z_e^{\mathrm{V}}) \; := \; \lambda x_b^{\mathrm{B}}.\ \lambda y_c^{\mathrm{B}}.\ \varepsilon z_e^{\mathrm{B}}.\ \neg\mathsf{P}(w_a^{\mathrm{V}}, x_b^{\mathrm{B}}, y_c^{\mathrm{B}}, z_e^{\mathrm{B}}) \tag{cf. (4.8.8)!}$$

$$C(w_a^{\mathrm{V}}) \; := \; \varepsilon w_a^{\mathrm{B}}.\ \mathsf{P}(w_a^{\mathrm{B}}, x_a^{\mathrm{V}}(w_a^{\mathrm{B}}), y_b^{\mathrm{V}}(w_a^{\mathrm{B}}), z_d^{\mathrm{V}}(w_a^{\mathrm{B}})) \tag{cf. (4.8.9)!}$$

$$C(z_d^{\mathrm{V}}) \; := \; \lambda w_a^{\mathrm{B}}.\ \varepsilon z_d^{\mathrm{B}}.\ \neg\mathsf{P}(w_a^{\mathrm{B}}, x_a^{\mathrm{V}}(w_a^{\mathrm{B}}), y_b^{\mathrm{V}}(w_a^{\mathrm{B}}), z_d^{\mathrm{B}}) \tag{cf. (4.8.10)!}$$

$$C(y_b^{\mathrm{V}}) \; := \; \lambda w_a^{\mathrm{B}}.\ \varepsilon y_b^{\mathrm{B}}.\ \mathsf{P}(w_a^{\mathrm{B}}, x_a^{\mathrm{V}}(w_a^{\mathrm{B}}), y_b^{\mathrm{B}}, z_c^{\mathrm{V}}(w_a^{\mathrm{B}})(y_b^{\mathrm{B}})) \tag{cf. (4.8.11)!}$$

$$C(z_c^{\mathrm{V}}) \; := \; \lambda w_a^{\mathrm{B}}.\ \lambda y_b^{\mathrm{B}}.\ \varepsilon z_c^{\mathrm{B}}.\ \neg\mathsf{P}(w_a^{\mathrm{B}}, x_a^{\mathrm{V}}(w_a^{\mathrm{B}}), y_b^{\mathrm{B}}, z_c^{\mathrm{B}}) \tag{cf. (4.8.12)!}$$

$$C(x_a^{\mathrm{V}}) \; := \; \lambda w_a^{\mathrm{B}}.\ \varepsilon x_a^{\mathrm{B}}.\ \neg\mathsf{P}(w_a^{\mathrm{B}}, x_a^{\mathrm{B}}, y_a^{\mathrm{V}}(w_a^{\mathrm{B}})(x_a^{\mathrm{B}}), z_b^{\mathrm{V}}(w_a^{\mathrm{B}})(x_a^{\mathrm{B}})) \tag{cf. (4.8.13)!}$$

$$C(z_b^{\mathrm{V}}) \; := \; \lambda w_a^{\mathrm{B}}.\ \lambda x_a^{\mathrm{B}}.\ \varepsilon z_b^{\mathrm{B}}.\ \neg\mathsf{P}(w_a^{\mathrm{B}}, x_a^{\mathrm{B}}, y_a^{\mathrm{V}}(w_a^{\mathrm{B}})(x_a^{\mathrm{B}}), z_b^{\mathrm{B}}) \tag{cf. (4.8.14)!}$$

$$C(y_a^{\mathrm{V}}) \; := \; \lambda w_a^{\mathrm{B}}.\ \lambda x_a^{\mathrm{B}}.\ \varepsilon y_a^{\mathrm{B}}.\ \mathsf{P}(w_a^{\mathrm{B}}, x_a^{\mathrm{B}}, y_a^{\mathrm{B}}, z_a^{\mathrm{V}}(w_a^{\mathrm{B}})(x_a^{\mathrm{B}})(y_a^{\mathrm{B}})) \tag{cf. (4.8.15)!}$$

$$C(z_a^{\mathrm{V}}) \; := \; \lambda w_a^{\mathrm{B}}.\ \lambda x_a^{\mathrm{B}}.\ \lambda y_a^{\mathrm{B}}.\ \varepsilon z_a^{\mathrm{B}}.\ \neg\mathsf{P}(w_a^{\mathrm{B}}, x_a^{\mathrm{B}}, y_a^{\mathrm{B}}, z_a^{\mathrm{B}}) \tag{cf. (4.8.16)!}$$

Note that this representation of (4.8.1) is smaller and easier to understand than all previous ones. Indeed, by combination of λ-abstraction and term sharing via free variables, in our framework the ε becomes practically feasible.

All in all, by this procedure we can replace all ε-terms in all formulas and sequents. The only place where the ε still occurs is the range of the choice-condition C; and also there it is not essential because, instead of

$$C(z^{\mathrm{V}}) \;\; = \;\; \lambda v_0^{\mathrm{B}}.\ \ldots \lambda v_{l-1}^{\mathrm{B}}.\ \varepsilon v_l^{\mathrm{B}}.\ A,$$

we could write

$$C(z^{\mathrm{V}}) \;\; = \;\; \lambda v_0^{\mathrm{B}}.\ \ldots \lambda v_{l-1}^{\mathrm{B}}.\ A\{v_l^{\mathrm{B}} \mapsto z^{\mathrm{V}}(v_0^{\mathrm{B}}) \cdots (v_{l-1}^{\mathrm{B}})\}$$

as we have actually done in [106; 107; 108; 110; 111].

4.12 Instantiating Free Variables ("ε-Substitution")

Having already realized Requirement I (Indication of Commitment) of § 4.2 in § 4.11, we are now going to explain how to satisfy Requirement II (Reasoning). To this end, we have to explain how to replace free variables with terms that satisfy their choice-conditions.

The first thing to know about free variables with choice-conditions is: Just like the free variables without choice-conditions (introduced by γ-rules e.g.) and contrary to free atoms, the free variables with choice-conditions (introduced by δ^+-rules e.g.) are *rigid* in the sense that the only way to replace a free variable is to do it *globally*, i.e. in all formulas and all choice-conditions with the same term in an atomic transaction.

In *reductive* theorem proving, such as in sequent, tableau, matrix, or indexed-formula-tree calculi, we are in the following situation: While a free variable without choice-condition can be replaced with nearly everything, the replacement of a free variable with a choice-condition requires some proof work, and a free atom cannot be instantiated at all.

Contrariwise, when formulas are used as tools instead of tasks, free atoms can indeed be replaced — and this even locally (i.e. non-rigidly) and repeatedly. This is the case not only for purely *generative* calculi (such as resolution and paramodulation calculi) and Hilbert-style calculi (such as the predicate calculus of Hilbert–Bernays [57; 58; 59; 60]), but also for the lemma and induction hypothesis application in the otherwise reductive calculi of [106], cf. [106, § 2.5.2].

More precisely — again considering *reductive* theorem proving, where formulas are proof tasks — a free variable without choice-condition may be instantiated with any term (of appropriate type) that does not violate the current variable-condition, cf. § 5.7 for details. The instantiation of a free variable with choice-condition additionally requires some proof work depending on the current choice-condition, cf. Definition 5.13 for the formal details. In general, if a substitution σ replaces the free variable y^{V} in the domain of the choice-condition C, then — to know that the global instantiation of the entire proof forest with σ is correct — we have to prove $(Q_C(y^{\mathrm{V}}))\sigma$, where Q_C is given as follows:

Definition 4.11 (Q_C)
Q_C is the function that maps every $z^{\mathrm{V}} \in \mathrm{dom}(C)$ with
$$C(z^{\mathrm{V}}) = \lambda v_0^{\mathrm{B}}. \ldots \lambda v_{l-1}^{\mathrm{B}}. \ \varepsilon v_l^{\mathrm{B}}. \ B$$
(for some bound atoms $v_0^{\mathrm{B}}, \ldots, v_l^{\mathrm{B}}$ and some formula B) to the single-formula sequent
$$\forall v_0^{\mathrm{B}}. \ldots \forall v_{l-1}^{\mathrm{B}}. \ (\ \exists v_l^{\mathrm{B}}. \ B \quad \Rightarrow \quad B\{v_l^{\mathrm{B}} \mapsto z^{\mathrm{V}}(v_0^{\mathrm{B}}) \cdots (v_{l-1}^{\mathrm{B}})\} \ ),$$
and is otherwise undefined.

Note that $Q_C(y^{\text{V}})$ is nothing but a formulation of Hilbert–Bernays' axiom (ε_0) in our framework. (See our § 4.5 for (ε_0)).

Moreover, Lemma 5.19 will state the validity of $Q_C(y^{\text{V}})$. Therefore, the commitment to a choice comes only with the substitution σ. Indeed, regarding the σ-instance of $Q_C(y^{\text{V}})$ whose provability is required, it is only the *arbitrariness* of the substitution σ that realizes the *indefiniteness* of the choice for the ε.

Now, as an example for Q_C, we can replay Example 3.1 and use it for a discussion of the δ^+-rule instead of the δ^--rule:

Example 4.12 (Soundness of δ^+-rule)
The formula $$\exists y^{\text{B}}.\ \forall x^{\text{B}}.\ (y^{\text{B}} = x^{\text{B}})$$
is not universally valid. We can start a reductive proof attempt as follows:

γ-step: $\forall x^{\text{B}}.\ (y^{\text{V}} = x^{\text{B}})$, $\quad \exists y^{\text{B}}.\ \forall x^{\text{B}}.\ (y^{\text{B}} = x^{\text{B}})$

δ^+-step: $\quad (y^{\text{V}} = x^{\text{V}})$, $\quad \exists y^{\text{B}}.\ \forall x^{\text{B}}.\ (y^{\text{B}} = x^{\text{B}})$

Now, if the free variable y^{V} could be replaced with the free variable x^{V}, then we would get the tautology $(x^{\text{V}} = x^{\text{V}})$, i.e. we would have proved an invalid formula. To prevent this, as indicated to the lower right of the bar of the first of the δ^+-rules in § 3.4 on Page 9, the δ^+-step has to record $\mathbb{VA}(\forall x^{\text{B}}.\ (y^{\text{V}} = x^{\text{B}})) \times \{x^{\text{V}}\} = \{(y^{\text{V}}, x^{\text{V}})\}$ in a positive variable-condition, where $(y^{\text{V}}, x^{\text{V}})$ means that "x^{V} positively depends on y^{V}" (or that "y^{V} is a subterm of the description of x^{V}"), so that we may never instantiate the free variable y^{V} with a term containing the free variable x^{V}, because this instantiation would result in cyclic dependencies (or in a cyclic term).

Contrary to Example 3.1, we have a further opportunity here to complete this proof attempt into a successful proof: If the the substitution $\sigma := \{x^{\text{V}} \mapsto y^{\text{V}}\}$ could be applied, then we would get the tautology $(y^{\text{V}} = y^{\text{V}})$, i.e. we would have proved an invalid formula. To prevent this — as indicated to the upper right of the bar of the first of the δ^+-rules in § 3.4 on Page 9 — the δ^+-step has to record
$$\left(\ x^{\text{V}},\quad \varepsilon x^{\text{B}}.\ \neg(y^{\text{V}} = x^{\text{B}})\ \right)$$
in the choice-condition C. If we take this pair as an equation, then the intuition behind the above statement that y^{V} is somehow a subterm of the description of x^{V} becomes immediately clear. If we take it as element of the graph of the function C, however, then we can compute $(Q_C(x^{\text{V}}))\sigma$ and try to prove it. $Q_C(x^{\text{V}})$ is

$\exists x^{\text{B}}.\ \neg(y^{\text{V}} = x^{\text{B}}) \ \Rightarrow\ \neg(y^{\text{V}} = x^{\text{V}})$; $\quad$ so $(Q_C(x^{\text{V}}))\sigma$ is

$\exists x^{\text{B}}.\ \neg(y^{\text{V}} = x^{\text{B}}) \ \Rightarrow\ \neg(y^{\text{V}} = y^{\text{V}})$. $\quad$ In classical logic with equality this is equivalent to $\forall x^{\text{B}}.\ (y^{\text{V}} = x^{\text{B}})$. If we were able to show the truth of this formula, then it would be sound to apply the substitution σ to prove the above sequent resulting from the γ-step. That sequent, however, already includes this formula as an element of its disjunction. Thus, no progress is possible by means of the δ^+-rules here; and so this example is not a counterexample to the soundness of the δ^+-rules.

Example 4.13 (Predecessor Function)
Suppose that our domain is natural numbers and that y^V has the choice-condition
$$C(y^\mathsf{V}) \;=\; \lambda v_0^\mathsf{B}.\; \varepsilon v_1^\mathsf{B}.\; \big(\; v_0^\mathsf{B} = v_1^\mathsf{B} + 1 \;\big).$$
Then, before we may instantiate y^V with the symbol p for the predecessor function
specified by
$$\forall x^\mathsf{B}.\; \big(\; \mathsf{p}(x^\mathsf{B}+1) = x^\mathsf{B} \;\big),$$
we have to prove the single-formula sequent $(Q(y^\mathsf{V}))\{y^\mathsf{V} \mapsto \mathsf{p}\}$, which reads
$$\forall v_0^\mathsf{B}.\; \Big(\; \exists v_1^\mathsf{B}.\; \big(\; v_0^\mathsf{B} = v_1^\mathsf{B} + 1 \;\big) \;\Rightarrow\; \big(\; v_0^\mathsf{B} = \mathsf{p}(v_0^\mathsf{B}) + 1 \;\big) \;\Big).$$
In fact, the single formula of this sequent immediately follows from the specification
of p. Note that the fact that $\mathsf{p}(0)$ is not specified here is no problem in this ε-
substitution because $\varepsilon v_1^\mathsf{B}.\; (0 = v_1^\mathsf{B} + 1)$ is not specified by (ε_0) either.

Example 4.14 (Canossa 1077) *(continuing Example 4.5)*
(See [26] if you want to look behind the omnipresent legend and find out what really
seems to have happened at Canossa in January 1077.)
The situation of Example 4.5 now reads
$$\mathsf{Holy\ Ghost} = z_0^\mathsf{V} \quad \wedge \quad \mathsf{Joseph} = z_1^\mathsf{V} \tag{4.14.1}$$
with
$$C(z_0^\mathsf{V}) = \varepsilon z_0^\mathsf{B}.\; \mathsf{Father}(z_0^\mathsf{B}, \mathsf{Jesus}),$$
and
$$C(z_1^\mathsf{V}) = \varepsilon z_1^\mathsf{B}.\; \mathsf{Father}(z_1^\mathsf{B}, \mathsf{Jesus}).$$

This does not bring us into the old trouble with the Pope because nobody knows
whether $z_0^\mathsf{V} = z_1^\mathsf{V}$ holds or not.
On the one hand, knowing (4.1.2) from Example 4.1 of § 4.4, we can prove (4.14.1)
as follows: Let us replace z_0^V with $\mathsf{Holy\ Ghost}$ because, for $\sigma_0 := \{z_0^\mathsf{V} \mapsto \mathsf{Holy\ Ghost}\}$,
from $\mathsf{Father}(\mathsf{Holy\ Ghost}, \mathsf{Jesus})$ we conclude
$$\exists z_0^\mathsf{B}.\; \mathsf{Father}(z_0^\mathsf{B}, \mathsf{Jesus}) \;\Rightarrow\; \mathsf{Father}(\mathsf{Holy\ Ghost}, \mathsf{Jesus}),$$
which is nothing but the required $(Q_C(z_0^\mathsf{V}))\sigma_0$.
Analogously, we replace z_1^V with Joseph because, for $\sigma_1 := \{z_1^\mathsf{V} \mapsto \mathsf{Joseph}\}$, from
(4.1.2) we conclude the required $(Q_C(z_1^\mathsf{V}))\sigma_1$. After these replacements, (4.14.1)
becomes the tautology
$$\mathsf{Holy\ Ghost} = \mathsf{Holy\ Ghost} \quad \wedge \quad \mathsf{Joseph} = \mathsf{Joseph}$$
On the other hand, if we want to have trouble, we can apply the substitution
$$\sigma' \;=\; \{z_0^\mathsf{V} \mapsto \mathsf{Joseph},\; z_1^\mathsf{V} \mapsto \mathsf{Joseph}\}$$
to (4.14.1) because both $(Q_C(z_0^\mathsf{V}))\sigma'$ and $(Q_C(z_1^\mathsf{V}))\sigma'$ are equal to $(Q_C(z_1^\mathsf{V}))\sigma_1$ up to
renaming of bound atoms. Then our task is to show
$$\mathsf{Holy\ Ghost} = \mathsf{Joseph} \quad \wedge \quad \mathsf{Joseph} = \mathsf{Joseph}.$$
Note that this course of action is stupid, even under the aspect of theorem proving
alone.

5 Formal Presentation of Our Semantics

To satisfy Requirement III (Semantics) of § 4.2, we will now present our novel semantics for Hilbert's ε formally. This is required for precision and consistency. As consistency of our new semantics is not trivial at all, technical rigor cannot be avoided. From §§ 2 and 4, the reader should have a good intuition of our intended representation and semantics of Hilbert's ε, free variables, atoms, and choice-conditions in our framework.

5.1 Organization

After some preliminary subsections, we formalize variable-conditions and their consistency (§ 5.5) and discuss alternatives to the design decisions in the formalization of variable-conditions (§ 5.6).

Moreover, we explain how to deal with free variables syntactically (§ 5.7) and semantically (§§ 5.8 and 5.9).

After formalizing choice-conditions and their compatibility (§ 5.10), we define our notion of validity and discuss some examples (§ 5.11). One of these examples is especially interesting because we show that — with our new more careful treatment of negative information in our *positive/negative* variable-conditions — we can now model Henkin quantification directly.

Our interest goes beyond soundness in that we want to have *"preservation of solutions"*. By this we mean the following: All *closing substitutions* for the free variables — i.e. all solutions that transform a proof attempt (to which a proposition has been reduced) into a closed proof — are also solutions of the original proposition. This is similar to a proof in PROLOG (cf. [69], [13]), computing answers to a query proposition that contains free variables. Therefore, we discuss this solution-preserving notion of *reduction* (§ 5.15), especially under the aspects of extensions of variable-conditions and choice-conditions (§ 5.12), and of global instantiation of free variables with choice-conditions ("ε-substitution") (§ 5.13).

Finally, in § 5.16, we show soundness, safeness, and solution-preservation for our γ-, δ^-, and δ^+-rules of §§ 3.2, 3.3, and 3.4.

All in all, we extend and simplify the presentation of [108], which already simplifies and extends the presentation of [106] and which is extended with additional linguistic applications in [111]. Note, however, that [106] additionally contains some comparative discussions and compatible extensions for *descente infinie*, which also apply to our new version here.

5.2 Basic Notions and Notation

'**N**' denotes the set of natural numbers and '$\prec$' the ordering on **N**. Let $\mathbf{N_+} := \{\, n \in \mathbf{N} \mid 0 \neq n \,\}$. We use '$\uplus$' for the union of disjoint classes and 'id' for the identity function. For classes R, A, and B we define:

$$\mathrm{dom}(R) := \{\, a \mid \exists b.\ (a, b) \in R \,\} \qquad \textit{domain}$$
$$_A\!\restriction\! R \quad := \{\, (a, b) \in R \mid a \in A \,\} \qquad \textit{(domain-) restriction to } A$$
$$\langle A \rangle R \quad := \{\, b \mid \exists a \in A.\ (a, b) \in R \,\} \quad \textit{image of } A, \text{ i.e. } \langle A \rangle R = \mathrm{ran}(_A\!\restriction\! R)$$

And the dual ones:

$$\mathrm{ran}(R) \ := \{\, b \mid \exists a.\ (a, b) \in R \,\} \qquad \textit{range}$$
$$R\!\restriction_B \quad := \{\, (a, b) \in R \mid b \in B \,\} \qquad \textit{range-restriction to } B$$
$$R \langle B \rangle \quad := \{\, a \mid \exists b \in B.\ (a, b) \in R \,\} \quad \textit{reverse-image of } B, \text{ i.e. } R \langle B \rangle = \mathrm{dom}(R\!\restriction_B)$$

We use '$\emptyset$' to denote the empty set as well as the empty function. Functions are (right-) unique relations, and so the meaning of "$f \circ g$" is extensionally given by $(f \circ g)(x) = g(f(x))$. The *class of total functions from A to B* is denoted as $A \to B$. The *class of (possibly) partial functions from A to B* is denoted as $A \rightsquigarrow B$. Both $\to$ and $\rightsquigarrow$ associate to the right, i.e. $A \rightsquigarrow B \to C$ reads $A \rightsquigarrow (B \to C)$.

Let R be a binary relation. R is said to be a relation *on A* if $\mathrm{dom}(R) \cup \mathrm{ran}(R) \subseteq A$. R is *irreflexive* if $\mathrm{id} \cap R = \emptyset$. It is *$A$-reflexive* if $_A\!\restriction\!\mathrm{id} \subseteq R$. Speaking of a *reflexive* relation we refer to the largest A that is appropriate in the local context, and referring to this A we write R^0 to ambiguously denote $_A\!\restriction\!\mathrm{id}$. With $R^1 := R$, and $R^{n+1} := R^n \circ R$ for $n \in \mathbf{N_+}$, R^m denotes the m-step relation for R. The *transitive closure* of R is $R^+ := \bigcup_{n \in \mathbf{N_+}} R^n$. The *reflexive transitive closure* of R is $R^* := \bigcup_{n \in \mathbf{N}} R^n$. A relation R (on A) is *well-founded* if any non-empty class B ($\subseteq A$) has an R-minimal element, i.e. $\exists a \in B.\ \neg \exists a' \in B.\ a' R a$.

5.3 Choice Functions

To be useful in context with Hilbert's ε, the notion of a "choice function" must be generalized: We need a *total* function on the power-set of any universe. Thus, a value must be supplied even for the empty set:

Definition 5.1 ([Generalized] [Function-] Choice Function)
f is a *choice function [on A]* if f is a function with [$A \subseteq \mathrm{dom}(f)$ and]
$$f : \mathrm{dom}(f) \to \bigcup(\mathrm{dom}(f)) \quad \text{and} \quad \forall Y \in \mathrm{dom}(f).\ (\ f(Y) \in Y \ ).$$
f is a *generalized choice function [on A]* if f is a function with [$A \subseteq \mathrm{dom}(f)$ and]
$$f : \mathrm{dom}(f) \to \bigcup(\mathrm{dom}(f)) \quad \text{and} \quad \forall Y \in \mathrm{dom}(f).\ (\ f(Y) \in Y \ \lor \ Y = \emptyset \ ).$$
f is a *function-choice function for a function F* if
$$f \text{ is a function with } \mathrm{dom}(F) \subseteq \mathrm{dom}(f) \text{ and } \forall x \in \mathrm{dom}(F).\ (\ f(x) \in F(x) \ ).$$

Corollary 5.2

The empty function $\emptyset$ is both a choice function and a generalized choice function.
If $\mathrm{dom}(f) = \{\emptyset\}$,

> *then f is neither a choice function nor a generalized choice function.*

If $\emptyset \notin \mathrm{dom}(f)$, then f is a generalized choice function if and only if

> *f is a choice function.*

If $\emptyset \in \mathrm{dom}(f)$, then f is a generalized choice function if and only if
there is a choice function f' and an $x \in \bigcup(\mathrm{dom}(f'))$ such that $f = f' \uplus \{(\emptyset, x)\}$.

5.4   Variables, Atoms, Constants, and Substitutions

We assume the following sets of symbols to be disjoint:

> $\mathbb{V}$ (free) (rigid) <u>v</u>ariables, which serve as unknowns or
> the free variables of Fitting's [20; 21]

> $\mathbb{A}$ *(free)* <u>a</u>toms, which serve as parameters and must not be bound

> $\mathbb{B}$ <u>b</u>ound atoms, which may be bound

> Σ *constants*, i.e. the function and predicate symbols from the <u>s</u>ignature

We define:

> $\mathbb{VA}$ $:=$ $\mathbb{V} \uplus \mathbb{A}$

> $\mathbb{VAB}$ $:=$ $\mathbb{V} \uplus \mathbb{A} \uplus \mathbb{B}$

By slight abuse of notation, for $S \in \{\mathbb{V}, \mathbb{A}, \mathbb{B}, \mathbb{VA}, \mathbb{VAB}\}$, we write "$S(\Gamma)$" to denote the set of symbols from S that have free occurrences in Γ.

Let σ be a substitution. σ is a *substitution on* V if $\mathrm{dom}(\sigma) \subseteq V$.

The following indented statement (as simple as it is) will require some discussion.

> We denote with "$\Gamma\sigma$" the result of replacing each (free) occurrence of a symbol $x \in \mathrm{dom}(\sigma)$ in Γ with $\sigma(x)$; possibly after renaming in Γ some symbols that are bound in Γ, in particular because a capture of their free occurrences in $\sigma(x)$ must be avoided.

Note that such a renaming of symbols that are bound in Γ will hardly be required for the following reason: We will bind only symbols from the set $\mathbb{B}$ of bound atoms. And — unless explicitly stated otherwise — we tacitly assume that all occurrences of bound atoms from $\mathbb{B}$ in a term or formula or in the range of a substitution are *bound occurrences* (i.e. that a bound atom $x^{\mathbb{B}} \in \mathbb{B}$ occurs only in the scope of a binder on $x^{\mathbb{B}}$). Thus, in standard situations, even without renaming, no additional occurrences can become bound (i.e. captured) when applying a substitution.

Only if we want to exclude the binding of a bound atom within the scope of another binding of the same bound atom (e.g. for the sake of readability and in the tradition of Hilbert–Bernays), then we may still have to rename some of the bound atoms in Γ. For example, for Γ being the formula $\forall x^{\mathtt{B}}.\,(x^{\mathtt{B}} = y^{\mathtt{V}})$ and σ being the substitution $\{y^{\mathtt{V}} \mapsto \varepsilon x^{\mathtt{B}}.\,(x^{\mathtt{B}} = x^{\mathtt{B}})\}$, we may want the result of $\Gamma\sigma$ to be something like $\forall z^{\mathtt{B}}.\,(z^{\mathtt{B}} = \varepsilon x^{\mathtt{B}}.\,(x^{\mathtt{B}} = x^{\mathtt{B}}))$ instead of $\forall x^{\mathtt{B}}.\,(x^{\mathtt{B}} = \varepsilon x^{\mathtt{B}}.\,(x^{\mathtt{B}} = x^{\mathtt{B}}))$.

Unless explicitly stated otherwise, in this paper we will use only substitutions on subsets of $\mathbb{V}$. Thus, also the occurrence of "(free)" in the statement indented above is hardly of any relevance here, because we will never bind elements of $\mathbb{V}$.

5.5 Consistent Positive/Negative Variable-Conditions

Variable-conditions are binary relations on free variables and free atoms. They put conditions on the possible instantiation of free variables, and on the dependencies of their valuations. In this paper, for clarity of presentation, a variable-condition is formalized as a pair (P, N) of binary relations, which we will call a "positive/negative variable-condition":

- The first component (P) of such a pair is a binary relation that is meant to express *positive* dependencies. It comes with the intention of transitivity, although it will typically not be closed up to transitivity for reasons of presentation and efficiency.

  The overall idea is that the occurrence of a pair $(x^{\mathtt{VA}}, y^{\mathtt{V}})$ in this positive relation means something like

  "the value of $y^{\mathtt{V}}$ may well depend on $x^{\mathtt{VA}}$"

  or

  "the value of $y^{\mathtt{V}}$ is described in terms of $x^{\mathtt{VA}}$".

- The second component (N), however, is meant to capture *negative* dependencies.

  The overall idea is that the occurrence of a pair $(x^{\mathtt{V}}, y^{\mathtt{A}})$ in this negative relation means something like

  "the value of $x^{\mathtt{V}}$ has to be fixed before the value of $y^{\mathtt{A}}$ can be determined"

  or

  "the value of $x^{\mathtt{V}}$ must not depend on $y^{\mathtt{A}}$"

  or

  "$y^{\mathtt{A}}$ is fresh for $x^{\mathtt{V}}$".

  Relations similar to this negative relation (N) already occurred as the only component of a variable-condition in [104], and later — with a completely different motivation — as *"freshness* conditions" also in [29].

Definition 5.3 (Positive/Negative Variable-Condition)
A *positive/negative variable-condition* is a pair (P, N) with
$$P \subseteq \mathbb{VA} \times \mathbb{V}$$
and
$$N \subseteq \mathbb{V} \times \mathbb{A}.$$

Note that, in a positive/negative variable-condition (P, N), the relations P and N are always disjoint because their ranges are subsets of the disjoint sets $\mathbb{V}$ and $\mathbb{A}$, respectively.

A relation exactly like this positive relation (P) was the only component of a variable-condition as defined and used identically throughout [105; 106; 107; 108; 110; 111]. Note, however, that, in these publications, we had to admit this single positive relation to be a subset of $\mathbb{VA} \times \mathbb{VA}$ (instead of the restriction to $\mathbb{VA} \times \mathbb{V}$ of Definition 5.3 in this paper), because it had to simulate the negative relation (N) in addition; thereby losing some expressive power as compared to our positive/negative variable-conditions here (cf. Example 5.20).

In the following definition, the well-foundedness guarantees that all dependencies can be traced back to independent symbols and that no variable may transitively depend on itself, whereas the irreflexivity makes sure that no contradictious dependencies can occur.

Definition 5.4 (Consistency)
A pair (P, N) is *consistent* if
$$P \text{ is well-founded}$$
and
$$P^+ \circ N \text{ is irreflexive.}$$

Let (P, N) be a positive/negative variable-condition. Let us think of our (binary) relations P and N as edges of a directed graph whose vertices are the symbols for atoms and variables currently in use. Then, $P^+ \circ N$ is irreflexive if and only if there is no cycle in $P \cup N$ that contains exactly one edge from N. Moreover, in practice, a positive/negative variable-condition (P, N) can always be chosen to be finite in both its components. In the case that P is finite, P is well-founded if and only if P is acyclic. Thus we get:

Corollary 5.5
Let (P, N) be a positive/negative variable-condition with $|P| \in \mathbf{N}$.
(P, N) is consistent if and only if
each cycle in the directed graph of $P \uplus N$ contains more than one edge from N.
In case of $|N| \in \mathbf{N}$, the right-hand side of this equivalence can be effectively tested with an asymptotic time complexity of $|P| + |N|$.

471

Note that, in the finite case, the test of Corollary 5.5 seems to be both the most efficient and the most human-oriented way to represent the question of consistency of positive/negative variable-conditions.

5.6 Further Discussion of our Formalization of Variable-Conditions

Let us recall that the two relations P and N of a positive/negative variable-condition (P, N) are always disjoint because their ranges must be disjoint according to Definition 5.3. Thus, from a technical point of view, we could merge P and N into a single relation, but we prefer to have two relations for the two different functions (the positive and the negative one) of the variable-conditions in this paper, instead of the one relation for one function of [105; 106; 107; 108; 110; 111], which realized the negative function only with a significant loss of relevant information.

Moreover, in Definition 5.3, we have excluded the possibility that two atoms $a^{\mathbb{A}}, b^{\mathbb{A}} \in \mathbb{A}$ may be related to each other in any of the two components of a positive/negative variable-condition (P, N):

- $y^{\mathbb{VA}} P a^{\mathbb{A}}$ is excluded for intentional reasons: An atom $a^{\mathbb{A}}$ cannot depend on any other symbol $y^{\mathbb{VA}}$. In this sense an atom is indeed atomic and can be seen as a black box.

- $b^{\mathbb{A}} N a^{\mathbb{A}}$, however, is excluded for technical reasons only.
  Two distinct atoms $a^{\mathbb{A}}, b^{\mathbb{A}}$ in nominal terms [101] are indeed always fresh for each other: $a^{\mathbb{A}} \# b^{\mathbb{A}}$. In our notation, this would read: $b^{\mathbb{A}} N a^{\mathbb{A}}$. ·
  The reason why we did not include $(\mathbb{A} \times \mathbb{A}) \setminus_{\mathbb{A}} \lceil \text{id}$ into the negative component N is simply that we want to be close to the data structures of a both efficient and human-oriented graph implementation.
  Furthermore, consistency of a positive/negative variable-condition (P, N) is equivalent to consistency of $\left(P, \quad N \uplus ((\mathbb{A} \times \mathbb{A}) \setminus _{\mathbb{A}} \lceil \text{id}) \right)$.
  Indeed, if we added $(\mathbb{A} \times \mathbb{A}) \setminus _{\mathbb{A}} \lceil \text{id}$ to N, the result of the acyclicity test of Corollary 5.5 would not be changed: If there were a cycle with a single edge from $(\mathbb{A} \times \mathbb{A}) \setminus _{\mathbb{A}} \lceil \text{id}$, then its previous edge would have to be one of the original edges of N; and so this cycle would have more than one edge from $N \uplus ((\mathbb{A} \times \mathbb{A}) \setminus _{\mathbb{A}} \lceil \text{id})$, and thus would not count as a counterexample to consistency.

Furthermore, we could remove the set $\mathbb{B}$ of bound atoms from our sets of symbols and consider its elements to be elements of the set $\mathbb{A}$ of (free) atoms. Besides some additional care on free occurrences of atoms in § 5.4, an additional price we would have to pay for this removal is that we would have to include $\mathbb{V} \times \mathbb{B}$ as a subset into the second component (N) of all our positive/negative variable-conditions (P, N). The reason for this is that we must guarantee that a bound atom $b^{\mathbb{B}}$ cannot be read by any variable x^{V}, especially not after an elimination of binders; then, by this inclusion, in case of $b^{\mathrm{B}}\ P^+\ x^{\mathrm{V}}$, we would get a cycle $b^{\mathrm{B}}\ P^+\ x^{\mathrm{V}}\ N\ b^{\mathrm{B}}$ with only one edge from N. Although, in practical contexts, we can always get along with a finite subset of $\mathbb{V}\times\mathbb{B}$, the essential pairs of this subset would still be quite many and would be most confusing already in small examples. For instance, for the higher-order choice-condition of Example 4.10, almost four dozen pairs from $\mathbb{V}\times\mathbb{B}$ are technically required, compared to only a good dozen pairs that are actually relevant to the problem (cf. Example 5.14(a)).

5.7   Extensions, σ-Updates, and (P,N)-Substitutions

Within a progressing reasoning process, positive/negative variable-conditions may be subject to only one kind of transformation, which we simply call an "extension".

Definition 5.6 ([Weak] Extension)
(P',N') is an [weak] *extension of* (P,N) if
(P',N') is a positive/negative variable-condition, $P\subseteq P'$ [or at least $P\subseteq (P')^+$],
and $N\subseteq N'$.

As an immediate corollary of Definitions 5.6 and 5.4 we get:

Corollary 5.7 *If (P',N') is a consistent positive/negative variable-condition and an [weak] extension of (P,N), then (P,N) is a consistent positive/negative variable-condition as well.*

A σ-update is a special form of an extension:

Definition 5.8 (σ-Update, Dependence Relation)
Let (P,N) be a positive/negative variable-condition and σ be a substitution on $\mathbb{V}$.
The *dependence relation of* σ is
$$D\ :=\ \{\ (z^{\mathrm{VA}},x^{\mathrm{V}})\ |\ x^{\mathrm{V}}\in\mathrm{dom}(\sigma)\wedge z^{\mathrm{VA}}\in\mathbb{VA}(\sigma(x^{\mathrm{V}}))\ \}.$$
The σ-*update of* (P,N) is $(P\cup D,N)$.[5]

Definition 5.9 ((P,N)-Substitution)
Let (P,N) be a positive/negative variable-condition. σ is a (P,N)-*substitution* if
$$\sigma\text{ is a substitution on }\mathbb{V}\text{ and the }\sigma\text{-update of }(P,N)\text{ is consistent.}$$

Syntactically, $(x^{\mathrm{V}},a^{\mathrm{A}})\in N$ is to express that a (P,N)-substitution σ must not replace x^{V} with a term in which a^{A} could ever occur; i.e. that a^{A} is fresh for x^{V}: $a^{\mathrm{A}}\ \#\ x^{\mathrm{V}}$.   This is indeed guaranteed if any σ-update (P',N') of (P,N) is again

---

[5]Cf. § C.2.

required to be consistent, and so on. We can see this as follows: For $z^V \in \mathbb{V}(\sigma(x^V))$, we get
$$z^V \ P' \ x^V \ N' \ a^A.$$
If we now try to apply a second substitution σ' with $a^A \in \mathbb{A}(\sigma'(z^V))$ (so that a^A occurs in $(x^V\sigma)\sigma'$, contrary to what we initially expressed as our freshness intention), then σ' is not a (P', N')-substitution because, for the σ'-update (P'', N'') of (P', N'), we have
$$a^A \ P'' \ z^V \ P'' \ x^V \ N'' \ a^A;$$
so $(P'')^+ \circ N''$ is not irreflexive. All in all, the positive/negative variable-condition

- (P', N') blocks any instantiation of $(x^V\sigma)$ resulting in a term containing a^A, just as

- (P, N) blocked x^V before the application of σ.

5.8 Semantic Presuppositions

Instead of defining truth from scratch, we require some abstract properties typically holding in two-valued model semantics.

Truth is given relative to a Σ-structure $\mathcal{S}$, which provides some *non-empty set* as the universe (or "carrier", "domain") (for each type). Moreover, we assume that every Σ-structure $\mathcal{S}$ is not only defined on the predicate and function symbols of the signature Σ, but is defined also on the symbols $\forall$ and $\exists$ such that $\mathcal{S}(\exists)$ serves as a function-choice function for the universe function $\mathcal{S}(\forall)$ in the sense that, for each type α of Σ, the universe for the type α is denoted by $\mathcal{S}(\forall)_\alpha$ and
$$\mathcal{S}(\exists)_\alpha \in \mathcal{S}(\forall)_\alpha.$$
For $\mathrm{X} \subseteq \mathbb{VAB}$, we denote the set of total $\mathcal{S}$-valuations of X (i.e. functions mapping atoms and variables in X to objects of the universe of $\mathcal{S}$) with
$$\mathrm{X} \to \mathcal{S},$$
and the set of (possibly) partial $\mathcal{S}$-valuations of X with
$$\mathrm{X} \rightsquigarrow \mathcal{S}.$$
Here we expect types to be respected in the sense that, for each $\delta : \mathrm{X} \to \mathcal{S}$ and for each $x^{VAB} \in \mathrm{X}$ with $x^{VAB} : \alpha$ (i.e. x^{VAB} has type α), we have $\delta(x^{VAB}) \in \mathcal{S}(\forall)_\alpha$.

For $\delta : \mathrm{X} \to \mathcal{S}$, we denote with "$\mathcal{S} \uplus \delta$" the extension of $\mathcal{S}$ to X. More precisely, we assume some evaluation function "eval" such that $\mathrm{eval}(\mathcal{S} \uplus \delta)$ maps every term whose free-occurring symbols are from $\Sigma \uplus \mathrm{X}$ into the universe of $\mathcal{S}$ (respecting types). Moreover, $\mathrm{eval}(\mathcal{S} \uplus \delta)$ maps every formula B whose free-occurring symbols are from $\Sigma \uplus \mathrm{X}$ to TRUE or FALSE, such that:
$$B \text{ is true in } \mathcal{S} \uplus \delta \quad \text{iff} \quad \mathrm{eval}(\mathcal{S} \uplus \delta)(B) = \mathsf{TRUE}.$$

We leave open what our formulas and what our Σ-structures exactly are. The latter can range from first-order Σ-structures to higher-order modal Σ-models; provided that the following three properties — which (explicitly or implicitly) belong to the standard of most logic textbooks — hold for every term or formula B, every Σ-structure $\mathcal{S}$, and every $\mathcal{S}$-valuation $\delta : \text{VAB} \rightsquigarrow \mathcal{S}$.

EXPLICITNESS LEMMA

The value of the evaluation of B depends only on the valuation of those variables and atoms that actually have free occurrences in B; i.e., for $X := \text{VAB}(B)$, if $X \subseteq \text{dom}(\delta)$, then:

$$\text{eval}(\mathcal{S} \uplus \delta)(B) = \text{eval}(\mathcal{S} \uplus {}_{X|}\delta)(B).$$

SUBSTITUTION [VALUE] LEMMA

Let σ be a substitution on VAB. If $\text{VAB}(B\sigma) \subseteq \text{dom}(\delta)$, then:

$$\text{eval}(\mathcal{S} \uplus \delta)(B\sigma) = \text{eval}\Big(\mathcal{S} \uplus \big((\sigma \uplus {}_{\text{VAB}\setminus\text{dom}(\sigma)|}\text{id}) \circ \text{eval}(\mathcal{S} \uplus \delta) \big) \Big)(B).$$

VALUATION LEMMA

The evaluation function treats application terms from VAB straightforwardly in the sense that for every $v_0^{\text{VAB}}, \ldots, v_{l-1}^{\text{VAB}}, y^{\text{VAB}} \in \text{dom}(\delta)$ with $v_0^{\text{VAB}} : \alpha_0, \ldots, v_{l-1}^{\text{VAB}} : \alpha_{l-1}$, $y^{\text{VAB}} : \alpha_0 \to \cdots \to \alpha_{l-1} \to \alpha_l$ for some types $\alpha_0, \ldots, \alpha_{l-1}, \alpha_l$, we have:

$$\text{eval}(\mathcal{S} \uplus \delta)(y^{\text{VAB}}(v_0^{\text{VAB}}) \cdots (v_{l-1}^{\text{VAB}})) = \delta(y^{\text{VAB}})(\delta(v_0^{\text{VAB}})) \cdots (\delta(v_{l-1}^{\text{VAB}})).$$

Note that we need the case of the VALUATION LEMMA where y^{VAB} is a higher-order symbol (i.e. the case of $l \succ 0$) only when higher-order choice-conditions are required. Besides this, the basic language of the general reasoning framework, however, may well be first-order and does not have to include function application.

Moreover, in the few cases where we explicitly refer to quantifiers, implication, or negation, such as in our inference rules of §§ 3.2, 3.3, and 3.4. or in our version of axiom (ε_0) (cf. Definition 4.11), and in the lemmas and theorems that refer to these (namely Lemmas 5.19 and 5.25, Theorem 5.27(6), and Theorem 5.28),[6] we have to know that the quantifiers, the implication, and the negation show the standard semantic behavior of classical logic:

∀-LEMMA

Assume $\text{VAB}(\forall x^{\text{B}}.\, A) \subseteq \text{dom}(\delta)$. The following two are equivalent:

- $\text{eval}(\mathcal{S} \uplus \delta)(\forall x^{\text{B}}.\, A) = \mathsf{TRUE}$

- $\text{eval}(\mathcal{S} \uplus {}_{\text{VAB}\setminus\{x^{\text{B}}\}|}\delta \uplus \chi)(A) = \mathsf{TRUE}$ for every $\chi : \{x^{\text{B}}\} \to \mathcal{S}$

---

[6]Lemma 5.19 depends on the backward directions of the ∀-LEMMA and the ⇒-LEMMA, and on the forward direction of the ∃-LEMMA. Lemma 5.25 and Theorem 5.27(6) depend on the forward directions of the ∀-LEMMA and the ⇒-LEMMA, and on the backward direction of the ∃-LEMMA. Theorem 5.28 depends on both directions of the ∀-LEMMA, of the ∃-LEMMA, and of the ¬-LEMMA.

∃-LEMMA
Assume $\mathbb{VAB}(\exists x^{\mathrm{B}}.\ A) \subseteq \mathrm{dom}(\delta)$. The following two are equivalent:

- $\mathrm{eval}(\mathcal{S} \uplus \delta)(\exists x^{\mathrm{B}}.\ A) = \mathsf{TRUE}$,

- $\mathrm{eval}(\mathcal{S} \uplus {}_{\mathbb{VAB}\backslash\{x^{\mathrm{B}}\}}|\delta \uplus \chi)(A) = \mathsf{TRUE}$ for some $\chi : \{x^{\mathrm{B}}\} \to \mathcal{S}$

⇒-LEMMA
Assume $\mathbb{VAB}(A{\Rightarrow}B) \subseteq \mathrm{dom}(\delta)$. The following two are equivalent:

- $\mathrm{eval}(\mathcal{S} \uplus \delta)(A{\Rightarrow}B) = \mathsf{TRUE}$

- $\mathrm{eval}(\mathcal{S} \uplus \delta)(A) = \mathsf{FALSE}$ or $\mathrm{eval}(\mathcal{S} \uplus \delta)(B) = \mathsf{TRUE}$

¬-LEMMA
Assume $\mathbb{VAB}(A) \subseteq \mathrm{dom}(\delta)$. The following two are equivalent:

- $\mathrm{eval}(\mathcal{S} \uplus \delta)(A) = \mathsf{TRUE}$

- $\mathrm{eval}(\mathcal{S} \uplus \delta)(\neg A) = \mathsf{FALSE}$

5.9   Semantic Relations and $\mathcal{S}$-Raising-Valuations

We now come to some technical definitions required for our semantic (model-theoretic) counterparts of our syntactic (P, N)-substitutions.

Let $\mathcal{S}$ be a Σ-structure. An $\mathcal{S}$-*raising-valuation* π plays the rôle of a *raising function*, a dual of a Skolem function as defined in [75]. This means that π does not simply map each variable directly to an object of $\mathcal{S}$ (of the same type), but may additionally read the values of some atoms under an $\mathcal{S}$-valuation $\tau : \mathbb{A} \to \mathcal{S}$. More precisely, we assume that π takes some restriction of τ as a second argument, say $\tau' : \mathbb{A} \rightsquigarrow \mathcal{S}$ with $\tau' \subseteq \tau$. In short:
$$\pi : \mathbb{V} \to (\mathbb{A} \rightsquigarrow \mathcal{S}) \rightsquigarrow \mathcal{S}.$$
Moreover, for each variable x^{V}, we require that the set $\mathrm{dom}(\tau')$ of atoms read by $\pi(x^{\mathrm{V}})$ is identical for all τ. This identical set will be denoted with $\mathcal{S}_\pi\langle\!\langle\{x^{\mathrm{V}}\}\rangle\!\rangle$ below. Technically, we require that there is some "semantic relation" $S_\pi \subseteq \mathbb{A}{\times}\mathbb{V}$ such that for all $x^{\mathrm{V}} \in \mathbb{V}$:
$$\pi(x^{\mathrm{V}}) \ : \ (S_\pi\langle\!\langle\{x^{\mathrm{V}}\}\rangle\!\rangle \to \mathcal{S}) \to \mathcal{S}.$$
This means that $\pi(x^{\mathrm{V}})$ can read the τ-value of y^{A} if and only if $(y^{\mathrm{A}}, x^{\mathrm{V}}) \in S_\pi$. Note that, for each $\pi : \mathbb{V} \to (\mathbb{A} \rightsquigarrow \mathcal{S}) \rightsquigarrow \mathcal{S}$, at most one such semantic relation exists, namely the one of the following definition.

Definition 5.10 (Semantic Relation (S_π))
The *semantic relation for* π is
$$S_\pi \ := \ \{\ (y^{\mathrm{A}}, x^{\mathrm{V}}) \mid x^{\mathrm{V}} \in \mathbb{V} \wedge y^{\mathrm{A}} \in \mathrm{dom}(\textstyle\bigcup(\mathrm{dom}(\pi(x^{\mathrm{V}})))) \ \}.$$

Definition 5.11 ($\mathcal{S}$-Raising-Valuation)
Let $\mathcal{S}$ be a Σ-structure. π is an $\mathcal{S}$-*raising-valuation* if
$$\pi : \mathbb{V} \to (\mathbb{A} \leadsto \mathcal{S}) \leadsto \mathcal{S}$$
and, for all $x^{\mathrm{V}} \in \mathrm{dom}(\pi)$:
$$\pi(x^{\mathrm{V}}) : (S_\pi \langle\!\langle x^{\mathrm{V}} \rangle\!\rangle \to \mathcal{S}) \to \mathcal{S}.$$

Finally, we need the technical means to turn an $\mathcal{S}$-raising-valuation π together with an $\mathcal{S}$-valuation τ of the atoms into an $\mathcal{S}$-valuation $\mathsf{e}(\pi)(\tau)$ of the variables:

Definition 5.12 (e)
We define the function $\quad \mathsf{e} : \quad (\mathbb{V} \to (\mathbb{A} \leadsto \mathcal{S}) \leadsto \mathcal{S}) \quad \to \quad (\mathbb{A} \to \mathcal{S}) \quad \to \quad \mathbb{V} \quad \leadsto \quad \mathcal{S}$
for $\quad\quad\quad\quad\quad \pi : \mathbb{V} \to (\mathbb{A} \leadsto \mathcal{S}) \leadsto \mathcal{S}, \quad\quad \tau : \mathbb{A} \to \mathcal{S}, \quad x^{\mathrm{V}} \in \mathbb{V}$
by $\quad\quad\quad\quad\quad \mathsf{e}(\pi)(\tau)(x^{\mathrm{V}}) := \pi(x^{\mathrm{V}})(_{S_\pi \langle\!\langle x^{\mathrm{V}} \rangle\!\rangle} | \tau).$

The "e" stands for "<u>e</u>valuation" and replaces an "ϵ" used in previous publications, which was too easily confused with Hilbert's ε.

5.10 Choice-Conditions and Compatibility

In the following definition, we define choice-conditions as syntactic objects. They influence our semantics by a compatibility requirement, which will be described in Definition 5.15.

Definition 5.13 (Choice-Condition, Return Type)
C is a (P, N)-*choice-condition* if

- (P, N) is a consistent positive/negative variable-condition and

- C is a partial function from $\mathbb{V}$ into the set of higher-order ε-terms

such that, for every $y^{\mathrm{V}} \in \mathrm{dom}(C)$, the following items hold for some types $\alpha_0, \ldots, \alpha_l$:

1. The value $C(y^{\mathrm{V}})$ is of the form
$$\lambda v_0^{\mathrm{B}}. \ \ldots \ \lambda v_{l-1}^{\mathrm{B}}. \ \varepsilon v_l^{\mathrm{B}}. \ B$$
for some formula B and for some mutually distinct bound atoms $v_0^{\mathrm{B}}, \ldots, v_l^{\mathrm{B}}$ $\in \mathbb{B}$ with $\mathbb{B}(B) \subseteq \{v_0^{\mathrm{B}}, \ldots, v_l^{\mathrm{B}}\}$ and $v_0^{\mathrm{B}} : \alpha_0, \ \ldots, \ v_l^{\mathrm{B}} : \alpha_l$.

2. $y^{\mathrm{V}} : \alpha_0 \to \cdots \to \alpha_{l-1} \to \alpha_l$.

3. $z^{\mathrm{VA}} \ P^+ \ y^{\mathrm{V}} \quad$ for all $\ z^{\mathrm{VA}} \in \mathbb{VA}(C(y^{\mathrm{V}}))$.

In the situation described, α_l is *the return type of* $C(y^{\mathrm{V}})$.
β is *a return type of* C if there is a $z^{\mathrm{V}} \in \mathrm{dom}(C)$ such that β is the return type of $C(z^{\mathrm{V}})$.

Example 5.14 (Choice-Condition)   *(continuing Example 4.10)*

(a) If (P,N) is a consistent positive/negative variable-condition that satisfies

$$z_a^\text{V}\ P\ y_a^\text{V}\ P\ z_b^\text{V}\ P\ x_a^\text{V}\ P\ z_c^\text{V}\ P\ y_b^\text{V}\ P\ z_d^\text{V}\ P\ w_a^\text{V}\ P\ z_e^\text{V}\ P\ y_c^\text{V}\ P\ z_f^\text{V}\ P\ x_b^\text{V}\ P\ z_g^\text{V}\ P\ y_d^\text{V}\ P\ z_h^\text{V},$$

then the C of Example 4.10 is a (P,N)-choice-condition, indeed.

(b) If some clever person tried to do the entire quantifier elimination of Example 4.10 by

$$
\begin{aligned}
C'(z_h^\text{V}) &:= \quad \varepsilon z_h^\text{B}.\ \neg\mathsf{P}(w_a^\text{V}, x_b^\text{V}, y_d^\text{V}, z_h^\text{B})\\
C'(y_d^\text{V}) &:= \quad \varepsilon y_d^\text{B}.\ \mathsf{P}(w_a^\text{V}, x_b^\text{V}, y_d^\text{B}, z_h^\text{V})\\
C'(x_b^\text{V}) &:= \quad \varepsilon x_b^\text{B}.\ \neg\mathsf{P}(w_a^\text{V}, x_b^\text{B}, y_d^\text{V}, z_h^\text{V})\\
C'(w_a^\text{V}) &:= \quad \varepsilon w_a^\text{B}.\ \mathsf{P}(w_a^\text{B}, x_b^\text{V}, y_d^\text{V}, z_h^\text{V})
\end{aligned}
$$

then he would — among other constraints — have to satisfy $z_h^\text{V}\ P^+\ y_d^\text{V}\ P^+\ z_h^\text{V}$, because of item 3 of Definition 5.13 and the values of C' at y_d^V and z_h^V. This would make P non-well-founded. Thus, this C' cannot be a (P,N)-choice-condition for any (P,N), because the consistency of (P,N) is required in Definition 5.13. Note that the choices required by C' for y_d^V and z_h^V are in an unsolvable conflict, indeed.

(c) For a more elementary example, take

$$C''(x^\text{V}) \quad := \quad \varepsilon x^\text{B}.\ (x^\text{B}=y^\text{V}) \qquad\qquad C''(y^\text{V}) \quad := \quad \varepsilon y^\text{B}.\ (x^\text{V}\neq y^\text{B})$$

Then x^V and y^V form a vicious circle of conflicting choices for which no valuation can be found that is compatible with C''. But C'' is no choice-condition at all because there is no *consistent* positive/negative variable-condition (P,N) such that C'' is a (P,N)-choice-condition.

Definition 5.15 (Compatibility)

Let C be a (P,N)-choice-condition. Let $\mathcal{S}$ be a Σ-structure.

π is $\mathcal{S}$-*compatible with* $(C,(P,N))$ if the following items hold:

1. π is an $\mathcal{S}$-raising-valuation (cf. Definition 5.11) and $(P\cup S_\pi, N)$ is consistent (cf. Definitions 5.4 and 5.10).

2. For every $y^\text{V}\in\text{dom}(C)$ with $C(y^\text{V})=\lambda v_0^\text{B}.\ \dots\ \lambda v_{l-1}^\text{B}.\ \varepsilon v_l^\text{B}.\ B$ for some formula B, and for every $\tau:\mathbb{A}\to\mathcal{S}$, and for every $\chi:\{v_0^\text{B},\dots,v_l^\text{B}\}\to\mathcal{S}$:

    If B is true in $\mathcal{S}\uplus\mathsf{e}(\pi)(\tau)\uplus\tau\uplus\chi$,

    then $B\{v_l^\text{B}\mapsto y^\text{V}(v_0^\text{B})\cdots(v_{l-1}^\text{B})\}$ is true in $\mathcal{S}\uplus\mathsf{e}(\pi)(\tau)\uplus\tau\uplus\chi$ as well.

    (For e, see Definition 5.12.)

To understand item 2 of Definition 5.15, let us consider a (P, N)-choice-condition

$$C := \{(y^{\vee}, \lambda v_0^{\text{B}}. \ldots \lambda v_{l-1}^{\text{B}}. \varepsilon v_l^{\text{B}}. B)\},$$

which restricts the value of $y^{\vee}$ according to the higher-order ε-term $\lambda v_0^{\text{B}}. \ldots \lambda v_{l-1}^{\text{B}}. \varepsilon v_l^{\text{B}}. B$. Then, roughly speaking, this choice-condition C requires that whenever there is a χ-value of v_l^{B} such that B is true in $\mathcal{S} \uplus \mathsf{e}(\pi)(\tau) \uplus \tau \uplus \chi$, the π-value of $y^{\vee}$ is chosen in such a way that $B\{v_l^{\text{B}} \mapsto y^{\vee}(v_0^{\text{B}}) \cdots (v_{l-1}^{\text{B}})\}$ becomes true in $\mathcal{S} \uplus \mathsf{e}(\pi)(\tau) \uplus \tau \uplus \chi$ as well. Note that the free variables of the formula $B\{v_l^{\text{B}} \mapsto y^{\vee}(v_0^{\text{B}}) \cdots (v_{l-1}^{\text{B}})\}$ cannot read the χ-value of any of the bound atoms $v_0^{\text{B}}, \ldots, v_l^{\text{B}}$, because free variables can never depend on the value of any bound atoms.

Moreover, item 2 of Definition 5.15 is closely related to Hilbert's ε-operator in the sense that — roughly speaking — $y^{\vee}$ must be given one of the values admissible for

$$\lambda v_0^{\text{B}}. \ldots \lambda v_{l-1}^{\text{B}}. \varepsilon v_l^{\text{B}}. B.$$

As the choice for $y^{\vee}$ depends on the symbols that have a free occurrence in that higher-order ε-term, we included these dependencies into the positive relation P of the consistent positive/negative variable-condition (P, N) in item 3 of Definition 5.13. By this inclusion, conflicts like the one shown in Example 5.14(c) are obviated.

Let (P, N) be a consistent positive/negative variable-condition. Then the empty function $\emptyset$ is a (P, N)-choice-condition. Moreover, each $\pi : \mathbb{V} \to \{\emptyset\} \to \mathcal{S}$ is $\mathcal{S}$-compatible with $(\emptyset, (P, N))$ because of $S_{\pi} = \emptyset$. Furthermore, assuming an adequate principle of choice on the meta level, a compatible π always exists according to the following lemma. This existence relies on item 3 of Definition 5.13 and on the well-foundedness of P.

Lemma 5.16 *Let C be a (P, N)-choice-condition. Let $\mathcal{S}$ be a Σ-structure.*
Assume that, for every return type α of C, there is a generalized choice function on the power-set of $\mathcal{S}(\forall)_{\alpha}$.
[Let ρ be an $\mathcal{S}$-raising-valuation with $S_{\rho} \subseteq P^{+}$.]
Then there is an $\mathcal{S}$-raising-valuation π such that the following hold:

- *π is $\mathcal{S}$-compatible with $(C, (P, N))$.*
- *$S_{\pi} = \mathbb{A}\!\downharpoonright(P^{+})$.*
- *[$\mathbb{V}\backslash\mathrm{dom}(C)\!\downharpoonright\pi = \mathbb{V}\backslash\mathrm{dom}(C)\!\downharpoonright\rho.$]*

5.11 $(C, (P, N))$-Validity

Definition 5.17 ($(C, (P, N))$-Validity, K)

Let C be a (P, N)-choice-condition. Let G be a set of sequents.
Let $\mathcal{S}$ be a Σ-structure. Let $\delta : \mathbb{VA} \rightsquigarrow \mathcal{S}$ be an $\mathcal{S}$-valuation.

G is $(C, (P, N))$-*valid in* $\mathcal{S}$ if
$\quad\quad\quad\quad$ G is $(\pi, \mathcal{S})$-valid for some π that is $\mathcal{S}$-compatible with $(C, (P, N))$.

G is $(\pi, \mathcal{S})$-*valid* if G is true in $\mathcal{S} \uplus \mathsf{e}(\pi)(\tau) \uplus \tau$ for every $\tau : \mathbb{A} \to \mathcal{S}$.

G is *true in* $\mathcal{S} \uplus \delta$ if Γ is true in $\mathcal{S} \uplus \delta$ for all $\Gamma \in G$.

A sequent Γ is *true in* $\mathcal{S} \uplus \delta$ if there is some formula listed in Γ that is true in $\mathcal{S} \uplus \delta$. Validity in a class of Σ-structures is understood as validity in each of the Σ-structures of that class. If we omit the reference to a special Σ-structure we mean validity in some fixed class K of Σ-structures, such as the class of all Σ-structures or the class of Herbrand Σ-structures.

Example 5.18 ($(\emptyset, (P, N))$-Validity)

For $x^{\mathrm{V}} \in \mathbb{V}$, $y^{\mathrm{A}} \in \mathbb{A}$, the single-formula sequent $x^{\mathrm{V}} = y^{\mathrm{A}}$ is $(\emptyset, (\emptyset, \emptyset))$-valid in any Σ-structure $\mathcal{S}$ because we can choose $S_\pi := \mathbb{A} \times \mathbb{V}$ and $\pi(x^{\mathrm{V}})(\tau) := \tau(y^{\mathrm{A}})$ for $\tau : \mathbb{A} \to \mathcal{S}$, resulting in

$$\mathsf{e}(\pi)(\tau)(x^{\mathrm{V}}) = \pi(x^{\mathrm{V}})(S_\pi \langle\!\langle x^{\mathrm{V}} \rangle\!\rangle | \tau) = \pi(x^{\mathrm{V}})(\mathbb{A} | \tau) = \pi(x^{\mathrm{V}})(\tau) = \tau(y^{\mathrm{A}}).$$

This means that $(\emptyset, (\emptyset, \emptyset))$-validity of $x^{\mathrm{V}} = y^{\mathrm{A}}$ is equivalent to validity of

$$\forall y_0^{\mathrm{B}}.\ \exists x_0^{\mathrm{B}}.\ (x_0^{\mathrm{B}} = y_0^{\mathrm{B}}). \tag{1}$$

Moreover, note that $\mathsf{e}(\pi)(\tau)$ has access to the τ-value of y^{A} just as a raising function x_1^{B} for x_0^{B} has access to y_0^{B} in the raised (i.e. dually Skolemized) form $\exists x_1^{\mathrm{B}}.\ \forall y_0^{\mathrm{B}}.$ $(x_1^{\mathrm{B}}(y_0^{\mathrm{B}}) = y_0^{\mathrm{B}})$ of (1).

Contrary to this, for $P := \emptyset$ and $N := \mathbb{V} \times \mathbb{A}$, the same single-formula sequent $x^{\mathrm{V}} = y^{\mathrm{A}}$ is not $(\emptyset, (P, N))$-valid in general, because then the required consistency of $(P \cup S_\pi, N)$ implies $S_\pi = \emptyset$; otherwise $P \cup S_\pi \cup N$ has a cycle of length 2 with exactly one edge from N. Thus, the value of x^{V} cannot depend on $\tau(y^{\mathrm{A}})$ anymore:

$$\pi(x^{\mathrm{V}})(S_\pi \langle\!\langle x^{\mathrm{V}} \rangle\!\rangle | \tau) = \pi(x^{\mathrm{V}})(\emptyset | \tau) = \pi(x^{\mathrm{V}})(\emptyset).$$

This means that $(\emptyset, (\emptyset, \mathbb{V} \times \mathbb{A}))$-validity of $x^{\mathrm{V}} = y^{\mathrm{A}}$ is equivalent to validity of

$$\exists x_0^{\mathrm{B}}.\ \forall y_0^{\mathrm{B}}.\ (x_0^{\mathrm{B}} = y_0^{\mathrm{B}}). \tag{2}$$

Moreover, note that $\mathsf{e}(\pi)(\tau)$ has no access to the τ-value of y^{A} just as a raising function x_1^{B} for x_0^{B} has no access to y_0^{B} in the raised form $\exists x_1^{\mathrm{B}}.\ \forall y_0^{\mathrm{B}}.\ (x_1^{\mathrm{B}}() = y_0^{\mathrm{B}})$ of (2).

For a more general example let $G = \{\ A_{i,0} \ldots A_{i,n_i-1} \mid i \in I\ \}$, where, for $i \in I$ and $j \prec n_i$, the $A_{i,j}$ are formulas with variables from $\boldsymbol{v}$ and atoms from $\boldsymbol{a}$. Then $(\emptyset, (\emptyset, \mathbb{V} \times \mathbb{A}))$-validity of G means validity of

$$\exists \boldsymbol{v}.\ \forall \boldsymbol{a}.\ \forall i \in I.\ \exists j \prec n_i.\ A_{i,j}$$

whereas $(\emptyset, (\emptyset, \emptyset))$-validity of G means validity of

$$\forall \boldsymbol{a}.\ \exists \boldsymbol{v}.\ \forall i \in I.\ \exists j \prec n_i.\ A_{i,j}$$

Ignoring the question of γ-multiplicity, also any other sequence of universal and existential quantifiers can be represented by a consistent positive/negative variable-condition (P, N), simply by starting from the consistent positive/negative variable-condition $(\emptyset, \emptyset)$ and applying the γ- and δ-rules from §§ 3.2, 3.3, and 3.4. A reverse translation of a positive/negative variable-condition (P, N) into a sequence of quantifiers, however, may require a strengthening of dependencies, in the sense that a subsequent backward translation would result in a *more restrictive* consistent positive/negative variable-condition (P', N') with $P \subseteq P'$ and $N \subseteq N'$. This means that our framework can express quantificational dependencies more fine-grained than standard quantifiers; cf. Example 5.20.

As already noted in § 4.12, the single-formula sequent $Q_C(y^{\mathrm{v}})$ of Definition 4.11 is a formulation of axiom (ε_0) of § 4.6 in our framework.

Lemma 5.19 ($((C, (P, N))$-Validity of $Q_C(y^{\mathrm{v}})$)
Let C be a (P, N)-choice-condition. Let $y^{\mathrm{v}} \in \mathrm{dom}(C)$. Let $\mathcal{S}$ be a Σ-structure.

1. *$Q_C(y^{\mathrm{v}})$ is $(\pi, \mathcal{S})$-valid for every π that is $\mathcal{S}$-compatible with $(C, (P, N))$.*

2. *$Q_C(y^{\mathrm{v}})$ is $(C, (P, N))$-valid in $\mathcal{S}$; provided that for every return type α of C (cf. Definition 5.13), there is a generalized choice function on the power-set of $\mathcal{S}(\forall)_\alpha$.*

In [111, § 6.4.1], we showed that Henkin quantification was problematic for the variable-conditions of that paper, which had only one component, namely the positive one of our positive/negative variable-conditions here: Indeed, there the only way to model an example of a Henkin quantification precisely was to increase the order of some variables by raising. Let us consider the same example here again and show that now we can model its Henkin quantification directly with a *consistent* positive/*negative* variable-condition, but *without raising*.

Example 5.20 (Henkin **Quantification**)

In Hintikka's [63] of 1974, quantifiers in first-order logic were found insufficient to give the precise semantics of some English sentences. In Hintikka's [64] of 1974, *IF logic*, i.e. Independence-Friendly logic — a first-order logic with more flexible quantifiers — was presented to overcome this weakness. In [63], we find the following sentence:

> Some relative of each villager and
> some relative of each townsman hate each other. (H0)

Let us first change to a lovelier subject:

> Some loved one of each woman and
> some loved one of each man love each other. (H1)

For our purposes here, we consider (H1) to be equivalent to the following sentence, which may be more meaningful and easier to understand:

> We can fix a loved one for each woman and a loved one for each man,
> such that for every pair of woman and man, these loved ones could love
> each other.

(H1) can be represented by the following Henkin-quantified IF-logic formula:

$$\forall x_0^{\mathbb{B}}.\ \forall y_0^{\mathbb{B}}.\left(\left(\begin{array}{c}\mathsf{Female}(x_0^{\mathbb{B}})\\ \wedge\ \ \mathsf{Male}(y_0^{\mathbb{B}})\end{array}\right) \Rightarrow \exists y_1^{\mathbb{B}}/y_0^{\mathbb{B}}.\ \exists x_1^{\mathbb{B}}/x_0^{\mathbb{B}}.\left(\begin{array}{c}\mathsf{Loves}(x_0^{\mathbb{B}}, y_1^{\mathbb{B}})\\ \wedge\ \ \mathsf{Loves}(y_0^{\mathbb{B}}, x_1^{\mathbb{B}})\\ \wedge\ \ \mathsf{Loves}(y_1^{\mathbb{B}}, x_1^{\mathbb{B}})\\ \wedge\ \ \mathsf{Loves}(x_1^{\mathbb{B}}, y_1^{\mathbb{B}})\end{array}\right)\right) \quad \text{(H2)}$$

Let us refer to the standard game-theoretic semantics for quantifiers (cf. e.g. [64]), which is defined as follows: Witnesses have to be picked for the quantified variables outside-in. We have to pick the witnesses for the γ-quantifiers (i.e., in (H2), for the existential quantifiers), and our opponent in the game picks the witnesses for the δ-quantifiers (i.e. for the universal quantifiers in (H2)). We win if the resulting quantifier-free formula evaluates to true. A formula is true if we have a winning strategy.

Then an IF-logic quantifier such as "$\exists y_1^{\mathbb{B}}/y_0^{\mathbb{B}}.$" in (H2) is a special quantifier, which is a bit different from "$\exists y_1^{\mathbb{B}}.$". Game-theoretically, it has the following semantics: It asks us to pick the loved one $y_1^{\mathbb{B}}$ independently from the choice of the man $y_0^{\mathbb{B}}$ (by our opponent in the game), although the IF-logic quantifier occurs in the scope of the quantifier "$\forall y_0^{\mathbb{B}}.$".

Note that Formula (H2) is already close to anti-prenex form. In fact, if we move its quantifiers closer toward the leaves of the formula tree, this does not admit us to reduce their dependencies. It is more interesting, however, to move the quantifiers of (H2) out — to obtain prenex form — and then to simplify the prenex form by using the equivalence of "$\forall y_0^{\mathbb{B}}.\ \exists y_1^{\mathbb{B}}/y_0^{\mathbb{B}}.$" and "$\exists y_1^{\mathbb{B}}.\ \forall y_0^{\mathbb{B}}.$", resulting in:

$$\forall x_0^{\mathbb{B}}.\ \exists y_1^{\mathbb{B}}.\ \forall y_0^{\mathbb{B}}.\ \exists x_1^{\mathbb{B}}/x_0^{\mathbb{B}}.\left(\left(\begin{array}{c}\mathsf{Female}(x_0^{\mathbb{B}})\\ \wedge\quad \mathsf{Male}(y_0^{\mathbb{B}})\end{array}\right)\Rightarrow\left(\begin{array}{c}\mathsf{Loves}(x_0^{\mathbb{B}},y_1^{\mathbb{B}})\\ \wedge\quad \mathsf{Loves}(y_0^{\mathbb{B}},x_1^{\mathbb{B}})\\ \wedge\quad \mathsf{Loves}(y_1^{\mathbb{B}},x_1^{\mathbb{B}})\\ \wedge\quad \mathsf{Loves}(x_1^{\mathbb{B}},y_1^{\mathbb{B}})\end{array}\right)\right)\qquad\text{(H2}')$$

Note that this formula has a semantics different from the following formula with standard quantifiers:

$$\forall x_0^{\mathbb{B}}.\ \exists y_1^{\mathbb{B}}.\ \forall y_0^{\mathbb{B}}.\ \exists x_1^{\mathbb{B}}.\left(\left(\begin{array}{c}\mathsf{Female}(x_0^{\mathbb{B}})\\ \wedge\quad \mathsf{Male}(y_0^{\mathbb{B}})\end{array}\right)\Rightarrow\left(\begin{array}{c}\mathsf{Loves}(x_0^{\mathbb{B}},y_1^{\mathbb{B}})\\ \wedge\quad \mathsf{Loves}(y_0^{\mathbb{B}},x_1^{\mathbb{B}})\\ \wedge\quad \mathsf{Loves}(y_1^{\mathbb{B}},x_1^{\mathbb{B}})\\ \wedge\quad \mathsf{Loves}(x_1^{\mathbb{B}},y_1^{\mathbb{B}})\end{array}\right)\right)\qquad\text{(S2}')$$

An alternative way to define the semantics of IF-logic quantifiers is by describing their effect on the equivalent *raised* forms of the formulas in which they occur. *Raising* is a dual of Skolemization, cf. [75]. The raised form of (S2$'$) is the following:

$$\exists y_1^{\mathbb{B}}.\ \exists x_1^{\mathbb{B}}.\ \forall x_0^{\mathbb{B}}.\ \forall y_0^{\mathbb{B}}.\left(\left(\begin{array}{c}\mathsf{Female}(x_0^{\mathbb{B}})\\ \wedge\quad \mathsf{Male}(y_0^{\mathbb{B}})\end{array}\right)\Rightarrow\left(\begin{array}{c}\mathsf{Loves}(x_0^{\mathbb{B}},y_1^{\mathbb{B}}(x_0^{\mathbb{B}}))\\ \wedge\quad \mathsf{Loves}(y_0^{\mathbb{B}},x_1^{\mathbb{B}}(y_0^{\mathbb{B}},x_0^{\mathbb{B}}))\\ \wedge\quad \mathsf{Loves}(y_1^{\mathbb{B}}(x_0^{\mathbb{B}}),x_1^{\mathbb{B}}(y_0^{\mathbb{B}},x_0^{\mathbb{B}}))\\ \wedge\quad \mathsf{Loves}(x_1^{\mathbb{B}}(y_0^{\mathbb{B}},x_0^{\mathbb{B}}),y_1^{\mathbb{B}}(x_0^{\mathbb{B}}))\end{array}\right)\right)\qquad\text{(S3)}$$

For Henkin-quantified IF-logic formulas, the raised form is defined as usual, besides that a γ-quantifier, say "$\exists x_1^{\mathbb{B}}.$", followed by a slash as in "$\exists x_1^{\mathbb{B}}/x_0^{\mathbb{B}}.$", is raised such that $x_0^{\mathbb{B}}$ does not appear as an argument to the raising function for $x_1^{\mathbb{B}}$. Accordingly, *mutatis mutandis*, (H2) as well as (H2$'$) are equivalent to their common raised form (H3) below, where $x_0^{\mathbb{B}}$ does not occur as an argument to the raising function $x_1^{\mathbb{B}}$ — contrary to (S3), which is strictly implied by (H3) because we can choose the loved one of the woman differently for different men.

$$\exists y_1^{\mathbb{B}}.\ \exists x_1^{\mathbb{B}}.\ \forall x_0^{\mathbb{B}}.\ \forall y_0^{\mathbb{B}}.\left(\left(\begin{array}{c}\mathsf{Female}(x_0^{\mathbb{B}})\\ \wedge\quad \mathsf{Male}(y_0^{\mathbb{B}})\end{array}\right)\Rightarrow\left(\begin{array}{c}\mathsf{Loves}(x_0^{\mathbb{B}},y_1^{\mathbb{B}}(x_0^{\mathbb{B}}))\\ \wedge\quad \mathsf{Loves}(y_0^{\mathbb{B}},x_1^{\mathbb{B}}(y_0^{\mathbb{B}}))\\ \wedge\quad \mathsf{Loves}(y_1^{\mathbb{B}}(x_0^{\mathbb{B}}),x_1^{\mathbb{B}}(y_0^{\mathbb{B}}))\\ \wedge\quad \mathsf{Loves}(x_1^{\mathbb{B}}(y_0^{\mathbb{B}}),y_1^{\mathbb{B}}(x_0^{\mathbb{B}}))\end{array}\right)\right)\qquad\text{(H3)}$$

Now, (H3) looks already very much like the following tentative representation of (H1) in our framework of free atoms and variables:

$$\left(\begin{array}{c}\mathsf{Female}(x_0^{\mathbb{A}})\\ \wedge\quad \mathsf{Male}(y_0^{\mathbb{A}})\end{array}\right)\Rightarrow\left(\begin{array}{c}\mathsf{Loves}(x_0^{\mathbb{A}},y_1^{\mathbb{V}})\\ \wedge\quad \mathsf{Loves}(y_0^{\mathbb{A}},x_1^{\mathbb{V}})\\ \wedge\quad \mathsf{Loves}(y_1^{\mathbb{V}},x_1^{\mathbb{V}})\\ \wedge\quad \mathsf{Loves}(x_1^{\mathbb{V}},y_1^{\mathbb{V}})\end{array}\right)\qquad\text{(H1}')$$

with choice-condition C given by

$$C(y_1^{\mathbb{V}})\quad :=\quad \varepsilon y_1^{\mathbb{B}}.\ (\mathsf{Female}(x_0^{\mathbb{A}})\Rightarrow\mathsf{Loves}(x_0^{\mathbb{A}},y_1^{\mathbb{B}}))$$
$$C(x_1^{\mathbb{V}})\quad :=\quad \varepsilon x_1^{\mathbb{B}}.\ (\mathsf{Male}(y_0^{\mathbb{A}})\Rightarrow\mathsf{Loves}(y_0^{\mathbb{A}},x_1^{\mathbb{B}}))$$

which requires our positive/negative variable-condition (P,N) to contain $(x_0^{\mathbb{A}},y_1^{\mathbb{V}})$ and $(y_0^{\mathbb{A}},x_1^{\mathbb{V}})$ in the positive relation P (by item 3 of Definition 5.13).

The concrete form of this choice-condition C was chosen to mirror the structure of the natural language sentence (H1) as close as possible. Actually, however, we do not need exactly this choice-condition here. Indeed, to find a representation in our framework, we could also work with an empty choice-condition. Crucial for our discussion, however, is that we can have

$$(x_0^{\mathbb{A}}, y_1^{\mathbb{V}}), (y_0^{\mathbb{A}}, x_1^{\mathbb{V}}) \in P;$$

otherwise the choice of the loved ones could not depend on their lovers.

In any case, we can add $(y_1^{\mathbb{V}}, y_0^{\mathbb{A}})$ to the negative relation N here, namely to express that $y_1^{\mathbb{V}}$ must not read $y_0^{\mathbb{A}}$. Then we obtain:

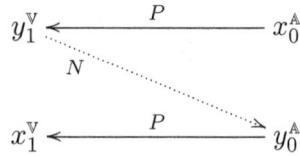

$$
\begin{array}{ccc}
y_1^{\mathbb{V}} & \xleftarrow{\quad P \quad} & x_0^{\mathbb{A}} \\[2pt]
& {\scriptstyle N} & \\[2pt]
x_1^{\mathbb{V}} & \xleftarrow{\quad P \quad} & y_0^{\mathbb{A}}
\end{array}
$$

The same variable-condition is also obtained if we start from the empty variable-condition $(\emptyset, \emptyset)$, remove all quantifiers with γ- and δ^--rules from (S2′), and then add $P := \{(x_0^{\mathbb{A}}, y_1^{\mathbb{V}}), (y_0^{\mathbb{A}}, x_1^{\mathbb{V}})\}$. The corresponding procedure for (H2′), however, has to add also $(x_1^{\mathbb{V}}, x_0^{\mathbb{A}})$ to N as part of the last γ-step that removes the IF-logic quantifier "$\exists x_1^{\mathbb{B}}/x_0^{\mathbb{A}}$." and replaces $x_1^{\mathbb{B}}$ with $x_1^{\mathbb{V}}$. After this procedure, our current positive/negative variable-condition is now given as (P, N) with $P = \{(x_0^{\mathbb{A}}, y_1^{\mathbb{V}}), (y_0^{\mathbb{A}}, x_1^{\mathbb{V}})\}$ and $N = \{(y_1^{\mathbb{V}}, y_0^{\mathbb{A}}), (x_1^{\mathbb{V}}, x_0^{\mathbb{A}})\}$. Thus, we have a single cycle in the graph, namely the following one:

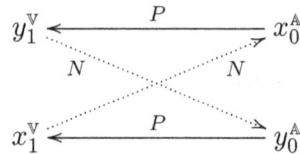

$$
\begin{array}{ccc}
y_1^{\mathbb{V}} & \xleftarrow{\quad P \quad} & x_0^{\mathbb{A}} \\[2pt]
& {\scriptstyle N} \quad {\scriptstyle N} & \\[2pt]
x_1^{\mathbb{V}} & \xleftarrow{\quad P \quad} & y_0^{\mathbb{A}}
\end{array}
$$

But this cycle necessarily has two edges from the negative relation N. Thus, in spite of this cycle, our positive/negative variable-condition (P, N) is consistent by Corollary 5.5.

With the variable-conditions of [105; 106; 107; 108; 110; 111], however, this cycle necessarily destroys the consistency, because they have no distinction between the edges of N and P.

Therefore — if the discussion in [111, § 6.4.1] is sound — our new framework of this paper with positive/*negative* variable-conditions is the only one among all approaches suitable for describing the semantics of noun phrases in natural languages that admits us to model IF logic and Henkin quantifiers without raising.

5.12 Extended Extensions

Just like the positive/negative variable-condition (P, N), the (P, N)-choice-condition C may be extended during proofs. This kind of extension together with a simple soundness condition plays an important rôle in inference:

Definition 5.21 (Extended Extension)
$(C', (P', N'))$ is an *extended extension of* $(C, (P, N))$ if
- C is a (P, N)-choice-condition (cf. Definition 5.13),
- C' is a (P', N')-choice-condition,
- (P', N') is an extension of (P, N) (cf. Definition 5.6),   and
- $C \subseteq C'$.

Lemma 5.22 (Extended Extension)
Let $(C', (P', N'))$ be an extended extension of $(C, (P, N))$.
If π is $\mathcal{S}$-compatible with $(C', (P', N'))$,   then π is $\mathcal{S}$-compatible with $(C, (P, N))$ as well.

5.13 Extended σ-Updates

After global application of a (P, N)-substitution σ, we now have to update both (P, N) and C:

Definition 5.23 (Extended σ-Update)
Let C be a (P, N)-choice-condition and let σ be a substitution on $\mathbb{V}$.
The *extended σ-update* $(C', (P', N'))$ *of* $(C, (P, N))$ is given as follows:
$$C' := \{ (x^{\vee}, B\sigma) \mid (x^{\vee}, B) \in C \wedge x^{\vee} \notin \mathrm{dom}(\sigma) \},$$
(P', N') is the σ-update of (P, N) (cf. Definition 5.8).

Note that a σ-update (cf. Definition 5.8) is an extension (cf. Definition 5.6), whereas an extended σ-update is not an extended extension in general, because entries of the choice-condition may be modified or even deleted, such that we may have $C \nsubseteq C'$. The remaining properties of an extended extension, however, are satisfied:

Lemma 5.24 (Extended σ-Update)   *Let C be a (P, N)-choice-condition.*
Let σ be a (P, N)-substitution.
Let $(C', (P', N'))$ be the extended σ-update of $(C, (P, N))$.
Then C' is a (P', N')-choice-condition.

5.14 The Main Lemma

Lemma 5.25 ((P, N)-Substitutions and $(C, (P, N))$-Validity)
Let (P, N) be a positive/negative variable-condition.
Let C be a (P, N)-choice-condition. Let σ be a (P, N)-substitution.
Let $(C', (P', N'))$ be the extended σ-update of $(C, (P, N))$. Let $\mathcal{S}$ be a Σ-structure.
Let π' be an $\mathcal{S}$-raising-valuation that is $\mathcal{S}$-compatible with $(C', (P', N'))$.
Let O and O' be two disjoint sets with $O \subseteq \mathrm{dom}(\sigma) \cap \mathrm{dom}(C)$ and $O' \subseteq \mathrm{dom}(C) \setminus O$.
Moreover, assume that σ respects C on O in the given semantic context in the sense
that $(\langle O \rangle Q_C)\sigma$ is $(\pi', \mathcal{S})$-valid (cf. Definition 4.11 for Q_C).
Furthermore, regarding the set O' (where σ may disrespect C), assume the following
items to hold:

- *O' covers the variables in $\mathrm{dom}(\sigma) \cap \mathrm{dom}(C)$ besides O in the sense of*

$$\mathrm{dom}(\sigma) \cap \mathrm{dom}(C) \;\subseteq\; O' \uplus O.$$

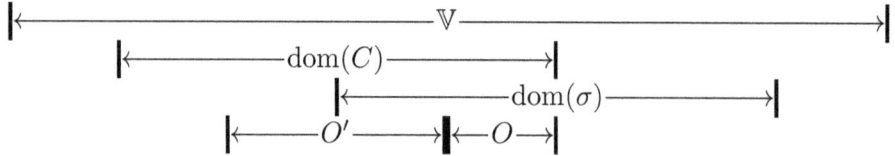

- *O' satisfies the closure condition $\langle O' \rangle P^+ \cap \mathrm{dom}(C) \subseteq O'$.*

- *For every $y^\forall \in O'$, for α being the return type of $C(y^\forall)$ (cf. Definition 5.13), there is a generalized choice function on the power-set of $\mathcal{S}(\forall)_\alpha$.*

Then there is an $\mathcal{S}$-raising-valuation π that is $\mathcal{S}$-compatible with $(C, (P, N))$ and that satisfies the following:

1. *For every term or formula B with $O' \cap \mathbb{V}(B) = \emptyset$ and possibly with some un-bound occurrences of bound atoms from a set $W \subseteq \mathbb{B}$, and for every $\tau : \mathbb{A} \to \mathcal{S}$ and every $\chi : W \to \mathcal{S}$:*

$$\mathrm{eval}(\mathcal{S} \uplus \mathrm{e}(\pi')(\tau) \uplus \tau \uplus \chi)(B\sigma) \;=\; \mathrm{eval}(\mathcal{S} \uplus \mathrm{e}(\pi)(\tau) \uplus \tau \uplus \chi)(B).$$

2. *For every set of sequents G with $O' \cap \mathbb{V}(G) = \emptyset$ we have:*

$$G\sigma \text{ is } (\pi', \mathcal{S})\text{-valid} \quad \text{iff} \quad G \text{ is } (\pi, \mathcal{S})\text{-valid}.$$

In Lemma 5.25, we illustrate the subclass relation with a Lambert diagram [71, Dianoiologie, §§ 173–194], similar to a Venn diagram. In general, a Lambert diagram expresses nothing but the following: If — in vertical projection — each point of the overlap of the lines for classes $A_1, \ldots, A_m$ is covered by a line of the classes $B_1, \ldots, B_n$ then $A_1 \cap \cdots \cap A_m \subseteq B_1 \cup \cdots \cup B_n$; moreover, the points not covered by a line for A are consider to be covered by a line for the complement $\overline{A}$.

Note that Lemma 5.25 gets a lot simpler when we require the entire (P, N)-substitution σ to respect the (P, N)-choice-condition C by setting $O := \mathrm{dom}(\sigma) \cap \mathrm{dom}(C)$ and $O' := \emptyset$; in particular all requirements on O' are trivially satisfied then. Moreover, note that the (still quite long) proof of Lemma 5.25 is more than a factor of 2 shorter than the proof of the analogous Lemma B.5 in [106] (together with Lemma B.1, its additionally required sub-lemma).

5.15 Reduction

Reduction is the reverse of consequence. It is the backbone of logical reasoning, especially of abduction and goal-directed deduction.

In our case, a reduction step does not only reduce a set of problems to another set of problems, but also guarantees that the solutions of the latter also solve the former; here "solutions" means those $\mathcal{S}$-raising-valuations of the (rigid) (free) variables from $\mathbb{V}$ which are $\mathcal{S}$-compatible with $(C, (P, N))$ for the positive/negative variable-condition (P, N) and the (P, N)-choice-condition C given by the context of the reduction step.

Definition 5.26 (Reduction)

Let (P, N) be a positive/negative variable-condition. Let C be a (P, N)-choice-condition. Let G_0 and G_1 be sets of sequents. Let $\mathcal{S}$ be a Σ-structure.

G_0 $(C, (P, N))$-*reduces to* G_1 *in* $\mathcal{S}$ if for every π that is $\mathcal{S}$-compatible with $(C, (P, N))$:

       If G_1 is $(\pi, \mathcal{S})$-valid, then G_0 is $(\pi, \mathcal{S})$-valid as well.

Theorem 5.27 (Reduction)

Let (P, N) be a positive/negative variable-condition.
Let C be a (P, N)-choice-condition.
Let G_0, G_1, G_2, and G_3 be sets of sequents. Let $\mathcal{S}$ be a Σ-structure.

1. **(Validity)** *If G_0 $(C, (P, N))$-reduces to G_1 in $\mathcal{S}$*
   and G_1 is $(C, (P, N))$-valid in $\mathcal{S}$,
   then G_0 is $(C, (P, N))$-valid in $\mathcal{S}$, too.

2. **(Reflexivity)** *In case of $G_0 \subseteq G_1$: G_0 $(C, (P, N))$-reduces to G_1 in $\mathcal{S}$.*

3. **(Transitivity)** *If G_0 $(C, (P, N))$-reduces to G_1 in $\mathcal{S}$*
   and G_1 $(C, (P, N))$-reduces to G_2 in $\mathcal{S}$,
   then G_0 $(C, (P, N))$-reduces to G_2 in $\mathcal{S}$.

4. **(Additivity)** *If G_0 $(C, (P, N))$-reduces to G_2 in $\mathcal{S}$*
   and G_1 $(C, (P, N))$-reduces to G_3 in $\mathcal{S}$,
   then $G_0 \cup G_1$ $(C, (P, N))$-reduces to $G_2 \cup G_3$ in $\mathcal{S}$.

5. **(Monotonicity)** *For $(C', (P', N'))$ being an extended extension of $(C, (P, N))$:*

   (a) *If G_0 is $(C', (P', N'))$-valid in $\mathcal{S}$, then G_0 is also $(C, (P, N))$-valid in $\mathcal{S}$.*

   (b) *If G_0 $(C, (P, N))$-reduces to G_1 in $\mathcal{S}$,*
       then G_0 also $(C', (P', N'))$-reduces to G_1 in $\mathcal{S}$.

6. **(Instantiation of Free Variables)**   *Let σ be a (P, N)-substitution.*
   Let $(C', (P', N'))$ be the extended σ-update of $(C, (P, N))$.
   Set $M := \mathrm{dom}(\sigma) \cap \mathrm{dom}(C)$. Choose some $V \subseteq \mathbb{V}$ with $\mathbb{V}(G_0, G_1) \subseteq V$.
   Set $O := M \cap P^* \langle V \rangle$.   Set $O' := \mathrm{dom}(C) \cap \langle M \backslash O \rangle P^*$.
   Assume that for every $y^{\mathrm{v}} \in O'$, for α being the return type of $C(y^{\mathrm{v}})$
   there is a generalized choice function on the power-set of $\mathcal{S}(\forall)_\alpha$.

   (a) *If $G_0 \sigma \cup (\langle O \rangle Q_C) \sigma$ is $(C', (P', N'))$-valid in $\mathcal{S}$,*
       then G_0 is $(C, (P, N))$-valid in $\mathcal{S}$.

   (b) *If G_0 $(C, (P, N))$-reduces to G_1 in $\mathcal{S}$,*
       then $G_0 \sigma$ $(C', (P', N'))$-reduces to $G_1 \sigma \cup (\langle O \rangle Q_C) \sigma$ in $\mathcal{S}$.

7. **(Instantiation of Free Atoms)**   *Let ν be a substitution on $\mathbb{A}$.*
   If $\mathbb{V}(G_0) \times \mathrm{dom}(\nu) \subseteq N$, then $G_0 \nu$ $(C, (P, N))$-reduces to G_0 in $\mathcal{S}$.

5.16   Soundness, Safeness, and Solution-Preservation

Soundness of inference rules has the global effect that if we reduce a set of sequents to an empty set, then we know that the original set is valid. Soundness is an essential property of inference rules.

Safeness of inference rules has the global effect that if we reduce a set of sequents to an invalid set, then we know that already the original set was invalid. Safeness is helpful in rejecting false assumptions and in patching failed proof attempts.

As explained before, for a reduction step in our framework, we are not contend with soundness: We want *solution-preservation* in the sense that an $\mathcal{S}$-raising-valuation π that makes the set of sequents of the reduced proof state $(\pi, \mathcal{S})$-valid is guaranteed to do the same for the original input proposition, provided that π is $\mathcal{S}$-compatible with $(C, (P, N))$ for the positive/negative variable-condition (P, N) and the (P, N)-choice-condition C of the reduced proof state.

All our inference rules of § 3 have all of these properties. This is obvious for the trivial α- and β-rules. For the inference rules where this is not obvious, i.e. our γ- and δ^-- and δ^+-rules of §§ 3.2, 3.3, and 3.4, we state these properties in the following theorem.

Theorem 5.28

Let (P, N) be a positive/negative variable-condition.
Let C be a (P, N)-choice-condition.
Let us consider any of the γ-, δ^--, and δ^+-rules of §§ 3.2, 3.3, and 3.4.
Let G_0 and G_1 be the sets of the sequent above and of the sequents below the bar of that rule, respectively.
Let C'' be the set of the pair indicated to the upper right of the bar if there is any (which is the case only for the δ^+-rules) or the empty set otherwise.
Let V be the relation indicated to the lower right of the bar if there is any (which is the case only for the δ^-- and δ^+-rules) or the empty set otherwise.
Let us weaken the informal requirement "Let $x^{\mathbb{A}}$ be a fresh free atom" of the δ^--rules to its technical essence "$x^{\mathbb{A}} \in \mathbb{A} \setminus (\operatorname{dom}(P) \cup \mathbb{A}(\Gamma, A, \Pi))$".
Let us weaken the informal statement "Let $x^{\mathbb{V}}$ be a fresh free variable" of the δ^+-rules to its technical essence "$x^{\mathbb{V}} \in \mathbb{V} \setminus (\operatorname{dom}(C \cup P \cup N) \cup \mathbb{V}(A))$".
Let us set $C' := C \cup C''$, $P' := P \cup V_{|\mathbb{V}}$, $N' := N \cup V_{|\mathbb{A}}$.

Then $(C', (P', N'))$ is an extended extension of $(C, (P, N))$ (cf. Definition 5.21).

Moreover, the considered inference rule is sound, safe, and solution-preserving in the sense that G_0 and G_1 mutually $(C', (P', N'))$-reduce to each other in every Σ-structure $\mathcal{S}$.

6 Summary and Discussion

6.1 Positive/Negative Variable-Conditions

We take a *sequent* to be a list of formulas which denotes the disjunction of these formulas. In addition to the standard frameworks of two-valued logics, our formulas may contain free atoms and variables with a context-independent semantics: While we admit explicit quantification to bind only *bound atoms* (written x^B), our *free atoms* (written x^A) are implicitly universally quantified. Moreover, free variables (written x^V) are implicitly existentially quantified. The structure of this implicit form of quantification without quantifiers and without binders is represented globally in a *positive/negative variable-condition* (P, N), which can be seen as a directed graph on free atoms and variables whose edges are elements of either P or N.

Without loss of generality in practice, let us assume that P is finite. Then, a positive/negative variable-condition (P, N) is *consistent* if each cycle of the directed graph has more than one edge from N.

Roughly speaking, on the one hand, a *free variable* y^V is put into the scope of another free variable or atom x^{VA} by an edge (x^{VA}, y^V) in P; and, on the other hand, a *free atom* y^A is put into the scope of another free variable or atom x^{VA} by an edge (x^{VA}, y^A) in N.

On the one hand, an edge (x^{VA}, y^V) *must* be put into P

- if y^V is introduced in a δ^+-step where x^{VA} occurs in the principal[2] formula, and also

- if y^V is globally replaced with a term in which x^{VA} occurs.

On the other hand, an edge (x^{VA}, y^A) *must* be put into N

- if x^{VA} is actually a free *variable*, and y^A is introduced in a δ^--step where x^{VA} occurs in the sequent (either in the principal formula or in the parametric formulas)[2]

Furthermore, such edges *may* always be added to the positive/negative variable-condition, as long as it remains consistent. Such an unforced addition of edges might be appropriate especially in the formulation of a new proposition:

- partly, because we may need this for modeling the intended semantics by representing the intended quantificational structure for the free variables and atoms of the new proposition;

- partly, because we may need this for enabling induction in the form of Fermat's *descente infinie* on the free atoms of the proposition; cf. [106, §§ 2.5.2 and 3.3]. (This is closely related to the satisfaction of the condition on N in Theorem 5.27(7).)

6.2   Semantics of Positive/Negative Variable-Conditions

The value assigned to a free variable y^{V} by an $\mathcal{S}$-*raising-valuation* π may depend on the value assigned to an atom x^{A} by an $\mathcal{S}$-*valuation*. In that case, the *semantic relation* S_π contains an edge $(x^{\mathrm{A}}, y^{\mathrm{V}})$. Moreover, π is enforced to obey the quantificational structure by the requirement that $(P \cup S_\pi, N)$ must be consistent; cf. Definitions 5.10 and 5.15.

6.3   Replacing ε-Terms with Free Variables

Suppose that an ε-term $\varepsilon z^{\mathrm{B}}.\, B$ has free occurrences of exactly the bound atoms $v_0^{\mathrm{B}}, \ldots, v_{l-1}^{\mathrm{B}}$ which are not free atoms of our framework, but are actually bound in the syntactic context in which this ε-term occurs. Then we can replace it in this context with the application term $z^{\mathrm{V}}(v_0^{\mathrm{B}}) \cdots (v_{l-1}^{\mathrm{B}})$ for a fresh free variable z^{V} and set the value of a global function C (called the *choice-condition*) at z^{V} according to

$$C(z^{\mathrm{V}}) \quad := \quad \lambda v_0^{\mathrm{B}}.\, \ldots \lambda v_{l-1}^{\mathrm{B}}.\, \varepsilon z^{\mathrm{B}}.\, B,$$

and augment P with an edge $(y^{\mathrm{VA}}, z^{\mathrm{V}})$ for each free variable or free atom y^{VA} occurring in B.

6.4   Semantics of Choice-Conditions

A free variable z^{V} in the domain of the global choice-condition C must take a value that makes $C(z^{\mathrm{V}})$ true — if such a choice is possible. This can be formalized as follows. Let "eval" be the standard evaluation function. Let $\mathcal{S}$ be any of the semantic structures (or models) under consideration. Let δ be a valuation of the free variables and free atoms (resulting from an $\mathcal{S}$-raising-valuation of the variables and an $\mathcal{S}$-valuation of the atoms). Let χ be an arbitrary $\mathcal{S}$-valuation of the bound atoms $v_0^{\mathrm{B}}, \ldots, v_{l-1}^{\mathrm{B}}, z^{\mathrm{B}}$. Then $\delta(z^{\mathrm{V}})$ must be a function that chooses a value that makes B true whenever possible, in the sense that $\mathrm{eval}(\mathcal{S} \uplus \delta \uplus \chi)(B) = \mathsf{TRUE}$ implies $\mathrm{eval}(\mathcal{S} \uplus \delta \uplus \chi)(B\mu) = \mathsf{TRUE}$ for

$$\mu := \{z^{\mathrm{B}} \mapsto z^{\mathrm{V}}(v_0^{\mathrm{B}}) \cdots (v_{l-1}^{\mathrm{B}})\}.$$

6.5 Substitution of Free Variables ("ε-Substitution")

The kind of logical inference we essentially need is (problem-) *reduction,* the backbone of abduction and goal-directed deduction; cf. § 5.15. In a tree of reduction steps our free variables and free atoms show the following behavior with respect to their instantiation:

Atoms behave as constant parameters. A free variable y^{v}, however, may be globally instantiated with any term by application of a substitution σ; unless, of course, in case y^{v} is in the domain of the global choice-condition C, in which case σ must additionally satisfy $C(y^{\mathsf{v}})$, in a sense to be explained below.

In addition, the applied substitution σ must always be an (P, N)-*substitution.* This means that the current positive/negative variable-condition (P, N) remains consistent when we extend it to its so-called σ-*update,* which augments P with the edges from the free variables and free atoms in $\sigma(z^{\mathsf{v}})$ to z^{v}, for each free variable z^{v} in the domain $\mathrm{dom}(\sigma)$.

Moreover, the global choice-condition C must be updated by removing z^{v} from its domain $\mathrm{dom}(C)$ and by applying σ to the C-values of the free variables remaining in $\mathrm{dom}(C)$.

Now, in case of a free variable $z^{\mathsf{v}} \in \mathrm{dom}(\sigma) \cap \mathrm{dom}(C)$, σ satisfies the current choice-condition C if $(Q_C(z^{\mathsf{v}}))\sigma$ is valid in the context of the updated variable-condition and choice-condition. Here, for a choice-condition $C(z^{\mathsf{v}})$ given as above, $Q_C(z^{\mathsf{v}})$ denotes the formula

$$\forall v_0^{\mathbb{B}}. \ldots \forall v_{l-1}^{\mathbb{B}}. \left(\exists z^{\mathbb{B}}. B \Rightarrow B\mu \right),$$

which is nothing but our version of Hilbert's axiom (ε_0); cf. Definition 4.11. Under these conditions, the invariance of reduction under substitution is stated in Theorem 5.27(6b).

Finally, note that $Q_C(z^{\mathsf{v}})$ itself is always valid in our framework; cf. Lemma 5.19.

6.6 Where have all the ε-Terms gone?

After the replacement described in § 6.3 and, in more detail, in § 4.11, the ε-symbol occurs neither in our terms, nor in our formulas, but only in the range of the current choice-condition, where its occurrences are inessential, as explained at the end of § 4.11.

As a consequence of this removal, our formulas are much more readable than in the standard approach of in-line presentation of ε-terms, which always was nothing but a theoretical presentation because in practical proofs the ε-terms would have grown so large that the mere size of them made them inaccessible to human inspection. To see this, compare our presentation in Example 4.10 to the one in

Example 4.8, and note that the latter is still hard to read although we have invested some efforts in finding a readable form of presentation.

From a mathematical point of view, however, the original ε-terms are still present in our approach; up to isomorphism and with the exception of some irrelevant term sharing. To make these ε-terms explicit in a formula A for a given (P, N)-choice-condition C, we just have to do the following:

Step 1: Let us consider the relation C not as a function, but as a ground term rewriting system: This means that we read $\left(z^{\mathrm{V}}, \; \lambda v_0^{\mathrm{B}}. \; \ldots \lambda v_{l-1}^{\mathrm{B}}. \; \varepsilon z^{\mathrm{B}} B \right) \in C$ as a rewrite rule saying that we may replace the free variable z^{V} (the left-hand side of the rule, which is not a variable but a constant w.r.t. the rewriting system) with the right-hand side $\lambda v_0^{\mathrm{B}}. \; \ldots \lambda v_{l-1}^{\mathrm{B}}. \; \varepsilon z^{\mathrm{B}}. \; B$ in any given context as long as we want.

By Definition 5.13(3), we know that all variables in B are smaller than z^{V} in P^+. By the consistency of our positive/negative variable-condition (P, N) (according to Definition 5.13), we know that P^+ is a well-founded ordering. Thus its multi-set extension is a well-founded ordering as well. Moreover, the multi-set of the free variable z^{V} of the left-hand side is bigger than the multi-set of the free-variable occurrences in the right-hand side in the well-founded multi-set extension of P^+. Thus, if we rewrite a formula, the multi-set of the free-variable occurrences in the rewritten formula is smaller than the multi-set of the free-variable occurrences in the original formula.

Therefore, normalization of any formula A with these rewrite rules terminates with a formula A'.

Step 2: As typed $\lambda\alpha\beta$-reduction is also terminating, we can apply it to remove the λ-terms introduced to A' by the rewriting of Step 1, resulting in a formula A''.

Then — with the proper semantics for the ε-binder — the formulas A' and A'' are equivalent to A, but do not contain any free variables that are in the domain of C. This means that A'' is equivalent to A, but does not contain ε-constrained free variables anymore.

Moreover, if the free variables in A resulted from the elimination of ε-terms as described in §§ 4.11 and 6.3, then all λ-terms that were not already present in A are provided with arguments and are removed by the rewriting of Step 2. Therefore, no λ-symbol occurs in the formula A'' if the formula A resulted from a first-order formula.

For example, if we normalize $\mathsf{P}(w_a^{\vee}, x_b^{\vee}, y_d^{\vee}, z_h^{\vee})$ with respect to the rewriting system given by the (P, N)-choice-condition C of of Example 4.10, and then by $\lambda\alpha\beta$-reduction, we end up in a normal form which is the first-order formula (4.8.1) of Example 4.8, with the exception of the renaming of some bound atoms that are bound by ε.

If each element $z^{\vee}$ in the domain of C binds a unique bound atom $z^{\mathbb{B}}$ by the ε in the higher-order ε-term $C(z^{\vee})$, then the normal form A'' can even preserve our information on committed choice when we consider any ε-term binding an occurrence of a bound atom of the same name to be committed to the same choice. In this sense, the representation given by the normal form is equivalent to our original one given by $\mathsf{P}(w_a^{\vee}, x_b^{\vee}, y_d^{\vee}, z_h^{\vee})$ and C.

6.7 Breaking with the Traditional Treatment of Hilbert's ε?

Our new semantic free-variable framework was actually developed to meet the requirements analysis for the combination of mathematical induction in the liberal style of Fermat's *descente infinie* with state-of-the-art logical deduction. The framework provides a formal system in which a working mathematician can straightforwardly develop his proofs supported by powerful automation; cf. [106].

If traditionalism meant restriction to the expressional means of the past — say the first half of the 20th century with its foundational crisis and special emphasis on constructivism, intuitionism, and finitism — then our approach would not classify as traditional. Although we offer the extras of non-committed choice and a model-theoretic notion of validity, we nevertheless see our framework based on Q_C as a form of (ε_0) (cf. § 4.12) as an upward-compatible extension of Hilbert–Bernays' original framework with (ε_0) as the only axiom for the ε. And with its equivalents for the traditional ε-terms (cf. § 6.6) and with its support for the global proof transformation given by the ε-substitution methods (cf. §§ 4.12, 5.15, and 6.5), our framework is indeed deeply rooted in the Hilbert–Bernays tradition.

Note that the fear of inconsistency should have been soothed anyway in the meantime by Wittgenstein, cf. e.g. [15]. The main disadvantage of an exclusively axiomatic framework as compared to one that also offers a model-theoretic semantics is the following: Constructive proofs of practically relevant theorems easily become too huge and too tedious, whereas semantic proofs are of a better manageable size. More important is the possibility to invent *new and more suitable logics for new applications* with semantic means, whereas proof transformations can refer only to already existing logics (cf. § 4.7).

We intend to pass the heritage of Hilbert's ε on to new generations interested in computational linguistics, automated theorem proving, and mathematics assistance systems; fields in which — with very few exceptions — the overall common opinion still is (the wrong one) that the ε hardly can be of any practical benefit.

The differences, however, between our free-variable framework for the ε and Hilbert's original underspecified ε-operator, in the order of increasing importance, are the following:

1. The term-sharing of ε-terms with the help of free variables improves the readability of our formulas considerably.

2. We do not have the requirement of globally committed choice for any ε-term: Different free variables with the same choice-condition may take different values. Nevertheless, *ε-substitution* works at least as well as in the original framework.

3. Opposed to all other classical validities for the ε (including the semantics of Asser's [4] of 1957, Hermes' [34] of 1965, and Leisenring's [72] of 1969), the implicit quantification over the choice of our free variables is existential instead of universal. This change simplifies formal reasoning in all relevant contexts, because we have to consider only an arbitrary single solution (or choice, substitution) instead of checking all of them.

7 Conclusion

Our more flexible semantics for Hilbert's ε and our novel free-variable framework presented in this paper were developed to solve the difficult soundness problems arising in the combination of mathematical induction in the liberal style of Fermat's *descente infinie* with state-of-the-art deduction.[7] Thereby, they had passed an evaluation of their usefulness even before they were recognized as a candidate for the semantics that Hilbert's school in logic may have had in mind for their ε. While this is a speculation, it is definite that the semantic framework for Hilbert's ε proposed in this paper has the following advantages:

---

[7] The well-foundedness required for the soundness of *descente infinie* gave rise to a notion of reduction which preserves solutions, cf. Definition 5.26. The liberalized δ-rules as found in [21] do not satisfy this notion. The addition of our choice-conditions finally turned out to be the only way to repair this defect of the liberalized δ-rules. See [106] for more details.

Indication of Commitment: The requirement of a commitment to a choice is expressed syntactically and most clearly by the sharing of a free variable; cf. § 4.11.

Semantics: The semantics of the ε is simple and straightforward in the sense that the ε-operator becomes similar to the referential use of indefinite articles and determiners in natural languages, cf. [111].

Our semantics for the ε is based on an abstract formal approach that extends a semantics for closed formulas (satisfying only very weak requirements, cf. § 5.8) to a semantics with existentially quantified "free variables" and universally quantified "free atoms", replacing the three kinds of free variables of [106; 107; 108; 110; 111], i.e. existential (free γ-variables), universal (free δ^--variables), and ε-constrained (free δ^+-variables). The simplification achieved by the reduction from three to two kinds of free variables results in a remarkable reduction of the complexity of our framework and will make its adaptation to applications much easier.

In spite of this simplification, we have enhanced the expressiveness of our framework by replacing the variable-conditions of [105; 106; 107; 108; 110; 111] with our *positive/negative* variable-conditions here, such that our framework now admits us to represent Henkin quantification directly; cf. Example 5.20. From a philosophical point of view, this clearer differentiation also provides a deep insight into the true nature and the relation of the δ^-- and the δ^+-rules.

Reasoning: Our representation of an ε-term $\varepsilon x^{\text{B}}. A$ can be replaced with *any* term t that satisfies the formula $\exists x^{\text{B}}. A \Rightarrow A\{x^{\text{B}} \mapsto t\}$, cf. § 4.12. Thus, the correctness of such a replacement is likely to be expressible and verifiable in the original calculus. Our free-variable framework for the ε is especially convenient for developing proofs in the style of a working mathematician, cf. [106; 107; 110]. Indeed, our approach makes proof work most simple because we do not have to consider all proper choices t for x (as in all other model-theoretic approaches) but only a single arbitrary one, which is fixed in a global proof transformation step.

Finally, we hope that our new semantic framework will help to solve further practical and theoretical problems with the ε and improve the applicability of the ε as a logic tool for description and reasoning. And already without the ε (i.e. for the case that the choice-condition is empty, cf. e.g. [109; 112]), our free-variable framework should find a multitude of applications in all areas of computer-supported reasoning.

A   Semantics for Hilbert's ε in the Literature

Here in § A of the appendix, we will review the literature on the ε's semantics with an emphasis on practical adequacy and the intentions of the Hilbert school in logic.

A.1   Right-Unique Semantics

In contrast to the indefiniteness we suggested in § 4.8, nearly all semantics for Hilbert's ε found elsewhere in the literature are functional, i.e. [right-] unique; cf. e.g. Leisenring's [72] and the references there.

A.1.1   Extensionality:
Ackermann's (II,4) = Bourbaki's (S7) = Leisenring's (E2)

In Ackermann's [2] of 1938 under the label (II,4), in Bourbaki's [11] of 1939ff. under the label (S7) (where a τ is written for the ε, which must not be confused with Hilbert's τ-operator[8]), and in Leisenring's [72] of 1969 under the label (E2), we find the following axiom scheme, which we presented already in § 4.10:

$$\forall x^{\mathtt{B}}.\,(A_0 \Leftrightarrow A_1) \quad \Rightarrow \quad \varepsilon x^{\mathtt{B}}.\,A_0 \;=\; \varepsilon x^{\mathtt{B}}.\,A_1 \tag{E2}$$

This axiom (E2) must not be confused with the similar formula (E2′) from [108, Lemma 31, § 5.6] and [111, Lemma 5.18, § 5.6], which reads in our new framework here as follows:

---

[8]Adding the ε either with (ε_0), with (ε_1), or with the ε-formula (cf. § 4.6) to intuitionist first-order logic is equivalent on the ε-free fragment to adding Plato's *Principle*, i.e.
$\exists y^{\mathtt{B}}.\,(\exists x^{\mathtt{B}}.\,A \;\Rightarrow\; A\{x^{\mathtt{B}}{\mapsto}y^{\mathtt{B}}\})$ with $y^{\mathtt{B}}$ not occurring in A, cf. Meyer-Viol's [74, § 3.3] of 1995.
Moreover, the non-trivial direction of (ε_2) is $\qquad \forall x^{\mathtt{B}}.\,A \;\Leftarrow\; A\{x^{\mathtt{B}} \mapsto \varepsilon x^{\mathtt{B}}.\,\neg A\}$.
Even intuitionistically, this entails its contrapositive $\quad \neg\forall x^{\mathtt{B}}.\,A \;\Rightarrow\; \neg A\{x^{\mathtt{B}} \mapsto \varepsilon x^{\mathtt{B}}.\,\neg A\}$,
and then, e.g. by the trivial direction of (ε_1) (when A is replaced with $\neg A$)

$$\neg\forall x^{\mathtt{B}}.\,A \;\Rightarrow\; \exists x^{\mathtt{B}}.\,\neg A \tag{Q2}$$

which is not valid in intuitionist logic in general. Thus, in intuitionist logic, the universal quantifier becomes strictly weaker by the inclusion of (ε_2) or anything similar for the universal quantifier, such as Hilbert's τ-operator (cf. Hilbert's [45] of 1923). More specifically, adding

$$\forall x^{\mathtt{B}}.\,A \;\Leftarrow\; A\{x^{\mathtt{B}} \mapsto \tau x^{\mathtt{B}}.\,A\} \tag{τ_0}$$

is equivalent on the τ-free theory to adding $\exists y^{\mathtt{B}}.\,(\forall x^{\mathtt{B}}.\,A \;\Leftarrow\; A\{x^{\mathtt{B}}{\mapsto}y^{\mathtt{B}}\})$ with $y^{\mathtt{B}}$ not occurring in A, which again implies (Q2), cf. Meyer-Viol's [74, § 3.4.2].
From a semantic point of view (cf. Gabbay's [27] of 1981), the intuitionist $\forall$ may be eliminated, however, by first applying the Gödel translation into the modal logic S4 with classical $\forall$ and $\neg$, cf. e.g. [22], and then adding the ε conservatively, e.g. by avoiding substitutions via λ-abstraction as in Fitting's [19] of 1975.

$$\forall x^{\mathbb{B}}. \, (A_0 \Leftrightarrow A_1) \qquad \Rightarrow \qquad x_0^{\vee} \; = \; x_1^{\vee} \qquad\qquad (\text{E2}')$$

for two different $x_0^{\vee}, x_1^{\vee} \in \mathbb{V} \backslash \mathbb{V}(A_1, A_2, \text{dom}(P \cup N))$ and for a (P, N)-choice-condition C with $C(x_i^{\vee}) = \varepsilon x^{\mathbb{B}}. \, A_i$ for $i \in \{0, 1\}$. Our (E2$'$) can be shown to be $(C, (P, N))$-valid by applying Theorem 5.27(1,5a,6a): Indeed, we can apply the substitution $\{x_1^{\vee} \mapsto y^{\vee}\}$ after an extended extension $(C', (P', N))$ for a fresh variable $y^{\vee} \in \mathbb{V} \backslash \mathbb{V}(A_1, A_2, x_0^{\vee}, x_1^{\vee}, \text{dom}(P \cup N))$ with[9]

$$C'(y^{\vee}) = \varepsilon y^{\mathbb{B}}. \left(\begin{array}{l} (\quad \forall x^{\mathbb{B}}. \, (A_0 \Leftrightarrow A_1) \;\; \Rightarrow \;\; y^{\mathbb{B}} = x_0^{\vee} \,) \\ \wedge \; (\;\; \neg \forall x^{\mathbb{B}}. \, (A_0 \Leftrightarrow A_1) \;\; \Rightarrow \;\; A_1 \{x^{\mathbb{B}} \mapsto y^{\mathbb{B}}\} \,) \end{array} \right).$$

Contrary to the valid proposition (E2$'$), however, (E2) is an axiom that imposes a right-unique behavior for the ε (in the standard framework), depending on the extension of the formula forming the scope of an ε-binder on $x^{\mathbb{B}}$, seen as a predicate on $x^{\mathbb{B}}$. Indeed — from a semantic point of view — the value of $\varepsilon x^{\mathbb{B}}. \, A$ in each Σ-structure $\mathcal{S}$ is functionally dependent on the extension of the formula A, i.e. on $\{ \, o \mid \text{eval}(\mathcal{S} \uplus \{x^{\mathbb{B}} \mapsto o\})(A) \, \}$.

Therefore, axiomatizations that have (E2) as an axiom or as a consequence of other axioms are called *extensional*.

---

[9]Let us give a formal proof of (E2$'$) in our framework on an abstract level by applying Theorem 5.27. We will reduce the set containing the single-formula sequent of the formula (E2$'$) to a valid set. Be aware of the requirements on occurrence of the variables as described in § A.1.1. We start with an extended extension $(C', (P', N))$ of the current $(C, (P, N))$ for a fresh variable $y^{\vee}$ with $C'(y^{\vee})$ as given § A.1.1. Of course, to satisfy Definition 5.13(3), here we set

$$P' \; := \; P \cup \mathbb{V\!A}(A_0, A_1, x_0^{\vee}) \times \{y^{\vee}\}.$$

Set $\sigma := \{x_1^{\vee} \mapsto y^{\vee}\}$. Let $(C'', (P'', N))$ be the extended σ-update of $(C', (P', N))$; then

$$\{y^{\vee}, x_0^{\vee}, x_1^{\vee}\} | C'' = \{y^{\vee}, x_0^{\vee}\} | C' \quad \text{and} \quad P'' = P' \cup \{(y^{\vee}, x_1^{\vee})\}.$$

Note that (P'', N) is consistent because every cycle not possible with (P, N) would have to run through the set $\{y^{\vee}, x_1^{\vee}\}$, which, however, is disjoint from $\text{dom}(N)$, closed under P'', and cycle-free.

Now we apply Theorem 5.27(6a). According to settings for the meta-variables given there, we have $O = M = \text{dom}(C') \cap \text{dom}(\sigma) = \{x_1^{\vee}\}$ and $O' = \emptyset$. Consider the set with the two single-formula sequents (E2)$'\sigma$ and $(Q_{C'}(x_1^{\vee}))\sigma$. The former sequent reads $\forall x^{\mathbb{B}}. \, (A_0 \Leftrightarrow A_1) \Rightarrow x_0^{\vee} = y^{\vee}$. According to Definition 4.11, the latter sequent reads $(\exists x^{\mathbb{B}}. \, A_1 \Rightarrow A_1 \{x^{\mathbb{B}} \mapsto x_1^{\vee}\})\sigma$, i.e. $\exists x^{\mathbb{B}}. \, A_1 \Rightarrow A_1 \{x^{\mathbb{B}} \mapsto y^{\vee}\}$. Now a simple case analysis on $\forall x^{\mathbb{B}}. (A_0 \Leftrightarrow A_1)$ shows that this two-element set $(C'', (P'', N))$-reduces to

$$\left\{ \exists x^{\mathbb{B}}. \, A_0 \Rightarrow A_0 \{x^{\mathbb{B}} \mapsto x_0^{\vee}\}; \quad \left(\exists y^{\mathbb{B}}. \left(\begin{array}{l} (\forall x^{\mathbb{B}}. \, (A_0 \Leftrightarrow A_1) \Rightarrow y^{\mathbb{B}} = x_0^{\vee}) \\ \wedge \; (\neg \forall x^{\mathbb{B}}. \, (A_0 \Leftrightarrow A_1) \Rightarrow A_1 \{x^{\mathbb{B}} \mapsto y^{\mathbb{B}}\}) \end{array} \right) \Rightarrow \left(\begin{array}{l} (\forall x^{\mathbb{B}}. \, (A_0 \Leftrightarrow A_1) \Rightarrow y^{\vee} = x_0^{\vee}) \\ \wedge \; (\neg \forall x^{\mathbb{B}}. \, (A_0 \Leftrightarrow A_1) \Rightarrow A_1 \{x^{\mathbb{B}} \mapsto y^{\vee}\}) \end{array} \right) \right) \right\},$$

i.e. to $\{Q_{C''}(x_0^{\vee}); \; Q_{C''}(y^{\vee})\}$, which is $(C'', (P'', N))$-valid by Lemma 5.19. Thus, (E2$'$)σ is $(C', (P'', N))$-valid. By (6a) this means that (E2$'$) is $(C', (P', N))$-valid, and by (5a) also $(C, (P, N))$-valid, as was to be shown.

Note that (E2) has a disastrous effect in intuitionist logic: The contrapositive of (E2) — together with (ε_0) and say "$0 \neq 1$" — turns every classical validity into an intuitionist one.[10] For the strong consequences of the ε-formula in intuitionist logic, see also Note 8.

A.1.2 Weaker than (E2), but still Right-Unique

To overcome this disastrous effect and to get more options for the definition of a semantics of the ε in general, in [4], [74], and [32] the value of $\varepsilon x^{\mathrm{B}}. A$ may additionally depend on the *syntax* besides the semantics of the formula in the scope of the ε. The semantics of the ε is then given as a function depending on a Σ-structure and on the syntactic details of the term $\varepsilon x^{\mathrm{B}}. A$.

In Giese & Ahrendt's [32, p.177] we read: "This definition contains no restriction whatsoever on the valuation of ε-terms." This claim, however, is not justified in its universality, because all considered options do still impose the restriction of a right-unique behavior; thereby the claim denies the possibility of an indefinite behavior as given in §§ 4.10 and 4.11. See also § A.2 for an alternative realization of an indefinite semantics.

---

[10]Let us prove $0 \neq 1$, $\varepsilon x^{\mathrm{B}}. A_0 \neq \varepsilon x^{\mathrm{B}}. A_1 \Rightarrow \neg(\forall x^{\mathrm{B}}. A_0 \wedge \forall x^{\mathrm{B}}. A_1)$ $\vdash$ $B \vee \neg B$ in intuitionist logic. (For the proof of the slightly weaker result $0 \neq 1$, (E2) $\vdash$ $B \vee \neg B$ for any formula B, cf. Bell's [7, Proof of Theorem 6.4], which already occurs in more detail in Bell's [5, § 3], and sketched in Bell's [6, § 7].)
Note that, for any implication $A \Rightarrow B$, its contrapositive $\neg B \Rightarrow \neg A$ is a consequence of it, and — in intuitionist logic — a *proper* consequence in general.
Let B be an arbitrary formula. By renaming we may w.l.o.g. assume that the free atom x^{A} of the ε-formula does not occur in B. We are going to show that $\vdash B \vee \neg B$ holds in intuitionist logic under the assumptions of reflexivity, symmetry, and transitivity of "$=$", the ε-formula (or (ε_0)), and of the formulas $0 \neq 1$ and $\varepsilon x^{\mathrm{B}}. A_0 \neq \varepsilon x^{\mathrm{B}}. A_1 \Rightarrow \neg(\forall x^{\mathrm{B}}. A_0 \wedge \forall x^{\mathrm{B}}. A_1)$.
Let x^{B} be a bound atom not occurring in B. Set $A_i := (B \vee x^{\mathrm{B}} = i)$ for $i \in \{0, 1\}$.
Now all that we have to show is a trivial consequence of the following Claims 1 and 2,
$$\varepsilon x^{\mathrm{B}}. A_0 \neq \varepsilon x^{\mathrm{B}}. A_1 \Rightarrow \neg(\forall x^{\mathrm{B}}. A_0 \wedge \forall x^{\mathrm{B}}. A_1), \quad \text{and Claim 3.}$$
Claim 1: $0 = 0$, $1 = 1$, $(\varepsilon\text{-formula})\{A \mapsto A_0\}\{x^{\mathrm{A}} \mapsto 0\}$, $(\varepsilon\text{-formula})\{A \mapsto A_1\}\{x^{\mathrm{A}} \mapsto 1\}$
$$\vdash \quad B \vee (\varepsilon x^{\mathrm{B}}. A_0 = 0 \wedge \varepsilon x^{\mathrm{B}}. A_1 = 1).$$
Claim 2: $\varepsilon x^{\mathrm{B}}. A_0 = 0 \wedge \varepsilon x^{\mathrm{B}}. A_1 = 1$, $0 \neq 1$, $\forall x^{\mathrm{B}}, y^{\mathrm{B}}, z^{\mathrm{B}}. (y^{\mathrm{B}} = x^{\mathrm{B}} \wedge y^{\mathrm{B}} = z^{\mathrm{B}} \Rightarrow x^{\mathrm{B}} = z^{\mathrm{B}})$
$$\vdash \quad \varepsilon x^{\mathrm{B}}. A_0 \neq \varepsilon x^{\mathrm{B}}. A_1.$$
Claim 3: $\neg(\forall x^{\mathrm{B}}. A_0 \wedge \forall x^{\mathrm{B}}. A_1)$ $\vdash$ $\neg B$.
Proof of Claim 1: Because neither x^{A} nor x^{B} occur in B, and because x^{A} does not occur in A_i, the instances of the ε-formulas read $(B \vee i = i) \Rightarrow (B \vee \varepsilon x^{\mathrm{B}}. A_i = i)$. Thus, from $i = i$, we get $B \vee \varepsilon x^{\mathrm{B}}. A_i = i$. Thus, we get $(B \vee \varepsilon x^{\mathrm{B}}. A_0 = 0) \wedge (B \vee \varepsilon x^{\mathrm{B}}. A_1 = 1)$, thus $B \vee (\varepsilon x^{\mathrm{B}}. A_0 = 0 \wedge \varepsilon x^{\mathrm{B}}. A_1 = 1)$ by distributivity.   Q.e.d. (Claim 1)
Proof of Claim 2: Trivial.   Q.e.d. (Claim 2)
Proof of Claim 3: As x^{B} does not occur in B, we get $B \vdash \forall x^{\mathrm{B}}. A_i$. The rest is trivial.
Q.e.d. (Claim 3)

A.1.3 Overspecification even beyond (E2)

In Hermes' [34, p.18] of 1965, the ε suffers further overspecification in addition to (E2):

$$\varepsilon x.\,\mathsf{false} \;=\; \varepsilon x.\,\mathsf{true} \qquad\qquad (\varepsilon_5)$$

Roughly speaking, this axiom sets the value of a generalized choice function on the empty set to its value on the whole universe. For classical logic, we can combine (E2) and (ε_5) into the following axiom of DeVidi's [14] of 1995 for "<u>v</u>ery <u>ext</u>ensional" semantics:

$$\forall x.\left(\begin{array}{c}(\exists y.\,A_0\{x\mapsto y\} \;\Rightarrow\; A_0)\\ \Leftrightarrow \quad (\exists y.\,A_1\{x\mapsto y\} \;\Rightarrow\; A_1)\end{array}\right) \;\Rightarrow\; \varepsilon x.\,A_0 = \varepsilon x.\,A_1 \qquad \text{(vext)}$$

Indeed, (vext) implies (E2) and (ε_5). The other direction, however, does not hold for intuitionist logic, where, roughly speaking, (vext) additionally implies that if the same elements make A_0 and A_1 as true as possible, then the ε-operator picks the same element of this set, even if the suprema $\exists y.\,A_0\{x\mapsto y\}$ and $\exists y.\,A_1\{x\mapsto y\}$ (in the complete Heyting algebra) are not equally true.

A.1.4 Strengthening Semantics to Turn Axiomatizations Complete

Although we have been concerned with soundness and safeness of our inference systems, we always accepted their incompleteness as the natural companion of semantics that are sufficiently weak to be useful in practice. Of course, completeness is the theoreticians' favorite puzzle because — as a global property of inference systems — it may be hard to prove, even for inconsistent systems. The objective of completeness gets particularly detached from practical usefulness, if a useful semantics is strengthened to obtain the completeness of a given inference system. Let us look at two examples for this procedure, resulting in practically useless semantics for the ε.

Different possible choices for the value of the generalized choice function on the empty set are discussed in Leisenring's [72] of 1969. As the consequences of any special choice are quite queer, the only solution that is found to be sufficiently adequate in [72] is validity in *all* models given by *all* generalized choice functions on the power-set of the universe. Note, however, that even in this case, in each model, the value of $\varepsilon x.\,A$ is *functionally* dependent on the extension of A.

Roughly speaking, in Leisenring's textbook [72], the axioms (ε_1) and (ε_2) from § 4.6 and (E2) from § 4.10 are shown to be complete w.r.t. this semantics of the ε in first-order logic.

This completeness makes it unlikely that extensional semantics matches the intentions of Hilbert's school in logic. Indeed, if their intended semantics for the ε could be completely captured by adding the single and straightforward axiom (E2), this axiom would not have been omitted in Hilbert–Bernays [58]; it would at least be possible to derive (E2) from some axiomatization in Hilbert–Bernays [58].

What makes Leisenring's notion of validity problematic for theorem proving is that a proof has to consider all appropriate choice functions and cannot just pick an advantageous single one of them. More specifically, when Leisenring does the step from satisfiability to validity he does the double duality switch from existence of a model and the existence of a choice function to all models and to all choice functions. Our notion of validity in Definition 5.17 does not switch the second duality, but stays with the *existence* of a choice function. Considering the influence that Leisenring's [72] of 1969 still has today, our avoidance of the universality requirement for choice functions in the definition of validity may be considered our practically most important conceptual contribution to the ε's semantics. If we stuck to Leisenring's definition of validity, then we would either have to give up the hope of finding proofs in practice, or have to avoid considering validity (beyond truth) in connection with Hilbert's ε, which is Hartley Slater' solution, carefully observed in Slater's [92; 93; 95; 96; 97].

This whole misleading procedure of strengthening semantics to obtain completeness for axiomatizations of the ε actually originates in Asser's [4] of 1957. The main objective of [4], however, is to find a semantics such that the basic ε-calculus of Hilbert–Bernays [58] — not containing (E2) — is sound and complete for it. This semantics, however, has to depend on the details of the syntactic form of the ε-terms and, moreover, turns out to be necessarily so artificial that Asser in [4] does not recommend it himself and admits that he thinks that it could not have been intended in Hilbert–Bernays [58].

"Allerdings ist dieser Begriff von Auswahlfunktion so kompliziert, daß sich seine Verwendung in der inhaltlichen Mathematik kaum empfiehlt."
[4, p. 59]

"This notion of a choice function, however," (i.e. the type-3 choice function, providing a semantics for the ε-operator) "is so intricate that its application in contentual mathematics is hardly to be recommended."

"Angesichts der Kompliziertheit des Begriffs der Auswahlfunktion dritter Art ergibt sich die Frage, ob bei Hilbert–Bernays (" ... ") wirklich beabsichtigt war, diesen Begriff von Auswahlfunktion axiomatisch zu beschreiben. Aus der Darstellung bei Hilbert–Bernays glaube ich entnehmen zu können, daß das nicht der Fall ist,"
[4, p. 65]

"The intricacy of the notion of the type-3 choice function puts up the
question whether the intention in Hilbert–Bernays [58] (" ... ") really
was to describe this notion of choice function axiomatically. I believe
I can draw from the presentation in Hilbert–Bernays [58] that that is
not the case,"

A.1.5 Roots of the Misunderstanding of Right-Uniqueness Requirement

The described prevalence of the right-uniqueness requirement may have its historical
justification in the fact that, if we expand the dots "..." in the quotation preceding
Example 4.2 in § 4.6, the full quotation on p.12 of Hilbert–Bernays [58; 60] reads:

"Das ε-Symbol bildet somit eine Art der Verallgemeinerung des
μ-Symbols für einen beliebigen Individuenbereich. Der Form nach stellt
es eine Funktion eines variablen Prädikates dar, welches außer dem-
jenigen Argument, auf welches sich die zu dem ε-Symbol gehörige gebundene Variable bezieht, noch freie Variable als Argumente ("Parameter") enthalten kann. Der Wert dieser Funktion für ein bestimmtes
Prädikat A (bei Festlegung der Parameter) ist ein Ding des Individuenbereichs, und zwar ist dieses Ding gemäß der inhaltlichen Übersetzung
der Formel (ε_0) *ein solches, auf das jenes Prädikat A zutrifft, voraus-
gesetzt, daß es überhaupt auf ein Ding des Individuenbereichs zutrifft.*"

"Thus, the ε-symbol forms a kind of generalization of the μ-symbol for
an arbitrary domain of individuals. According to its form, it constitues
a function of a variable predicate, which may contain free variables as
arguments ("parameters") in addition to the argument to which the
bound variable of the ε-symbol refers. The value of this function for
a given predicate A (for fixed parameters) is a thing of the domain of
individuals for which — according to the contentual translation of the
formula (ε_0) — *the predicate A holds, provided that A holds for any thing
of the domain of individuals at all.*"

Here the word "function" could be misunderstood in its narrower mathematical
sense, namely to denote a (right-) unique relation. It is stated to be a function,
however, only "according to its form", which — in the vernacular that becomes
obvious from reading [62] — means nothing but "with respect to the process of
the formation of formulas".

Thus, Hilbert–Bernays' notation of the ε takes the syntactic form of a function. This syntactic weakness was not bothering the work of the Hilbert school in the field of proof theory. With our more practical intentions, the ε's form of a function turns out as a problem even regarding syntax alone, cf. §§ 4.10 and 4.11. And we are not the only ones who have seen this applicational problem: For instance, in Heusinger's [35] of 1997, an index was introduced to the ε to overcome right-uniqueness.

If we nevertheless read "function" as a right-unique relation in the above quotation, what kind of function could be meant but a choice function, choosing an element from the set of objects that satisfy A, i.e. from its extension
$$\{\, o \mid \mathrm{eval}(\mathcal{S} \uplus \{x^{\mathbb{B}} \mapsto o\})(A) \,\}.$$
Accordingly, in Hilbert's earlier publication [48], we read (p. 68):

> "Darüber hinaus hat das ε die Rolle der Auswahlfunktion, d. h. im Falle, wo $A\,a$ auf mehrere Dinge zutreffen kann, ist εA *irgendeines* von den Dingen a, auf welche $A\,a$ zutrifft."

> "Beyond that, the ε has the rôle of the choice function, i.e., if $A\,a$ may hold for several objects, εA is *an arbitrary one* of the things a for which $A\,a$ holds."

Regarding the notation in this quotation, the syntax of the ε is not that of a binder here, but a functional $\varepsilon : (i \to o) \to i$, applied to $A : i \to o$.

The meaning of having "the rôle of the choice function" is defined by the text that follows in the quotation. Thus, it is obvious that Hilbert wants to state the arbitrariness of choice as given by an arbitrary choice function, and that the word "function" does not refer to a requirement of right-uniqueness here.

Moreover, note that the definite article in "the choice function" (instead of the indefinite one) is in conflict with an interpretation as a mathematical function in the narrower sense as well.

Furthermore, David Hilbert was sometimes pretty sloppy with the usage of choice functions in general: For instance, he may well have misinterpreted the consequences of the ε on the Axiom of Choice (cf. [87], [65]) in the one but last paragraph of [45]. Let us therefore point out the following: Although the ε supplies us with a syntactic means for expressing an *indefinite universal (generalized) choice function* (cf. § 5.2), the axioms (E2), (ε_0), (ε_1), and (ε_2) do not imply the Axiom of Choice in set theories, unless the axiom schemes of Replacement (Collection) and Comprehension (Separation, Subset) also range over expressions containing the ε; cf. [72, § IV 4.4].

Hilbert's school in logic may well have wanted to express what we call *"committed choice"* today, but they simply used the word "function" for the following three reasons:

1. They were not too much interested in semantics anyway.

2. The technical term "committed choice" did not exist at their time.

3. Last but not least, right-uniqueness conveniently serves as a global commitment to any choice and thereby avoids the problem illustrated in Example 4.6 of § 4.8.

A.2   Indefinite Semantics in the Literature

The only occurrence of an indefinite semantics for Hilbert's ε in the literature seems to be Blass & Gurevich's [10] of 2000 (and the references there), unless we count the indexed ε of Heusinger's [35] of 1997 for indefinite indices as such a semantics as well. The right-uniqueness is actually so prevalent in the literature that a "δ" is written instead of an "ε" in [10], because there the right-unique behavior is considered to be essential for the ε.

Consider the formula $\varepsilon x.\,(x\!=\!x) = \varepsilon x.\,(x\!=\!x)$ from [10] or the even simpler $\varepsilon x.\,\mathsf{true} = \varepsilon x.\,\mathsf{true}$ (discussed already in § 4.10), which may be valid or not, depending on the question whether the same object is taken on both sides of the equation or not. In natural language this like "Something is equal to something.", whose truth is indefinite. If you do not think so, consider $\varepsilon x.\,\mathsf{true} \neq \varepsilon x.\,\mathsf{true}$ in addition, i.e. "Something is unequal to something.", and notice that the two sentences seem to be contradictory.

In [10], Kleene's strong three-valued logic is taken as a mathematically elegant means to solve the problems with indefiniteness. In spite of the theoretical significance of this solution, however, Kleene's strong three-valued logic severely restricts its applicability from a practical point of view: In applications, a logic is not an object of investigation but a meta-logical tool, and logical arguments are never made explicit because the presence of logic is either not realized at all or taken to be trivial, even by academics (unless they are formalists); see, for instance, [83, p.14f.] for Wizard of Oz studies with young students.

Therefore, regarding applications, we had better stick to our common meta-logic, which in the western world is a subset of (modal) classical logic: A western court may accept that Lee Harvey Oswald killed John F. Kennedy as well as that he did not — but cannot accept a third possibility, a *tertium,* as required for Kleene's strong three-valued logic, and especially not the interpretation given in [10], namely that he *both* did and did not kill him, which contradicts any common sense.

B   The Proofs

Proof of Lemma 5.16

Under the given assumptions, set $\lhd := P^+$ and $S_\pi := {}_{\mathbb{A}}\!\lhd$.

<u>Claim A:</u> $\lhd = P^+ = (P \cup S_\pi)^+$ is a well-founded ordering.

<u>Claim B:</u> $(P \cup S_\pi, N)$ is a consistent positive/negative variable-condition.

<u>Claim C:</u> $S_\rho \subseteq {}_{\mathbb{A}}\!\lhd = S_\pi \subseteq \lhd$.

<u>Claim D:</u> $S_\pi \circ \lhd \subseteq S_\pi$.

<u>Proof of Claims A, B, C, and D:</u> (P, N) is consistent because C is a (P, N)-choice-condition. Thus, P is well-founded and $\lhd = P^+ = (P \cup S_\pi)^+$ is a well-founded ordering. Moreover, we have $S_\rho, S_\pi, P \subseteq \lhd$. Thus, (P, N) is a weak extension of $(P \cup S_\pi, N)$. Thus, by Corollary 5.7, $(P \cup S_\pi, N)$ is a consistent positive/negative variable-condition. Finally, $S_\pi \circ \lhd = {}_{\mathbb{A}}\!\lhd \circ \lhd \subseteq {}_{\mathbb{A}}\!\lhd = S_\pi$.

Q.e.d. (Claims A, B, C, and D)

By recursion on $y^{\mathrm{V}} \in \mathbb{V}$ in $\lhd$, we can define $\pi(y^{\mathrm{V}}) : (S_\pi \langle\!\langle y^{\mathrm{V}} \rangle\!\rangle \to \mathcal{S}) \to \mathcal{S}$ as follows. Let $\tau' : S_\pi \langle\!\langle y^{\mathrm{V}} \rangle\!\rangle \to \mathcal{S}$ be arbitrary.

<u>$y^{\mathrm{V}} \in \mathbb{V} \backslash \mathrm{dom}(C)$:</u> If an $\mathcal{S}$-raising-valuation ρ is given, then we set
$$\pi(y^{\mathrm{V}})(\tau') := \rho(y^{\mathrm{V}})(S_\rho \langle\!\langle y^{\mathrm{V}} \rangle\!\rangle | \tau');$$
which is well-defined according to Claim C. Otherwise, we choose an arbitrary value for $\pi(y^{\mathrm{V}})(\tau')$ from the universe of $\mathcal{S}$ (of the appropriate type). Note that $\mathcal{S}$ is assumed to provide some choice function $\mathcal{S}(\exists)$ for the universe function $\mathcal{S}(\forall)$ according to § 5.8.

<u>$y^{\mathrm{V}} \in \mathrm{dom}(C)$:</u> In   this   case,   we   have   the   following   situation:
$$C(y^{\mathrm{V}}) = \lambda v_0^{\mathrm{B}}. \ \ldots \lambda v_{l-1}^{\mathrm{B}}. \ \varepsilon v_l^{\mathrm{B}}. \ B \quad \text{for some formula } B \text{ and}$$
some $v_0^{\mathrm{B}}, \ldots, v_l^{\mathrm{B}} \in \mathbb{B}$ with $v_0^{\mathrm{B}} : \alpha_0, \ \ldots, \ v_l^{\mathrm{B}} : \alpha_l, \ y^{\mathrm{V}} : \alpha_0 \to \ldots \to \alpha_{l-1} \to \alpha_l$, and $z^{\mathrm{VA}} \lhd y^{\mathrm{V}}$ for all $z^{\mathrm{VA}} \in \mathbb{VA}(B)$, because C is a (P, N)-choice-condition. In particular, by Claim A, $y^{\mathrm{V}} \notin \mathbb{V}(B)$.

In this case, with the help of the assumed generalized choice function on the power-set of the universe of $\mathcal{S}$ of the sort α_l, we let $\pi(y^{\mathrm{V}})(\tau')$ be the function f that for $\chi : \{v_0^{\mathrm{B}}, \ldots, v_{l-1}^{\mathrm{B}}\} \to \mathcal{S}$ chooses a value from the universe of $\mathcal{S}$ of type α_l for $f(\chi(v_0^{\mathrm{B}})) \cdots (\chi(v_{l-1}^{\mathrm{B}}))$, such that,

if possible, B is true in $\mathcal{S} \uplus \delta' \uplus \chi'$,

for $\delta' := \mathrm{e}(\pi)(\tau' \uplus \tau'') \uplus \tau' \uplus \tau'' \uplus \chi$ for an arbitrary $\tau'' : (\mathbb{A} \backslash \mathrm{dom}(\tau')) \to \mathcal{S}$, and for $\chi' := \{v_l^{\mathrm{B}} \mapsto f(\chi(v_0^{\mathrm{B}})) \cdots (\chi(v_{l-1}^{\mathrm{B}}))\}$.

Note that the point-wise definition of f is correct: by the Explicitness Lemma and because of $y^{\mathrm{V}} \notin \mathbb{V}(B)$, the definition of the value of $f(\chi(v_0^{\mathrm{B}})) \cdots (\chi(v_{l-1}^{\mathrm{B}}))$ does not depend on the values of $f(\chi''(v_0^{\mathrm{B}})) \cdots (\chi''(v_{l-1}^{\mathrm{B}}))$ for a different $\chi'' : \{v_0^{\mathrm{B}}, \ldots, v_{l-1}^{\mathrm{B}}\} \to \mathcal{S}$. Therefore, the function f is well-defined, because it also

505

does not depend on τ'' according to the EXPLICITNESS LEMMA and Claim 1 below. Finally, π is well-defined by induction on $\lhd$ according to Claim 2 below.

<u>Claim 1:</u> For $z^{\mathbb{VA}} \lhd y^{\mathbb{V}}$, the application term $(\delta' \uplus \chi')(z^{\mathbb{VA}})$ has the the value $\tau'(z^{\mathbb{VA}})$ in case of $z^{\mathbb{VA}} \in \mathbb{A}$, and the value $\pi(z^{\mathbb{VA}})(_{S_\pi \langle\!\langle z^{\mathbb{VA}} \rangle\!\rangle}|\tau')$ in case of $z^{\mathbb{VA}} \in \mathbb{V}$.

<u>Claim 2:</u> The definition of $\pi(y^{\mathbb{V}})(\tau')$ depends only on such values of $\pi(v^{\mathbb{V}})$ with $v^{\mathbb{V}} \lhd y^{\mathbb{V}}$, and does not depend on τ'' at all.

<u>Proof of Claim 1:</u> For $z^{\mathbb{VA}} \in \mathbb{A}$ the application term has the value $\tau'(z^{\mathbb{VA}})$ because of $z^{\mathbb{VA}} \in S_\pi \langle\!\langle y^{\mathbb{V}} \rangle\!\rangle$. Moreover, for $z^{\mathbb{VA}} \in \mathbb{V}$, we have $S_\pi \langle\!\langle z^{\mathbb{VA}} \rangle\!\rangle \subseteq S_\pi \langle\!\langle y^{\mathbb{V}} \rangle\!\rangle$ by Claim D, and therefore the applicative term has the value $\pi(z^{\mathbb{VA}})(_{S_\pi \langle\!\langle z^{\mathbb{VA}} \rangle\!\rangle}|(\tau' \uplus \tau''))$ $=\pi(z^{\mathbb{VA}})(_{S_\pi \langle\!\langle z^{\mathbb{VA}} \rangle\!\rangle}|\tau')$. $\hfill$ Q.e.d. (Claim 1)

<u>Proof of Claim 2:</u> In case of $y^{\mathbb{V}} \notin \mathrm{dom}(C)$, the definition of $\pi(y^{\mathbb{V}})(\tau')$ is immediate and independent. Otherwise, we have $z^{\mathbb{VA}} \lhd y^{\mathbb{V}}$ for all $z^{\mathbb{VA}} \in \mathbb{VA}(C(y^{\mathbb{V}}))$. Thus, Claim 2 follows from the EXPLICITNESS LEMMA and Claim 1. $\hfill$ Q.e.d. (Claim 2)

Moreover, $\pi : \mathbb{V} \to (\mathbb{A} \rightsquigarrow \mathcal{S}) \rightsquigarrow \mathcal{S}$ is obviously an $\mathcal{S}$-raising-valuation. Thus, item 1 of Definition 5.15 is satisfied for π by Claim B.

To show that also item 2 of Definition 5.15 is satisfied, let us assume $y^{\mathbb{V}} \in \mathrm{dom}(C)$ and $\tau : \mathbb{A} \to \mathcal{S}$ to be arbitrary with $C(y^{\mathbb{V}}) = \lambda v_0^{\mathbb{B}}. \ \ldots \lambda v_{l-1}^{\mathbb{B}}. \ \varepsilon v_l^{\mathbb{B}}. \ B$, and let us then assume to the contrary of item 2 that, for some $\chi : \{v_0^{\mathbb{B}}, \ldots, v_l^{\mathbb{B}}\} \to \mathcal{S}$ and for $\delta := \mathsf{e}(\pi)(\tau) \uplus \tau \uplus \chi$ and $\sigma := \{v_l^{\mathbb{B}} \mapsto y^{\mathbb{V}}(v_0^{\mathbb{B}}) \cdots (v_{l-1}^{\mathbb{B}})\}$, we have $\mathrm{eval}(\mathcal{S} \uplus \delta)(B) = \mathsf{TRUE}$ and $\mathrm{eval}(\mathcal{S} \uplus \delta)(B\sigma) = \mathsf{FALSE}$.

Set $\tau' := {}_{S_\pi \langle\!\langle y^{\mathbb{V}} \rangle\!\rangle}|\tau$ and $\tau'' := {}_{\mathbb{A} \backslash \mathrm{dom}(\tau')}|\tau$.

Set $\delta' := {}_{\mathbb{VAB} \backslash \{v_l^{\mathbb{B}}\}}|\delta$ and $f := \pi(y^{\mathbb{V}})(\tau')$.

Set $\chi' := \{v_l^{\mathbb{B}} \mapsto f(\chi(v_0^{\mathbb{B}})) \cdots (\chi(v_{l-1}^{\mathbb{B}}))\}$.

Then $\delta' = \mathsf{e}(\pi)(\tau' \uplus \tau'') \uplus \tau' \uplus \tau'' \uplus {}_{\{v_0, \ldots, v_{l-1}\}}|\chi$. Moreover, by the EXPLICITNESS LEMMA, we have $\delta' = {}_{\mathbb{VAB} \backslash \{v_l^{\mathbb{B}}\}}|\mathrm{id} \circ \mathrm{eval}(\mathcal{S} \uplus \delta)$.

By the VALUATION LEMMA we have

$$
\begin{aligned}
& \mathrm{eval}(\mathcal{S} \uplus \delta)(y^{\mathbb{V}}(v_0^{\mathbb{B}}) \cdots (v_{l-1}^{\mathbb{B}})) \\
= \ & \delta(y^{\mathbb{V}})(\delta(v_0^{\mathbb{B}})) \cdots (\delta(v_{l-1}^{\mathbb{B}})) \\
= \ & \mathsf{e}(\pi)(\tau)(y^{\mathbb{V}})(\chi(v_0^{\mathbb{B}})) \cdots (\chi(v_{l-1}^{\mathbb{B}})) \\
= \ & \pi(y^{\mathbb{V}})(\tau')(\chi(v_0^{\mathbb{B}})) \cdots (\chi(v_{l-1}^{\mathbb{B}})) \\
= \ & f(\chi(v_0^{\mathbb{B}})) \cdots (\chi(v_{l-1}^{\mathbb{B}})).
\end{aligned}
$$

Thus, $\chi' = \sigma \circ \mathrm{eval}(\mathcal{S} \uplus \delta)$.

Thus, $\delta' \uplus \chi' = ({}_{\mathbb{VAB} \backslash \{v_l^{\mathbb{B}}\}}|\mathrm{id} \uplus \sigma) \circ \mathrm{eval}(\mathcal{S} \uplus \delta)$.

Thus, we have, on the one hand,

$$
\begin{aligned}
& \mathrm{eval}(\mathcal{S} \uplus \delta' \uplus \chi')(B) \\
= \ & \mathrm{eval}(\mathcal{S} \uplus (({}_{\mathbb{VAB} \backslash \{v_l^{\mathbb{B}}\}}|\mathrm{id} \uplus \sigma) \circ \mathrm{eval}(\mathcal{S} \uplus \delta)))(B) \\
= \ & \mathrm{eval}(\mathcal{S} \uplus \delta)(B\sigma) \\
= \ & \mathsf{FALSE},
\end{aligned}
$$

where the second equation holds by the Substitution [Value] Lemma.
Moreover, on the other hand, we have

$$\mathsf{eval}(\mathcal{S} \uplus \delta' \uplus_{\{v_l^{\mathbb{B}}\}} | \chi)(B)$$
$$= \mathsf{eval}(\mathcal{S} \uplus \delta)(B)$$
$$= \mathsf{TRUE}.$$

This means that a value (such as $\chi(v_l^{\mathbb{B}})$) could have been chosen for $f(\chi(v_0^{\mathbb{B}})) \cdots (\chi(v_{l-1}^{\mathbb{B}}))$ to make B true in $\mathcal{S} \uplus \delta' \uplus \chi'$, but it was not. This contradicts the definition of f. **Q.e.d. (Lemma 5.16)**

Proof of Lemma 5.19

Let $C(y^{\mathrm{v}}) = \lambda v_0^{\mathbb{B}}. \dots \lambda v_{l-1}^{\mathbb{B}}. \, \varepsilon v_l^{\mathbb{B}}. \, B$ for a formula B. Set $\sigma := \{v_l^{\mathbb{B}} \mapsto y^{\mathrm{v}}(v_0^{\mathbb{B}}) \cdots (v_{l-1}^{\mathbb{B}})\}$. Then we have $Q_C(y^{\mathrm{v}}) = \forall v_0^{\mathbb{B}}. \dots \forall v_{l-1}^{\mathbb{B}}.$ $\big(\exists v_l^{\mathbb{B}}. \, B \; \Rightarrow \; B\sigma \big)$. Let π be $\mathcal{S}$-compatible with $(C, (P, N))$; namely, in the case of item 1, the π mentioned in the lemma, or, in the case of item 2, the π that exists according to Lemma 5.16. Let $\tau : \mathbb{A} \to \mathcal{S}$ be arbitrary. It now suffices to show $\mathsf{eval}(\mathcal{S} \uplus \mathsf{e}(\pi)(\tau) \uplus \tau)(Q_C(y^{\mathrm{v}})) = \mathsf{TRUE}$. By the backward direction of the $\forall$-Lemma, it suffices to show $\mathsf{eval}(\mathcal{S} \uplus \delta)(\exists v_l^{\mathbb{B}}. \, B \; \Rightarrow \; B\sigma) = \mathsf{TRUE}$ for an arbitrary $\chi : \{v_0^{\mathbb{B}}, \dots, v_{l-1}^{\mathbb{B}}\} \to \mathcal{S}$, setting $\delta := \mathsf{e}(\pi)(\tau) \uplus \tau \uplus \chi$. By the backward direction of the $\Rightarrow$-Lemma, it suffices to show $\mathsf{eval}(\mathcal{S} \uplus \delta)(B\sigma) = \mathsf{TRUE}$ under the assumption of $\mathsf{eval}(\mathcal{S} \uplus \delta)(\exists v_l^{\mathbb{B}}. \, B) = \mathsf{TRUE}$. From the latter, by the forward direction of the $\exists$-Lemma, there is a $\chi' : \{v_l^{\mathbb{B}}\} \to \mathcal{S}$ such that $\mathsf{eval}(\mathcal{S} \uplus \delta \uplus \chi')(B) = \mathsf{TRUE}$. By item 2 of Definition 5.15, we get $\mathsf{eval}(\mathcal{S} \uplus \delta \uplus \chi')(B\sigma) = \mathsf{TRUE}$. By the Explicitness Lemma, we get $\mathsf{eval}(\mathcal{S} \uplus \delta)(B\sigma) = \mathsf{TRUE}$. **Q.e.d. (Lemma 5.19)**

Proof of Lemma 5.22

Let us assume that π is $\mathcal{S}$-compatible with $(C', (P', N'))$. Then, by item 1 of Definition 5.15, $\pi : \mathbb{V} \to (\mathbb{A} \rightsquigarrow \mathcal{S}) \rightsquigarrow \mathcal{S}$ is an $\mathcal{S}$-raising-valuation and $(P' \cup S_\pi, N')$ is consistent. As (P', N') is an extension of (P, N), we have $P \subseteq P'$ and $N \subseteq N'$. Thus, $(P' \cup S_\pi, N')$ is an extension of $(P \cup S_\pi, N)$. Thus, $(P \cup S_\pi, N)$ is consistent by Corollary 5.7. For π to be $\mathcal{S}$-compatible with $(C, (P, N))$, it now suffices to show item 2 of Definition 5.15. As this property does not depend on the positive/negative variable-conditions anymore, it suffices to note that it actually holds because it holds for C' by assumption and we also have $C \subseteq C'$ by assumption. **Q.e.d. (Lemma 5.22)**

Proof of Lemma 5.24

By assumption, $(C', (P', N'))$ is the extended σ-update of $(C, (P, N))$. Thus, (P', N') is the σ-update of (P, N). Thus, because σ is a (P, N)-substitution, (P', N') is a consistent positive/negative variable-condition by Definition 5.9. Moreover, C is a (P, N)-choice-condition. Thus, C is a partial function from $\mathbb{V}$ into the set of higher-order ε-terms, such that Items 1, 2, and 3 of Definition 5.13 hold. Thus, C' is a partial function from $\mathbb{V}$ into the set of higher-order ε-terms satisfying items 1 and 2 of Definition 5.13 as well. For C' to satisfy also item 3 of Definition 5.13, it now suffices to show the following Claim 1.

<u>Claim 1:</u> Let $y^{\vee} \in \mathrm{dom}(C')$ and $z^{\vee\!\mathbb{A}} \in \mathbb{VA}(C'(y^{\vee}))$. Then we have $z^{\vee\!\mathbb{A}}\,(P')^+ y^{\vee}$.

<u>Proof of Claim 1:</u> By the definition of C', we have $z^{\vee\!\mathbb{A}} \in \mathbb{VA}(C(y^{\vee}))$ or else there is some $x^{\vee} \in \mathrm{dom}(\sigma) \cap \mathbb{V}(C(y^{\vee}))$ with $z^{\vee\!\mathbb{A}} \in \mathbb{VA}(\sigma(x^{\vee}))$. Thus, as C is a (P, N)-choice-condition, we have either $z^{\vee\!\mathbb{A}}\, P^+ y^{\vee}$ or else $x^{\vee}\, P^+ y^{\vee}$ and $z^{\vee\!\mathbb{A}} \in \mathbb{VA}(\sigma(x^{\vee}))$. Then, as (P', N') is the σ-update of (P, N), by Definition 5.8, we have either $z^{\vee\!\mathbb{A}}\,(P')^+ y^{\vee}$ or else $x^{\vee}\,(P')^+ y^{\vee}$ and $z^{\vee\!\mathbb{A}}\, P' x^{\vee}$. Thus, in any case, $z^{\vee\!\mathbb{A}}\,(P')^+ y^{\vee}$. Q.e.d. (Claim 1)

Q.e.d. (Lemma 5.24)

Proof of Lemma 5.25

Let us assume the situation described in the lemma.

We set $A := \mathrm{dom}(\sigma) \setminus (O' \uplus O)$. As σ is a substitution on $\mathbb{V}$, we have $\mathrm{dom}(\sigma) \subseteq O' \uplus O \uplus A \subseteq \mathbb{V}$.

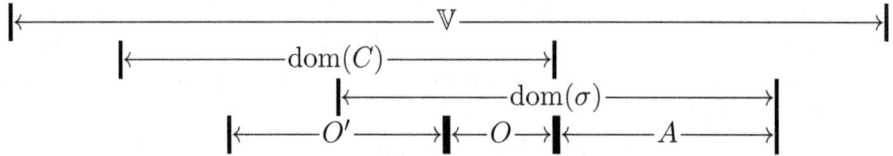

Note that C' is a (P', N')-choice-condition by Lemma 5.24.

As π' is $\mathcal{S}$-compatible with $(C', (P', N'))$, we know that $(P' \cup S_{\pi'}, N')$ s a consistent positive/negative variable-condition. Thus, $\lhd := (P' \cup S_{\pi'})^+$ is a well-founded ordering.

Let D be the dependence relation of σ. Set $S_{\pi} := {}_{\mathbb{A}}\!\!\restriction\lhd$.

<u>Claim 1:</u> We have $P', S_{\pi'}, P, D, S_{\pi} \subseteq\, \lhd$ and

$\qquad (P' \cup S_{\pi'}, N')$ is a weak extension of $(P \cup S_{\pi}, N)$ and of $(\lhd, N)$ (cf. Definition 5.6).

<u>Proof of Claim 1:</u> As (P', N') is the σ-update of (P, N), we have $P' = P \cup D$ and $N' = N$. Thus, $P', S_{\pi'}, P, D, S_{\pi} \subseteq (P' \cup S_{\pi'})^+ =\, \lhd$. Q.e.d. (Claim 1)

<u>Claim 2:</u> $(P \cup S_{\pi}, N)$ and $(\lhd, N)$ are consistent positive/negative variable-conditions.
<u>Proof of Claim 2:</u> This follows from Claim 1 by Corollary 5.7. Q.e.d. (Claim 2)

<u>Claim 3:</u> ${}_{O'}\!\!\restriction C$ is an $(\lhd, N)$-choice-condition.
<u>Proof of Claim 3:</u> By Claims 1 and 2 and the assumption that C is a (P, N)-choice-condition. Q.e.d. (Claim 3)

The plan for defining the $\mathcal{S}$-raising-valuation π (which we have to find) is to give $\pi(y^{\vee})(S_{\pi}\langle\!\langle\{y^{\vee}\}\rangle\!\rangle\restriction\tau)$ a value as follows:

508

(α) For $y^\vee \in \mathbb{V} \backslash (O' \uplus O \uplus A)$, we take this value to be

$$\pi'(y^\vee)(S_{\pi'}\langle\!\langle y^\vee\rangle\!\rangle|\tau).$$

This is indeed possible because of $S_{\pi'} \subseteq \mathbb{A}|\lhd = S_\pi$, so $S_{\pi'}\langle\!\langle y^\vee\rangle\!\rangle|\tau \subseteq S_\pi\langle\!\langle y^\vee\rangle\!\rangle|\tau$.

(β) For $y^\vee \in O \uplus A$, we take this value to be

$$\mathrm{eval}(\mathcal{S} \uplus \mathsf{e}(\pi')(\tau) \uplus \tau)(\sigma(y^\vee)).$$

Note that, in case of $y^\vee \in O$, we know that $(Q_C(y^\vee))\sigma$ is $(\pi', \mathcal{S})$-valid by assumption of the lemma. Moreover, the case of $y^\vee \in A$ is unproblematic because of $y^\vee \notin \mathrm{dom}(C)$. Again, π is well-defined in this case because the only part of τ that is accessed by the given value is $S_\pi\langle\!\langle y^\vee\rangle\!\rangle|\tau$. Indeed, this can be seen as follows: By Claim 1 and the transitivity of $\lhd$, we have: $\mathbb{A}|D \cup S_{\pi'} \circ D \subseteq \mathbb{A}|\lhd = S_\pi$.

(γ) For $y^\vee \in O'$, however, we have to take care of $\mathcal{S}$-compatibility with $(C, (P, N))$ explicitly in an $\lhd$-recursive definition on the basis a function ρ implementing (α) and (β). This disturbance does not interfere with the semantic invariance stated in the lemma because occurrences of variables from O' are explicitly excluded in the relevant terms and formulas and, according to the statement of lemma, O' satisfies the appropriate closure condition.

Set $S_\rho := S_\pi$. Let ρ be defined by $(y^\vee \in \mathbb{V},\ \tau : \mathbb{A} \to \mathcal{S})$

$$\rho(y^\vee)(S_\pi\langle\!\langle y^\vee\rangle\!\rangle|\tau) := \begin{cases} \pi'(y^\vee)(S_{\pi'}\langle\!\langle y^\vee\rangle\!\rangle|\tau) & \text{if } y^\vee \in \mathbb{V}\backslash(O \uplus A) \\ \mathrm{eval}(\mathcal{S} \uplus \mathsf{e}(\pi')(\tau) \uplus \tau)(\sigma(y^\vee)) & \text{if } y^\vee \in O \uplus A \end{cases}$$

Let π be the $\mathcal{S}$-raising-valuation that exists according to Lemma 5.16 for the $\mathcal{S}$-raising-valuation ρ and the $(\lhd, N)$-choice-condition $_{O'}|C$ (cf. Claim 3). Note that the assumptions of Lemma 5.16 are indeed satisfied here and that the resulting semantic relation S_π of Lemma 5.16 is indeed identical to our pre-defined relation of the same name, thereby justifying our abuse of notation: Indeed, by assumption of Lemma 5.25, for every return type α of $_{O'}|C$, there is a generalized choice function on the power-set of the universe of $\mathcal{S}$ for the type α; and we have

$$S_\rho = S_\pi = \mathbb{A}|\lhd = \mathbb{A}|(\lhd^+).$$

Because of $\mathrm{dom}(_{O'}|C) = O'$, according to Lemma 5.16, we then have

$$_{\mathbb{V}\backslash O'}|\pi = _{\mathbb{V}\backslash O'}|\rho$$

and π is $\mathcal{S}$-compatible with $(_{O'}|C, (\lhd, N))$.

<u>Claim 4:</u> For all $y^\vee \in O \uplus A$ and $\tau : \mathbb{A} \to \mathcal{S}$, when we set $\delta' := \mathsf{e}(\pi')(\tau) \uplus \tau$:
$$\mathsf{e}(\pi)(\tau)(y^\vee) = \mathrm{eval}(\mathcal{S} \uplus \delta')(\sigma(y^\vee)).$$

<u>Proof of Claim 4:</u> We have $O \uplus A \subseteq \mathbb{V}\backslash O'$. Thus, Claim 4 follows immediately from the definition of ρ. $\hspace{2cm}$ Q.e.d. (Claim 4)

<u>Claim 5:</u> For all $y^{\mathrm{v}} \in \mathbb{V} \backslash (O' \uplus O \uplus A)$ and $\tau : \mathbb{A} \to \mathcal{S}$: $\mathsf{e}(\pi)(\tau)(y^{\mathrm{v}}) = \mathsf{e}(\pi')(\tau)(y^{\mathrm{v}})$.

<u>Proof of Claim 5:</u> For $y^{\mathrm{v}} \in \mathbb{V} \backslash (O' \uplus O \uplus A)$, we have $y^{\mathrm{v}} \in \mathbb{V} \backslash O'$ and $y^{\mathrm{v}} \in \mathbb{V} \backslash (O \uplus A)$. Thus, $\mathsf{e}(\pi)(\tau)(y^{\mathrm{v}}) = \pi(y^{\mathrm{v}})(s_{\pi \langle\!\langle y^{\mathrm{v}} \rangle\!\rangle} | \tau) = \rho(y^{\mathrm{v}})(s_{\pi \langle\!\langle y^{\mathrm{v}} \rangle\!\rangle} | \tau) = \pi'(y^{\mathrm{v}})(s_{\pi'\langle\!\langle y^{\mathrm{v}} \rangle\!\rangle} | \tau) = \mathsf{e}(\pi')(\tau)(y^{\mathrm{v}})$.

Q.e.d. (Claim 5)

<u>Claim 6:</u> For any term or formula B (possibly with some unbound occurrences of bound atoms from the set $W \subseteq \mathbb{B}$) with $O' \cap \mathbb{V}(B) = \emptyset$, and for every $\tau : \mathbb{A} \to \mathcal{S}$ and every $\chi : W \to \mathcal{S}$, when we set $\delta := \mathsf{e}(\pi)(\tau) \uplus \tau$ and $\delta' := \mathsf{e}(\pi')(\tau) \uplus \tau$, we have $\mathrm{eval}(\mathcal{S} \uplus \delta' \uplus \chi)(B\sigma) = \mathrm{eval}(\mathcal{S} \uplus \delta \uplus \chi)(B)$.

<u>Proof of Claim 6:</u> $\mathrm{eval}(\mathcal{S} \uplus \delta' \uplus \chi)(B\sigma) =$

(by the SUBSTITUTION [VALUE] LEMMA)

$\mathrm{eval}(\mathcal{S} \uplus (\sigma \uplus {}_{\mathbb{VAB} \backslash \mathrm{dom}(\sigma)} | \mathrm{id}) \circ \mathrm{eval}(\mathcal{S} \uplus \delta' \uplus \chi))(B) =$

(by the EXPLICITNESS LEMMA and the VALUATION LEMMA (for the case of $l = 0$))

$\mathrm{eval}(\mathcal{S} \uplus (\sigma \circ \mathrm{eval}(\mathcal{S} \uplus \delta')) \uplus {}_{\mathbb{VA} \backslash \mathrm{dom}(\sigma)} | \delta' \uplus \chi)(B) =$

(by $O \uplus A \subseteq \mathrm{dom}(\sigma) \subseteq O' \uplus O \uplus A$, $O' \cap \mathbb{V}(B) = \emptyset$, and the EXPLICITNESS LEMMA)

$\mathrm{eval}(\mathcal{S} \uplus {}_{O \uplus A} | \sigma \circ \mathrm{eval}(\mathcal{S} \uplus \delta') \uplus {}_{\mathbb{VA} \backslash (O' \uplus O \uplus A)} | \delta' \uplus \chi)(B) =$

(by Claim 4 and Claim 5)

$\mathrm{eval}(\mathcal{S} \uplus {}_{O \uplus A} | \delta \uplus {}_{\mathbb{VA} \backslash (O' \uplus O \uplus A)} | \delta \uplus \chi)(B) =$

(by $O' \cap \mathbb{V}(B) = \emptyset$ and the EXPLICITNESS LEMMA)

$\mathrm{eval}(\mathcal{S} \uplus \delta \uplus \chi)(B)$.

Q.e.d. (Claim 6)

<u>Claim 7:</u> For every set of sequents G' (possibly with some unbound occurrences of bound atoms from the set $W \subseteq \mathbb{B}$) with $O' \cap \mathbb{V}(G') = \emptyset$, and for every $\tau : \mathbb{A} \to \mathcal{S}$ and for every $\chi : W \to \mathcal{S}$: Truth of G' in $\mathcal{S} \uplus \mathsf{e}(\pi)(\tau) \uplus \tau \uplus \chi$ is equivalent to

truth of $G'\sigma$ in $\mathcal{S} \uplus \mathsf{e}(\pi')(\tau) \uplus \tau \uplus \chi$.

<u>Proof of Claim 7:</u> This is a trivial consequence of Claim 6.

Q.e.d. (Claim 7)

<u>Claim 8:</u> For $y^{\mathrm{v}} \in \mathrm{dom}(C) \setminus O'$, we have $O' \cap \mathbb{V}(C(y^{\mathrm{v}})) = \emptyset$.

<u>Proof of Claim 8:</u> Otherwise there is some $y^{\mathrm{v}} \in \mathrm{dom}(C) \setminus O'$ and some $z^{\mathrm{v}} \in O' \cap \mathbb{V}(C(y^{\mathrm{v}}))$. Then $z^{\mathrm{v}} P^+ y^{\mathrm{v}}$ because C is a (P, N)-choice-condition, and then, as $\langle O' \rangle P^+ \cap \mathrm{dom}(C) \subseteq O'$ by assumption of the lemma, we have the contradicting $y^{\mathrm{v}} \in O'$.

Q.e.d. (Claim 8)

<u>Claim 9:</u> Let $y^{\mathrm{v}} \in \mathrm{dom}(C)$ and $C(y^{\mathrm{v}}) = \lambda v_0^{\mathbb{B}}. \ldots \lambda v_{l-1}^{\mathbb{B}}. \varepsilon v_l^{\mathbb{B}}. B$. Let $\tau : \mathbb{A} \to \mathcal{S}$ and $\chi : \{v_0^{\mathbb{B}}, \ldots, v_l^{\mathbb{B}}\} \to \mathcal{S}$. Set $\delta := \mathsf{e}(\pi)(\tau) \uplus \tau \uplus \chi$. Set $\mu := \{v_l^{\mathbb{B}} \mapsto y^{\mathrm{v}}(v_0^{\mathbb{B}}) \cdots (v_{l-1}^{\mathbb{B}})\}$. If B is true in $\mathcal{S} \uplus \delta$, then $B\mu$ is true in $\mathcal{S} \uplus \delta$ as well.

<u>Proof of Claim 9:</u> Set $\delta' := \mathsf{e}(\pi')(\tau) \uplus \tau \uplus \chi$.

$y^{\mathrm{v}} \notin O' \uplus O$: In this case, because of $\mathrm{dom}(\sigma) \cap \mathrm{dom}(C) \subseteq O' \uplus O$, we have $y^{\mathrm{v}} \notin \mathrm{dom}(\sigma)$. Thus, as $(C', (P', N'))$ is the extended σ-update of

510

$(C, (P, N))$, we have $C'(y^\vee) = (C(y^\vee))\sigma$. By Claim 8, we have $O' \cap \mathbb{V}(B) = \emptyset$. And then, by our case assumption, also $O' \cap \mathbb{V}(B\mu) = \emptyset$. By assumption of Claim 9, B is true in $\mathcal{S} \uplus \delta$. Thus, by Claim 7, $B\sigma$ is true in $\mathcal{S} \uplus \delta'$. Thus, as π' is $\mathcal{S}$-compatible with $(C', (P', N'))$, we know that $(B\sigma)\mu$ is true in $\mathcal{S} \uplus \delta'$. Because of $y^\vee \notin \mathrm{dom}(\sigma)$, this means that $(B\mu)\sigma$ is true in $\mathcal{S} \uplus \delta'$. Thus, by Claim 7, $B\mu$ is true in $\mathcal{S} \uplus \delta$.

$\underline{y^\vee \in O}$: By Claim 8, we have $O' \cap \mathbb{V}(B) = \emptyset$.

And then, by our case assumption, also $O' \cap \mathbb{V}(B\mu) = \emptyset$. Moreover, $(Q_C(y^\vee))\sigma$ is equal to $\forall v_0^\mathrm{B}. \ldots \forall v_{l-1}^\mathrm{B}. \left(\exists v_l^\mathrm{B}. B \Rightarrow B\mu \right)\sigma$ and $(\pi', \mathcal{S})$-valid by assumption of the lemma. Thus, by the forward direction of the $\forall$-LEMMA, $\left(\exists v_l^\mathrm{B}. B \Rightarrow B\mu \right)\sigma$ is true in $\mathcal{S} \uplus \delta'$. Thus, by Claim 7, $\exists v_l^\mathrm{B}. B \Rightarrow B\mu$ is true in $\mathcal{S} \uplus \delta$. As, by assumption of Claim 9, B is true in $\mathcal{S} \uplus \delta$, by the backward direction of the $\exists$-LEMMA, $\exists v_l^\mathrm{B}. B$ is true in $\mathcal{S} \uplus \delta$ as well. Thus, by the forward direction of the $\Rightarrow$-LEMMA, $B\mu$ is true in $\mathcal{S} \uplus \delta$ as well.

$\underline{y^\vee \in O'}$: π is $\mathcal{S}$-compatible with $(_O\langle C, (\lhd, N))$ by definition, as explicitly stated before Claim 4. $\hspace{2cm}$ Q.e.d. (Claim 9)

By Claims 2 and 9, π is $\mathcal{S}$-compatible with $(C, (P, N))$. And then items 1 and 2 of the lemma are trivial consequences of Claims 6 and 7, respectively.

$$\textbf{Q.e.d. (Lemma 5.25)}$$

Proof of Theorem 5.27

The first four items are trivial (Validity, Reflexivity, Transitivity, Additivity).

$\underline{(5a)}$: If G_0 is $(C', (P', N'))$-valid in $\mathcal{S}$, then there is some π that is $\mathcal{S}$-compatible with $(C', (P', N'))$ such that G_0 is $(\pi, \mathcal{S})$-valid. By Lemma 5.22, π is also $\mathcal{S}$-compatible with $(C, (P, N))$. Thus, G_0 is $(C, (P, N))$-valid, in $\mathcal{S}$.

$\underline{(5b)}$: Suppose that π is $\mathcal{S}$-compatible with $(C', (P', N'))$, and that G_1 is $(\pi, \mathcal{S})$-valid. By Lemma 5.22, π is also $\mathcal{S}$-compatible with $(C, (P, N))$. Thus, since G_0 $(C, (P, N))$-reduces to G_1, also G_0 is $(\pi, \mathcal{S})$-valid as was to be shown.

$\underline{(6)}$: Assume the situation described in the lemma.

$\underline{\text{Claim 1:}}$ $O' \subseteq \mathrm{dom}(C) \setminus O$.

$\underline{\text{Proof of Claim 1:}}$ By definition, $O' \subseteq \mathrm{dom}(C)$. It remains to show $O' \cap O = \emptyset$. To the contrary, suppose that there is some $y^\vee \in O' \cap O$. Then, by the definition of O', there is some $z^\vee \in M \setminus O$ with $z^\vee P^* y^\vee$. By definition of O, however, we have $y^\vee \in P^*\langle V \rangle$. Thus, $z^\vee \in P^*\langle V \rangle$. Thus, $z^\vee \in O$, a contradiction. $\hspace{0.5cm}$ Q.e.d. (Claim 1)

511

<u>Claim 2:</u> $\langle O' \rangle P^+ \cap \mathrm{dom}(C) \subseteq O'$.

<u>Proof of Claim 2:</u> Assume $y^\mathrm{v} \in O'$ and $z^\mathrm{v} \in \mathrm{dom}(C)$ with $y^\mathrm{v} \, P^+ \, z^\mathrm{v}$. It now suffices to show $z^\mathrm{v} \in O'$. Because of $y^\mathrm{v} \in O'$, there is some $x^\mathrm{v} \in M\backslash O$ with $x^\mathrm{v} \, P^* \, y^\mathrm{v}$. Thus, $x^\mathrm{v} \, P^* \, z^\mathrm{v}$. Thus, $z^\mathrm{v} \in O'$. Q.e.d. (Claim 2)

<u>Claim 3:</u> $\mathrm{dom}(\sigma) \cap \mathrm{dom}(C) \subseteq O' \cup O$.

<u>Proof of Claim 3:</u> $\mathrm{dom}(\sigma) \cap \mathrm{dom}(C) = \mathrm{dom}(C) \cap M \subseteq O \cup (\mathrm{dom}(C) \cap (M\backslash O))$ $\subseteq O \cup (\mathrm{dom}(C) \cap \langle M\backslash O \rangle P^*) = O \cup O'$. Q.e.d. (Claim 3)

<u>Claim 4:</u> $O' \cap \mathbb{V}(G_0, G_1) = O' \cap V = \emptyset$.

<u>Proof of Claim 4:</u> Because of $\mathbb{V}(G_0, G_1) \subseteq V$, it suffices to show the second equality. To the contrary of the second equality, suppose that there is some $y^\mathrm{v} \in O' \cap V$. Then, by the definition of O', there is some $z^\mathrm{v} \in M\backslash O$ with $z^\mathrm{v} \, P^* \, y^\mathrm{v}$. By definition of O, however, we have $z^\mathrm{v} \in O$, a contradiction. Q.e.d. (Claim 4)

<u>(6a):</u> In case that $G_0\sigma \cup (\langle O \rangle Q_C)\sigma$ is $(C', (P', N'))$-valid in $\mathcal{S}$, there is some π' that is $\mathcal{S}$-compatible with $(C', (P', N'))$ such that $G_0\sigma \cup (\langle O \rangle Q_C)\sigma$ is $(\pi', \mathcal{S})$-valid. Then both $G_0\sigma$ and $(\langle O \rangle Q_C)\sigma$ are $(\pi', \mathcal{S})$-valid. By Claims 1, 2, 3, and 4, let π be given as in Lemma 5.25. Then G_0 is $(\pi, \mathcal{S})$-valid. Moreover, as π is $\mathcal{S}$-compatible with $(C, (P, N))$, G_0 is $(C, (P, N))$-valid in $\mathcal{S}$.

<u>(6b):</u> Let π' be $\mathcal{S}$-compatible with $(C', (P', N'))$, and suppose that $G_1\sigma \cup (\langle O \rangle Q_C)\sigma$ is $(\pi', \mathcal{S})$-valid. Then both $G_1\sigma$ and $(\langle O \rangle Q_C)\sigma$ are $(\pi', \mathcal{S})$-valid. By Claims 1, 2, 3, and 4, let π be given as in Lemma 5.25. Then π is $\mathcal{S}$-compatible with $(C, (P, N))$, and G_1 is $(\pi, \mathcal{S})$-valid. By assumption, G_0 $(C, (P, N))$-reduces to G_1. Thus, G_0 is $(\pi, \mathcal{S})$-valid, too. Thus, by Lemma 5.25, $G_0\sigma$ is $(\pi', \mathcal{S})$-valid as was to be shown.

<u>(7):</u> Let π be $\mathcal{S}$-compatible with $(C, (P, N))$, and suppose that G_0 is $(\pi, \mathcal{S})$-valid. Let $\tau : \mathbb{A} \to \mathcal{S}$ be an arbitrary $\mathcal{S}$-valuation. Set $\delta := \mathsf{e}(\pi)(\tau) \uplus \tau$. It suffices to show $\mathrm{eval}(\mathcal{S} \uplus \delta)(G_0\nu) = \mathsf{TRUE}$.

Define $\tau' : \mathbb{A} \to \mathcal{S}$ via $\tau'(y^\mathbb{A}) := \left\{ \begin{array}{ll} \tau(y^\mathbb{A}) & \text{for } y^\mathbb{A} \in \mathbb{A}\backslash\mathrm{dom}(\nu) \\ \mathrm{eval}(\mathcal{S} \uplus \delta)(\nu(y^\mathbb{A})) & \text{for } y^\mathbb{A} \in \mathrm{dom}(\nu) \end{array} \right\}$.

<u>Claim 5:</u> For $v^\mathrm{v} \in \mathbb{V}(G_0)$ we have $\mathsf{e}(\pi)(\tau)(v^\mathrm{v}) = \mathsf{e}(\pi)(\tau')(v^\mathrm{v})$.

<u>Proof of Claim 5:</u> Otherwise there must be some $y^\mathbb{A} \in \mathrm{dom}(\nu)$ with $y^\mathbb{A} \, S_\pi \, v^\mathrm{v}$. Because of $v^\mathrm{v} \in \mathbb{V}(G_0)$ and $\mathbb{V}(G_0) \times \mathrm{dom}(\nu) \subseteq N$, we have $v^\mathrm{v} \, N \, y^\mathbb{A}$. But then $(P \cup S_\pi, N)$ is not consistent, which contradicts π being $\mathcal{S}$-compatible with $(C, (P, N))$. Q.e.d. (Claim 5)

Then we get by the SUBSTITUTION [VALUE] LEMMA (1st equation), the VALUATION LEMMA (for the case of $l=0$) (2nd equation), by definition of τ' and δ (3rd equation), by the EXPLICITNESS LEMMA and Claim 5 (4th equation), and by the $(\pi, \mathcal{S})$-validity of G_0 (5th equation):

$$
\begin{aligned}
\mathrm{eval}(\mathcal{S} \uplus \delta)(G_0\nu) \; &= \; \mathrm{eval}\Big(\mathcal{S} \uplus \big((\nu \uplus {}_{\mathbb{VA}\backslash\mathrm{dom}(\nu)}\!\upharpoonright\mathrm{id}) \circ \mathrm{eval}(\mathcal{S} \uplus \delta) \big)\Big)\Big(G_0\Big) \\
&= \; \mathrm{eval}\Big(\mathcal{S} \uplus \big(\nu \circ \mathrm{eval}(\mathcal{S} \uplus \delta) \big) \uplus {}_{\mathbb{VA}\backslash\mathrm{dom}(\nu)}\!\upharpoonright\delta\Big)\Big(G_0\Big) \\
&= \; \mathrm{eval}\Big(\mathcal{S} \uplus \tau' \uplus \mathsf{e}(\pi)(\tau)\Big)\Big(G_0\Big) \\
&= \; \mathrm{eval}\Big(\mathcal{S} \uplus \tau' \uplus \mathsf{e}(\pi)(\tau')\Big)\Big(G_0\Big) \\
&= \; \mathsf{TRUE}
\end{aligned}
$$

Q.e.d. (Theorem 5.27)

Proof of Theorem 5.28

To illustrate our techniques, we only treat the first rule of each kind; the other rules can be treated most similarly. In the situation described in the theorem, it suffices to show that C' is a (P', N')-choice-condition (because the other properties of an extended extension are trivial), and that, for every $\mathcal{S}$-raising-valuation π that is $\mathcal{S}$-compatible with $(C', (P', N'))$, the sets G_0 and G_1 of the upper and lower sequents of the inference rule are equivalent w.r.t. their $(\pi, \mathcal{S})$-validity.

$\underline{\gamma\text{-rule:}}$ In this case we have $(C', (P', N')) = (C, (P, N))$. Thus, C' is a (P', N')-choice-condition by assumption of the theorem. Moreover, for every $\mathcal{S}$-valuation $\tau : \mathbb{A} \to \mathcal{S}$, and for $\delta := \mathsf{e}(\pi)(\tau) \uplus \tau$, the truths of
$$ \{ \Gamma \quad \exists y^{\mathbb{B}}.\, A \quad \Pi \} \quad \text{and} \quad \{ A\{y^{\mathbb{B}} \mapsto t\} \quad \Gamma \quad \exists y^{\mathbb{B}}.\, A \quad \Pi \} $$
in $\mathcal{S}\uplus\delta$ are indeed equivalent. The implication from left to right is trivial because the former sequent is a sub-sequent of the latter.

For the other direction, assume that $A\{y^{\mathbb{B}} \mapsto t\}$ is true in $\mathcal{S}\uplus\delta$. Thus, by the SUB-STITUTION [VALUE] LEMMA (second equation) and the VALUATION LEMMA for $l = 0$ (third equation):
$$
\begin{aligned}
\mathsf{TRUE} \; &= \; \mathrm{eval}(\mathcal{S} \uplus \delta)(A\{y^{\mathbb{B}} \mapsto t\}) \\
&= \; \mathrm{eval}(\mathcal{S} \uplus ((\{y^{\mathbb{B}} \mapsto t\} \uplus {}_{\mathbb{VAB}\backslash\{y^{\mathbb{B}}\}}\!\upharpoonright\mathrm{id}) \circ \mathrm{eval}(\mathcal{S}\uplus\delta)))(A) \\
&= \; \mathrm{eval}(\mathcal{S} \uplus \{y^{\mathbb{B}} \mapsto \mathrm{eval}(\mathcal{S}\uplus\delta)(t)\} \uplus \delta)(A)
\end{aligned}
$$

Thus, by the backward direction of the $\exists$-LEMMA, $\exists y^{\mathbb{B}}.\, A$ is true in $\mathcal{S}\uplus\delta$. Thus, the upper sequent is true $\mathcal{S}\uplus\delta$.

$\underline{\delta^-\text{-rule:}}$ In this case, we have $x^{\mathbb{A}} \in \mathbb{A} \setminus (\mathrm{dom}(P) \cup \mathbb{A}(\Gamma, A, \Pi))$, $C'' = \emptyset$, and $V = \mathbb{V}(\Gamma \quad \forall x^{\mathbb{B}}.\, A \quad \Pi) \times \{x^{\mathbb{A}}\}$. Thus, $C' = C$, $P' = P$, and $N' = N \cup V$.

<u>Claim 1:</u> C' is a (P', N')-choice-condition.

<u>Proof of Claim 1:</u> By assumption of the theorem, C is a (P, N)-choice-condition. Thus, (P, N) is a consistent positive/negative variable-condition. By Definition 5.4, P is well-founded and $P^+ \circ N$ is irreflexive. Since $x^A \notin \mathrm{dom}(P)$, we have $x^A \notin \mathrm{dom}(P^+)$. Thus, because of $\mathrm{ran}(V) = \{x^A\}$, also $P^+ \circ N'$ is irreflexive. Thus, (P', N') is a consistent positive/negative variable-condition, and C' is a (P', N')-choice-condition. $\hspace{2cm}$ Q.e.d. (Claim 1)

Now, for the soundness direction, it suffices to show the contrapositive, namely to assume that there is an $\mathcal{S}$-valuation $\tau : \mathbb{A} \to \mathcal{S}$ such that $\{\Gamma \ \ \forall x^B.\ A \ \ \Pi\}$ is false in $\mathcal{S} \uplus \mathrm{e}(\pi)(\tau) \uplus \tau$, and to show that there is an $\mathcal{S}$-valuation $\tau' : \mathbb{A} \to \mathcal{S}$ such that $\{A\{x^B \mapsto x^A\} \ \ \Gamma \ \ \Pi\}$ is false in $\mathcal{S} \uplus \mathrm{e}(\pi)(\tau') \uplus \tau'$. Under this assumption, the sequent $\Gamma \Pi$ is false in $\mathcal{S} \uplus \mathrm{e}(\pi)(\tau) \uplus \tau$.

<u>Claim 2:</u> $\Gamma \Pi$ is false in $\mathcal{S} \uplus \mathrm{e}(\pi)(\tau') \uplus \tau'$ for all $\tau' : \mathbb{A} \to \mathcal{S}$ with $_{\mathbb{A} \backslash \{x^A\}}|\tau' = {}_{\mathbb{A} \backslash \{x^A\}}|\tau$.

<u>Proof of Claim 2:</u> Because of $x^A \notin \mathbb{A}(\Gamma \Pi)$, by the Explicitness Lemma, if Claim 2 did not hold, there would have to be some $u^V \in \mathbb{V}(\Gamma \Pi)$ with $x^A \ S_\pi \ u^V$. Then we have $u^V \ N' \ x^A$. Thus, we know that $(P' \cup S_\pi)^+ \circ N'$ is not irreflexive, which contradicts π being $\mathcal{S}$-compatible with $(C', (P', N'))$. $\hspace{1cm}$ Q.e.d. (Claim 2)

Moreover, under the above assumption, also $\forall x^B.\ A$ is false in $\mathcal{S} \uplus \mathrm{e}(\pi)(\tau) \uplus \tau$. By the backward direction of the $\forall$-Lemma, this means that there is some object o such that A is false in $\mathcal{S} \uplus \{x^B \mapsto o\} \uplus \mathrm{e}(\pi)(\tau) \uplus \tau$. Set $\tau' := {}_{\mathbb{A} \backslash \{x^A\}}|\tau \uplus \{x^A \mapsto o\}$. Then, by the Substitution [Value] Lemma (1st equation), by the Valuation Lemma (for $l = 0$) (2nd equation), and by the Explicitness Lemma and $x^A \notin \mathbb{A}(A)$ (3rd equation), we have:

$$\mathrm{eval}(\mathcal{S} \uplus \mathrm{e}(\pi)(\tau) \uplus \tau')(A\{x^B \mapsto x^A\}) \ =$$
$$\mathrm{eval}(\mathcal{S} \uplus ((\{x^B \mapsto x^A\} \uplus {}_{\mathbb{V} \mathbb{A} \mathbb{B} \backslash \{x^B\}}|\mathrm{id}) \circ \mathrm{eval}(\mathcal{S} \uplus \mathrm{e}(\pi)(\tau) \uplus \tau')))(A) \ =$$
$$\mathrm{eval}(\mathcal{S} \uplus \{x^B \mapsto o\} \uplus \mathrm{e}(\pi)(\tau) \uplus \tau')(A) \ =$$
$$\mathrm{eval}(\mathcal{S} \uplus \{x^B \mapsto o\} \uplus \mathrm{e}(\pi)(\tau) \uplus \tau)(A) \ = \ \mathsf{FALSE}.$$

<u>Claim 4:</u> $A\{x^B \mapsto x^A\}$ is false in $\mathcal{S} \uplus \mathrm{e}(\pi)(\tau') \uplus \tau'$.

<u>Proof of Claim 4:</u> Otherwise, there must be some $u^V \in \mathbb{V}(A\{x^B \mapsto x^A\})$ with $x^A \ S_\pi \ u^V$. Then we have $u^V \ N' \ x^A$. Thus, we know that $(P' \cup S_\pi)^+ \circ N'$ is not irreflexive, which contradicts π being $\mathcal{S}$-compatible with $(C', (P', N'))$. $\hspace{1cm}$ Q.e.d. (Claim 4)

By the Claims 4 and 2, $\{A\{x^B \mapsto x^A\} \ \ \Gamma \ \ \Pi\}$ is false in $\mathcal{S} \uplus \mathrm{e}(\pi)(\tau') \uplus \tau'$, as was to be show for the soundness direction of the proof.

Finally, for the safeness direction, assume that the sequent $\Gamma \ \forall x^B.\ A \ \Pi$ is $(\pi, \mathcal{S})$-valid. For arbitrary $\tau : \mathbb{A} \to \mathcal{S}$, we have to show that the lower sequent $A\{x^B \mapsto x^A\} \ \Gamma \ \Pi$ is true in $\mathcal{S} \uplus \delta$ for $\delta := \mathrm{e}(\pi)(\tau) \uplus \tau$. If some formula in $\Gamma \Pi$ is true in $\mathcal{S} \uplus \delta$, then the lower sequent is true in $\mathcal{S} \uplus \delta$ as well. Otherwise, $\forall x^B.\ A$ is true in $\mathcal{S} \uplus \delta$. Then, by the forward direction of the $\forall$-Lemma, this means that A is true in $\mathcal{S} \uplus \chi \uplus \delta$ for all $\mathcal{S}$-valuations $\chi : \{x^B\} \to \mathcal{S}$. Then, by the Substitution [Value] Lemma (1st equation), and by the Valuation Lemma (for $l = 0$) (2nd equation), we have:

$$\text{eval}(\mathcal{S} \uplus \delta)(A\{x^{\mathbb{B}} \mapsto x^{\mathbb{A}}\}) \;=\;$$
$$\text{eval}(\mathcal{S} \uplus ((\{x^{\mathbb{B}} \mapsto x^{\mathbb{A}}\} \uplus_{\text{\tiny VAB} \setminus \{x^{\mathbb{B}}\}} | \text{id}) \circ \text{eval}(\mathcal{S} \uplus \delta)))(A) \;=\;$$
$$\text{eval}(\mathcal{S} \uplus \{x^{\mathbb{B}} \mapsto \delta(x^{\mathbb{A}})\} \uplus \delta)(A) \;=\; \text{TRUE}.$$

$\underline{\delta^+\text{-rule:}}$ In this case, we have $x^{\mathbb{V}} \in \mathbb{V} \setminus (\text{dom}(C \cup P \cup N) \cup \mathbb{V}(A))$,
$\qquad C'' = \{(x^{\mathbb{V}}, \varepsilon x^{\mathbb{B}}. \neg A)\}$, and $V = \mathbb{V}\mathbb{A}(\forall x^{\mathbb{B}}. A) \times \{x^{\mathbb{V}}\} = \mathbb{V}\mathbb{A}(A) \times \{x^{\mathbb{V}}\}$.
Thus, $C' = C \cup \{(x^{\mathbb{V}}, \varepsilon x^{\mathbb{B}}. \neg A)\}$, $P' = P \cup V$, and $N' = N$.
By assumption of the theorem, C is a (P,N)-choice-condition. Thus, (P,N) is a consistent positive/negative variable-condition. Thus, by Definition 5.4, P is well-founded and $P^+ \circ N$ is irreflexive.

Claim 5: P' is well-founded.

Proof of Claim 5: Let B be a non-empty class. We have to show that there is a P'-minimal element in B. Because P is well-founded, there is some P-minimal element in B. If this element is V-minimal in B, then it is a P'-minimal element in B. Otherwise, this element must be $x^{\mathbb{V}}$ and there is an element $n^{\mathbb{V}\mathbb{A}} \in B \cap \mathbb{V}\mathbb{A}(A)$. Set $B' := \{ b^{\mathbb{V}\mathbb{A}} \in B \mid b^{\mathbb{V}\mathbb{A}} P^* n^{\mathbb{V}\mathbb{A}} \}$. Because of $n^{\mathbb{V}\mathbb{A}} \in B'$, we know that B' is a non-empty subset of B. Because P is well-founded, there is some P-minimal element $m^{\mathbb{V}\mathbb{A}}$ in B'. Then $m^{\mathbb{V}\mathbb{A}}$ is also a P-minimal element in B. Because of $x^{\mathbb{V}} \notin \mathbb{V}\mathbb{A}(A) \cup \text{dom}(P)$, we know that $x^{\mathbb{V}} \notin B'$. Thus, $m^{\mathbb{V}\mathbb{A}} \neq x^{\mathbb{V}}$. Thus, $m^{\mathbb{V}\mathbb{A}}$ is also a V-minimal element of B. Thus, $m^{\mathbb{V}\mathbb{A}}$ is also a P'-minimal element of B. Q.e.d. (Claim 5)

Claim 6: $(P')^+ \circ N'$ is irreflexive.

Proof of Claim 6: Suppose the contrary. Because $P^+ \circ N$ is irreflexive, $P^* \circ (V \circ P^*)^+ \circ N$ must be reflexive. Because of $\text{ran}(V) = \{x^{\mathbb{V}}\}$ and $\{x^{\mathbb{V}}\} \cap \text{dom}(P \cup N) = \emptyset$, we have $V \circ P = \emptyset$ and $V \circ N = \emptyset$. Thus, $P^* \circ (V \circ P^*)^+ \circ N = P^* \circ V^+ \circ N = \emptyset$. Q.e.d. (Claim 6)

Claim 7: C' is a (P', N')-choice-condition.

Proof of Claim 7: By Claims 5 and 6, (P', N') is a consistent positive/negative variable-condition. As $x^{\mathbb{V}} \in \mathbb{V} \setminus \text{dom}(C)$, we know that C' is a partial function on $\mathbb{V}$ just as C. Moreover, for $y^{\mathbb{V}} \in \text{dom}(C')$, we either have $y^{\mathbb{V}} \in \text{dom}(C)$ and then $\mathbb{V}\mathbb{A}(C'(y^{\mathbb{V}})) \times \{y^{\mathbb{V}}\} = \mathbb{V}\mathbb{A}(C(y^{\mathbb{V}})) \times \{y^{\mathbb{V}}\} \subseteq P^+ \subseteq (P')^+$, or $y^{\mathbb{V}} = x^{\mathbb{V}}$ and then $\mathbb{V}\mathbb{A}(C'(y^{\mathbb{V}})) \times \{y^{\mathbb{V}}\} = \mathbb{V}\mathbb{A}(\varepsilon x^{\mathbb{B}}. \neg A) \times \{x^{\mathbb{V}}\} = V \subseteq P' \subseteq (P')^+$. Q.e.d. (Claim 7)

Now it suffices to show that, for each $\tau : \mathbb{A} \to \mathcal{S}$, and for $\delta := \mathsf{e}(\pi)(\tau) \uplus \tau$, the truth of $\{\Gamma \quad \forall x^{\mathbb{B}}. A \quad \Pi\}$ in $\mathcal{S} \uplus \delta$ is equivalent that of $\{A\{x^{\mathbb{B}} \mapsto x^{\mathbb{V}}\} \quad \Gamma \quad \Pi\}$. For the soundness direction, it suffices to show that the former sequent is true in $\mathcal{S} \uplus \delta$ under the assumption that the latter is. If some formula in $\Gamma \Pi$ is true in $\mathcal{S} \uplus \delta$, then the former sequent is true in $\mathcal{S} \uplus \delta$ as well. Otherwise, this means that $A\{x^{\mathbb{B}} \mapsto x^{\mathbb{V}}\}$ is true in $\mathcal{S} \uplus \delta$. Then, by the forward direction of the ¬-Lemma, $\neg A\{x^{\mathbb{B}} \mapsto x^{\mathbb{V}}\}$ is false in $\mathcal{S} \uplus \delta$. By the Explicitness Lemma, $\neg A\{x^{\mathbb{B}} \mapsto x^{\mathbb{V}}\}$ is false in $\mathcal{S} \uplus \delta \uplus \chi$ for all $\chi : \{x^{\mathbb{B}}\} \to \mathcal{S}$. Because π is $\mathcal{S}$-compatible with $(C', (P', N'))$ and because of $C'(x^{\vee}) = \varepsilon x^{\text{\tiny B}}.\ \neg A$, by Item 2 of Definition 5.15, $\neg A$ is false in $\mathcal{S} \uplus \delta \uplus \chi$ for all $\chi : \{x^{\text{\tiny B}}\} \to \mathcal{S}$. Then, by the backward direction of the $\neg$-LEMMA, A is true in $\mathcal{S} \uplus \delta \uplus \chi$ for all $\chi : \{x^{\text{\tiny B}}\} \to \mathcal{S}$. Then, by the backward direction of the $\forall$-LEMMA, $\forall x^{\text{\tiny B}}.\ A$ is true in $\mathcal{S} \uplus \delta$.

The safeness direction is perfectly analogous to the case of the δ^--rule.

Q.e.d. (Theorem 5.28)

C   Notes

Finally, in § C.1 and § C.2, come two notes that were to big to fit into the footnotes 3 and 5, respectively.

C.1   Are Liberalized δ-Rules Really More Liberal?

We could object with the following two points to the classification of the δ^+-rules as being more "liberal" than the δ^--rules:

- $\mathbb{V}\!\mathbb{A}(\forall x^{\text{\tiny B}}.\ A)$ is not necessarily a subset of $\mathbb{V}(\Gamma\ \forall x^{\text{\tiny B}}.\ A\ \Pi)$, because $\mathbb{V}\!\mathbb{A}(\forall x^{\text{\tiny B}}.\ A)$ may include some additional free atoms.
  First note that δ^--rules and the free atoms did not occur in inference systems with δ^+-rules before the publication of [106]; so in the earlier systems with free δ^+-rules only, $\mathbb{V}\!\mathbb{A}(\forall x^{\text{\tiny B}}.\ A)$ was indeed a subset of $\mathbb{V}(\Gamma\ \forall x^{\text{\tiny B}}.\ A\ \Pi)$.
  Moreover, the additional atoms blocked by the δ^+-rules (as compared to the δ^--rules) can hardly block any reductive proofs of formulas without free atoms and variables. This has following reason.
  If a proof uses only δ^+-reductions, then there will be no (free) atoms around and the critical subset relation holds anyway. So a critical variable-condition can only arise if a δ^+-step follows a δ^--step on the same branch. With a reasonably minimal positive/negative variable-condition (P, N), the only additional cycles that could occur by the δ^+-rule as compared to the alternative application of a δ^--rules are of the form
  $$y^{\vee}\ N\ z^{\text{\tiny A}}\ P\ x^{\vee}\ P^*\ w^{\vee}\ P\ y^{\vee},$$
  resulting from the following scenario: $y^{\vee}\ N\ z^{\text{\tiny A}}$ results from a δ^--step, $z^{\text{\tiny A}}\ P\ x^{\vee}$ results from a subsequent δ^+-step on the same branch, $x^{\vee}\ P^*\ w^{\vee}$ results from possible further δ^+-steps (δ^--steps cannot produce a relevant cycle!) and instantiations of free variables, and $w^{\vee}\ P\ y^{\vee}$ finally results from an instantiation of $y^{\vee}$.
  Let us now see what happens if we replace the δ^+-step with a δ^--step with $x^{\text{\tiny A}}$ replacing $x^{\vee}$, *ceteris paribus*. Note that this is only possible if $x^{\vee}$ was never

instantiated, which again explains why there must be at least one step of P between x^V and y^V. If the free variable y^V occurs in the upper sequent of this changed step, then new proof immediately fails due to the new cycle
$$y^\mathrm{V} \ N \ x^\mathbb{A} \ P^* \ w^\mathrm{V} \ P \ y^\mathrm{V}.$$
Otherwise, y^V was lost on this branch; but then we must ask ourselves why we instantiated it with a term containing w^V. If w^V is essentially shared with another branch, on which y^V has survived, then it must occur in the sequent before the original δ^+-step, and so we get the cycle
$$w^\mathrm{V} \ N \ x^\mathbb{A} \ P^* \ w^\mathrm{V}.$$
Otherwise, if w^V is not shared with another branch, we do not see any reason to instantiate y^V with a term containing w^V. Indeed, if w^V is only this branch, then there is no reason; if w^V occurs only on another branch, then a good reason for $x^\mathrm{V} \ P^* \ w^\mathrm{V}$ can be rejected just as for y^V before.

- The δ^+-rule may contribute an P-edge to a cycle with exactly one edge from N, whereas the analogous δ^--rule would contribute an N-edge instead, so the analogous cycle would then not count as counterexample to the consistency of the positive/negative variable-condition because it has two edges from N.

  Also in this case we conjecture that δ^--rules do not admit any successful proofs that are not possible with the analogous δ^+-rules. A proof of this conjecture, however, is not easy: First, it is a global property which requires us to consider the entire inference system. Second, δ^--rules indeed admit some extra (P, N)-substitutions, which have to be shown not to generate essentially additional proofs. E.g., if we want to prove $\forall y^\mathbb{B}. \ \mathsf{Q}(a^\mathrm{V}, y^\mathbb{B}) \wedge \forall x^\mathbb{B}. \ \mathsf{Q}(x^\mathbb{B}, b^\mathrm{V})$, which is true for a reflexive ordering Q with a minimal and a maximal element, β- and δ^--rules reduce this to the two goals $\mathsf{Q}(a^\mathrm{V}, y^\mathbb{A})$ and $\mathsf{Q}(x^\mathbb{A}, b^\mathrm{V})$ with positive/negative variable-condition (P, N) given by $P = \emptyset$ and $N = \{(a^\mathrm{V}, y^\mathbb{A}), \ (b^\mathrm{V}, x^\mathbb{A})\}$. Then $\sigma_\mathbb{A} := \{a^\mathrm{V}{\mapsto}x^\mathbb{A}, \ b^\mathrm{V}{\mapsto}y^\mathbb{A}\}$ is a (P, N)-substitution. The analogous δ^+-rules would have resulted in the positive/negative variable-condition (P', N') given by $P' = \{(a^\mathrm{V}, y^\mathrm{V}), \ (b^\mathrm{V}, x^\mathrm{V})\}$ and $N' = \emptyset$. But $\sigma_\mathbb{V} := \{a^\mathrm{V}{\mapsto}x^\mathrm{V}, \ b^\mathrm{V}{\mapsto}y^\mathrm{V}\}$ is not a (P', N')-substitution!

C.2 σ-Updates Admitting Variable-Reuse and -Permutation

For a version of σ-updates that admits variable-reuse and -permutation as explained in Note 10 of [106] and executed in Notes 26–30 of [106], the σ-update has to forget about the old meaning of the variables in $\mathrm{dom}(\sigma)$. To this end — instead of the simpler $(P \cup D, N)$ — we have to chose a σ-update admitting variable-reuse and -permutation to be

$$\left(\left({}_{\mathbb{VA}\backslash\mathrm{dom}(\sigma)}| P \,\cup\, P' \circ P \right) |_{\mathbb{V}\backslash\mathrm{dom}(\sigma)}, \qquad {}_{\mathbb{V}\backslash\mathrm{dom}(\sigma)}| N \,\cup\, {}_{\mathbb{V}}| P' \circ N \right)$$

for $P' := D \,\cup\, {}_{\mathbb{VA}\backslash\mathrm{dom}(\sigma)}|(P|_{\mathrm{dom}(\sigma)})^{+}$.

Note that P' can be simplified to D here by taking as the σ-*update admitting* V_γ-*reuse and -permutation:*

$$\left(\begin{array}{l} {}_{\mathbb{A}\cup\mathrm{V}_{\delta+}\cup(\mathrm{V}_\gamma\backslash\mathrm{dom}(\sigma))}| P \,\cup\, D \circ P \,\cup\, D|_{\mathrm{V}_{\delta+}}, \\ {}_{\mathrm{V}_{\delta+}\cup(\mathrm{V}_\gamma\backslash\mathrm{dom}(\sigma))}| N \,\cup\, {}_{\mathbb{V}}| D|_{\mathrm{V}_\gamma\cap\mathrm{dom}(\sigma)} \circ N \end{array} \right),$$

provided that we partition $\mathbb{V}$ into two sets $\mathrm{V}_{\delta+} \uplus \mathrm{V}_\gamma$, use $\mathrm{V}_{\delta+}$ as the possible domain of the choice-conditions, and admit variable-reuse and -permutation only on V_γ, similar to what we already did in Note 10 of [106]. (The crucial restriction becomes here the following: For a (positive/negative) σ-update (P'', N'') admitting V_γ-reuse and -permutation we have $P'' \subseteq \mathbb{VA} \times \mathrm{V}_{\delta+}$ and $N'' \subseteq \mathbb{V} \times \mathbb{A}$).

Note, however, that it is actually better to work with the more complicated P', simply because it is more general and because the transitive closure will not be computed in practice, but a graph will be updated just as exemplified in Note 10 of [106].

References

[1] Wilhelm Ackermann. Begründung des „tertium non datur" mittels der Hilbertschen Theorie der Widerspruchsfreiheit. *Mathematische Annalen*, 93:1–36, 1925. Received March 30, 1924. Inauguraldissertation, 1924, Göttingen.

[2] Wilhelm Ackermann. Mengentheoretische Begründung der Logik. *Mathematische Annalen*, 115:1–22, 1938. Received April 23, 1937.

[3] Anon, editor. *Festschrift zur Feier der Enthüllung des* Gauß-Weber-*Denkmals in Göttingen, herausgegeben von dem Fest-Comitee.* Verlag von B. G. Teubner, Leipzig, 1899.

[4] Günter Asser. Theorie der logischen Auswahlfunktionen. *Zeitschrift für math. Logik und Grundlagen der Mathematik*, 3:30–68, 1957.

[5] John Lane Bell. Hilbert's ε-Operator and Classical Logic. *J. Philosophical Logic*, 22:1–18, 1993.

[6] John Lane Bell. Hilbert's ε-Operator in Intuitionistic Type Theories. *Math. Logic Quart. (until 1993: Zeitschrift für math. Logik und Grundlagen der Mathematik)*, 39:323–337, 1993.

[7] John Lane Bell, David DeVidi, and Graham Solomon. *Logical Options: An Introduction to Classical and Alternative Logics.* Broadview Press, 2001.

[8] Karel Berka and Lothar Kreiser, editors. *Logik-Texte – Kommentierte Auswahl zur Geschichte der modernen Logik.* Akademie Verlag GmbH, Berlin, 1973. 2nd rev. edn. (1st edn. 1971; 4th rev. rev. edn. 1986).

[9]   Wolfgang Bibel and Peter H. Schmitt, editors. *Automated Deduction — A Basis for Applications*. Number 8–10 in Applied Logic Series. Kluwer (Springer Science+Business Media), 1998. Vol. I: Foundations — Calculi and Methods. Vol. II: Systems and Implementation Techniques. Vol. III: Applications.

[10] Andreas Blass and Yuri Gurevich. The logic of choice. *J. Symbolic Logic*, 65:1264–1310, 2000.

[11] Nicolas Bourbaki. *Éléments des Mathématique — Livres 1–9*. Actualités Scientifiques et Industrielles. Hermann, Paris, 1939ff..

[12] Ricardo Caferra and Gernot Salzer, editors. *Automated Deduction in Classical and Non-Classical Logics*. Number 1761 in Lecture Notes in Artificial Intelligence. Springer, 2000.

[13] William F. Clocksin and Christopher S. Mellish. *Programming in* PROLOG. Springer, 2003. 5th edn. (1st edn. 1981).

[14] David DeVidi. Intuitionistic ε- and τ-calculi. *Math. Logic Quart. (until 1993: Zeitschrift für math. Logik und Grundlagen der Mathematik)*, 41:523–546, 1995.

[15] Cora Diamond, editor. *Wittgenstein's Lectures on the Foundations of Mathematics, Cambridge, 1939*. Cornell Univ., Ithaca (NY), 1976. From the notes of R. G. Bosanquet, Norman Malcolm, Rush Rhees, and Yorick Smythies. German translation is [16].

[16] Cora Diamond, editor. *Wittgensteins Vorlesungen über die Grundlagen der Mathematik, Cambridge, 1939*. Suhrkamp Verlag, Frankfurt am Main, 1978. German translation of [15] by Joachim Schulte. 1st edn. as "Ludwig Wittgenstein, Schriften 7" (ISBN 3518072471). Note that this volume was excluded from the later Suhrkamp edn. "Ludwig Wittgenstein, Werkausgabe".

[17] Uwe Egly and Christian G. Fermüller, editors. *11th Int. Conf. on Tableaux and Related Methods, København, 2002*, number 2381 in Lecture Notes in Artificial Intelligence. Springer, 2002.

[18] William Ewald, editor. *From* Kant *to* Hilbert *— A source book in the foundations of mathematics*. Oxford Univ. Press, 1996.

[19] Melvin Fitting. A modal logic ε-calculus. *Notre Dame J. of Formal Logic*, XVI:1–16, 1975.

[20] Melvin Fitting. *First-order logic and automated theorem proving*. Springer, 1990. 1st edn. (2nd rev. edn. is [21]).

[21] Melvin Fitting. *First-order logic and automated theorem proving*. Springer, 1996. 2nd rev. edn. (1st edn. is [20]).

[22] Melvin Fitting. On quantified modal logic. *Fundamenta Informaticae*, 39:105–121, 1999.

[23] Gottlob Frege. *Grundgesetze der Arithmetik — Begriffsschriftlich abgeleitet*. Verlag von Hermann Pohle, Jena, 1893/1903. Vol. I/II. As facsimile with corrigenda by Christian Thiel: Georg Olms Verlag, Hildesheim (Germany), 1998. English translations: [24], [25].

[24] Gottlob Frege. *The Basic Laws of Arithmetic — Exposition of the System.* Univ. of California Press, 1964. English translation of [23, Vol I], with an introduction, by Montgomery Furth.

[25] Gottlob Frege. *Basic Laws of Arithmetic.* Oxford Univ. Press, 2013. English translation of [23] by Philip Ebert and Marcus Rossberg, with an Introduction by Crispin Wright and an appendix by Roy T. Cook.

[26] Johannes Fried. *Canossa – Entlarvung einer Legende. Eine Streitschrift.* Akademie Verlag GmbH, Berlin, 2012.

[27] Dov Gabbay. *Semantical Investigations in* Heyting*'s Intuitionistic Logic.* Kluwer (Springer Science+Business Media), 1981.

[28] Dov Gabbay and John Woods, editors. *Handbook of the History of Logic.* North-Holland (Elsevier), 2004ff..

[29] Murdoch J. Gabbay and Andrew M. Pitts. A new approach to abstract syntax with variable binding. *Formal Asp. Comput.*, 13:341–363, 2002.

[30] Gerhard Gentzen. Untersuchungen über das logische Schließen. *Mathematische Zeitschrift*, 39:176–210,405–431, 1935. Also in [8, pp. 192–253]. English translation in [31].

[31] Gerhard Gentzen. *The Collected Papers of* Gerhard Gentzen. North-Holland (Elsevier), 1969. Ed. by Manfred E. Szabo.

[32] Martin Giese and Wolfgang Ahrendt. Hilbert's ε-terms in automated theorem proving. 1999. In [76, pp. 171–185].

[33] Jean van Heijenoort. *From* Frege *to* Gödel*: A Source Book in Mathematical Logic, 1879–1931.* Harvard Univ. Press, 1971. 2nd rev. edn. (1st edn. 1967).

[34] Hans Hermes. *Eine Termlogik mit Auswahloperator.* Number 6 in LNM. Springer, 1965.

[35] Klaus von Heusinger. *Salienz und Referenz — Der Epsilonoperator in der Semantik der Nominalphrase und anaphorischer Pronomen.* Number 43 in Studia grammatica. Akademie Verlag GmbH, Berlin, 1997.

[36] David Hilbert. Grundlagen der Geometrie. 1899. In [3, pp. 1–92]. 1st edn. without appendixes. Reprinted in [56, pp. 436–525]. *(Last edition of "Grundlagen der Geometrie" by Hilbert is [50], which is also most complete regarding the appendixes. Last three editions by Paul Bernays are [52; 53; 55], which are also most complete regarding supplements and figures. Its first appearance as a separate book was the French translation [38]. Two substantially different English translations are [39] and [54]).*

[37] David Hilbert. Über den Zahlbegriff. *Jahresbericht der Deutschen Mathematiker-Vereinigung*, 8:180–184, 1900. Received Dec. 1899. Reprinted as Appendix VI of [42; 43; 44; 46; 50].

[38] David Hilbert. Les principes fondamentaux de la géométrie. *Annales Scientifiques de l'École Normale Supérieure*, Série 3, 17:103–209, 1900. French translation by Léonce Laugel of special version of [36], revised and authorized by Hilbert. Also in published

as a separate book by the same publisher (Gauthier-Villars, Paris).

[39] David Hilbert. *The Foundations of Geometry*. Open Court, Chicago, 1902. English translation by E. J. Townsend of special version of [36], revised and authorized by Hilbert, http://www.gutenberg.org/etext/17384.

[40] David Hilbert. *Grundlagen der Geometrie. — Zweite, durch Zusätze vermehrte und mit fünf Anhängen versehene Auflage. Mit zahlreichen in den Text gedruckten Figuren.* Druck und Verlag von B. G. Teubner, Leipzig, 1903. 2nd rev. extd. edn. of [36], rev. and extd. with five appendixes, newly added figures, and an index of notion names.

[41] David Hilbert. Über die Grundlagen der Logik und der Arithmetik. 1905. In [70, pp. 174–185]. Reprinted as Appendix VII of [42; 43; 44; 46; 50]. English translation *On the foundations of logic and arithmetic* by Beverly Woodward with an introduction by Jean van Heijenoort in [33, pp. 129–138].

[42] David Hilbert. *Grundlagen der Geometrie. — Dritte, durch Zusätze und Literaturhinweise von neuem vermehrte und mit sieben Anhängen versehene Auflage. Mit zahlreichen in den Text gedruckten Figuren.* Number VII in Wissenschaft und Hypothese. Druck und Verlag von B. G. Teubner, Leipzig, Berlin, 1909. 3rd rev. extd. edn. of [36], rev. edn. of [40], extd. with a bibliography and two additional appendixes (now seven in total) (Appendix VI: [37]) (Appendix VII: [41]).

[43] David Hilbert. *Grundlagen der Geometrie. — Vierte, durch Zusätze und Literaturhinweise von neuem vermehrte und mit sieben Anhängen versehene Auflage. Mit zahlreichen in den Text gedruckten Figuren.* Druck und Verlag von B. G. Teubner, Leipzig, Berlin, 1913. 4th rev. extd. edn. of [36], rev. edn. of [42].

[44] David Hilbert. *Grundlagen der Geometrie. — Fünfte, durch Zusätze und Literaturhinweise von neuem vermehrte und mit sieben Anhängen versehene Auflage. Mit zahlreichen in den Text gedruckten Figuren.* Verlag und Druck von B. G. Teubner, Leipzig, Berlin, 1922. 5th extd. edn. of [36]. Contrary to what the sub-title may suggest, this is an anastatic reprint of [43], extended by a very short preface on the changes w.r.t. [43], and with augmentations to Appendix II, Appendix III, and Chapter IV, § 21.

[45] David Hilbert. Die logischen Grundlagen der Mathematik. *Mathematische Annalen*, 88:151–165, 1923. Received Sept. 29, 1922. Talk given at the Deutsche Naturforschergesellschaft in Leipzig, Sept. 1922. English translation in [18, pp. 1134–1148].

[46] David Hilbert. *Grundlagen der Geometrie. — Sechste unveränderte Auflage. Anastatischer Nachdruck. Mit zahlreichen in den Text gedruckten Figuren.* Verlag und Druck von B. G. Teubner, Leipzig, Berlin, 1923. 6th rev. extd. edn. of [36], anastatic reprint of [44].

[47] David Hilbert. Über das Unendliche — Vortrag, gehalten am 4. Juni 1925 gelegentlich einer zur Ehrung des Andenkens an Weierstraß von der Westfälischen Math. Ges. veranstalteten Mathematiker-Zusammenkunft in Münster i. W. *Mathematische Annalen*, 95:161–190, 1926. Received June 24, 1925. Reprinted as Appendix VIII of [50]. English translation *On the infinite* by Stefan Bauer-Mengelberg with an introduction by Jean van Heijenoort in [33, pp. 367–392].

[48] David Hilbert. Die Grundlagen der Mathematik — Vortrag, gehalten auf Einladung

des Mathematischen Seminars im Juli 1927 in Hamburg. *Abhandlungen aus dem mathematischen Seminar der Univ. Hamburg*, 6:65–85, 1928. Reprinted as Appendix IX of [50]. English translation *The foundations of mathematics* by Stefan Bauer-Mengelberg and Dagfinn Føllesdal with a short introduction by Jean van Heijenoort in [33, pp. 464–479].

[49] David Hilbert. Probleme der Grundlegung der Mathematik. *Mathematische Annalen*, 102:1–9, 1930. Vortrag gehalten auf dem Internationalen Mathematiker-Kongreß in Bologna, Sept. 3, 1928. Received March 25, 1929. Reprinted as Appendix X of [50]. Short version in *Atti del congresso internationale dei matematici, Bologna, 3–10 settembre 1928*, Vol. 1, pp. 135–141, Bologna, 1929.

[50] David Hilbert. *Grundlagen der Geometrie. — Siebente umgearbeitete und vermehrte Auflage. Mit 100 in den Text gedruckten Figuren.* Verlag und Druck von B. G. Teubner, Leipzig, Berlin, 1930. 7th rev. extd. edn. of [36], thoroughly revised edition of [46], extd. with three new appendixes (now ten in total) (Appendix VIII: [47]) (Appendix IX: [48]) (Appendix X: [49]).

[51] David Hilbert. *Grundlagen der Geometrie. — Achte Auflage, mit Revisionen und Ergänzungen von Dr. Paul Bernays. Mit 124 Abbildungen.* B. G. Teubner Verlagsgesellschaft, Stuttgart, 1956. 8th rev. extd. edn. of [36], rev. edn. of [50], omitting appendixes VI–X, extd. by Paul Bernays, now with 24 additional figures and 3 additional supplements.

[52] David Hilbert. *Grundlagen der Geometrie. — Neunte Auflage, revidiert und ergänzt von Dr. Paul Bernays. Mit 129 Abbildungen.* B. G. Teubner Verlagsgesellschaft, Stuttgart, 1962. 9th rev. extd. edn. of [36], rev. edn. of [51], extd. by Paul Bernays, now with 129 figures, 5 appendixes, and 8 supplements (I 1, I 2, II, III, IV 1, IV 2, V 1, V 2).

[53] David Hilbert. *Grundlagen der Geometrie. — Zehnte Auflage, revidiert und ergänzt von Dr. Paul Bernays. Mit 124 Abbildungen.* B. G. Teubner Verlagsgesellschaft, Stuttgart, 1968. 10th rev. extd. edn. of [36], rev. edn. of [52] by Paul Bernays.

[54] David Hilbert. *The Foundations of Geometry.* Open Court, Chicago and La Salle (IL), 1971. Newly translated and fundamentally different 2nd edn. of [39], actually an English translation of [53] by Leo Unger.

[55] David Hilbert. *Grundlagen der Geometrie. — 11. Auflage. Mit Supplementen von Dr. Paul Bernays.* B. G. Teubner Verlagsgesellschaft, Stuttgart, 1972. 11th rev. extd. edn. of [36], rev. edn. of [53] by Paul Bernays.

[56] David Hilbert. *David Hilbert's Lectures on the Foundations of Geometry, 1891–1902.* Springer, 2004. Ed. by Michael Hallett and Ulrich Majer.

[57] David Hilbert and Paul Bernays. *Grundlagen der Mathematik — Erster Band.* Number XL in Grundlehren der mathematischen Wissenschaften. Springer, 1934. 1st edn. (2nd edn. is [59]). Reprinted by J. W. Edwards Publ., Ann Arbor (MI), 1944. English translation is [61; 62].

[58] David Hilbert and Paul Bernays. *Grundlagen der Mathematik — Zweiter Band.* Number L in Grundlehren der mathematischen Wissenschaften. Springer, 1939. 1st edn. (2nd edn. is [60]). Reprinted by J. W. Edwards Publ., Ann Arbor (MI), 1944.

[59]   David Hilbert and Paul Bernays. *Grundlagen der Mathematik I*. Number 40 in Grundlehren der mathematischen Wissenschaften. Springer, 1968. 2nd rev. edn. of [57]. English translation is [61; 62].

[60]   David Hilbert and Paul Bernays. *Grundlagen der Mathematik II*. Number 50 in Grundlehren der mathematischen Wissenschaften. Springer, 1970. 2nd rev. extd. edn. of [58].

[61]   David Hilbert and Paul Bernays. *Grundlagen der Mathematik I — Foundations of Mathematics I, Part A: Title Pages, Prefaces, and §§ 1–2*. Springer, 2017. First English translation and bilingual facsimile edn. of the 2nd German edn. [59], incl. the annotation and translation of all differences of the 1st German edn. [57]. Ed. by Claus-Peter Wirth, Jörg Siekmann, Volker Peckhaus, Michael Gabbay, Dov Gabbay. Translated and commented by Claus-Peter Wirth &al. Thoroughly rev. 3rd edn. (1st edn. College Publications, London, 2011; 2nd edn. `http://wirth.bplaced.net/p/hilbertbernays`, 2013).

[62]   David Hilbert and Paul Bernays. *Grundlagen der Mathematik I — Foundations of Mathematics I, Part B: §§ 3–5 and Deleted Part I of the 1st Edn.*. Springer, 2017. First English translation and bilingual facsimile edn. of the 2nd German edn. [59], incl. the annotation and translation of all deleted texts of the 1st German edn. [57]. Ed. by Claus-Peter Wirth, Jörg Siekmann, Volker Peckhaus, Michael Gabbay, Dov Gabbay. Translated and commented by Claus-Peter Wirth &al. Thoroughly rev. 3rd edn. (1st edn. College Publications, London, 2012; 2nd edn. `http://wirth.bplaced.net/p/hilbertbernays`, 2013).

[63]   K. Jaakko J. Hintikka. Quantifiers vs. quantification theory. *Linguistic Inquiry*, V(2):153–177, 1974.

[64]   K. Jaakko J. Hintikka. *The Principles of Mathematics Revisited*. Cambridge Univ. Press, 1996.

[65]   Paul Howard and Jean E. Rubin. *Consequences of the Axiom of Choice*. American Math. Soc., 1998. `http://consequences.emich.edu/conseq.htm`.

[66]   Stig Kanger. A simplified proof method for elementary logic. 1963. In [91, Vol. 1, pp. 364–371].

[67]   Hubert C. Kennedy. *Selected works of* Guiseppe Peano. George Allen & Unwin, London, 1973.

[68]   Michael Kohlhase. Higher-order automated theorem proving. 1998. In [9, Vol. 1, pp. 431–462].

[69]   Robert A. Kowalski. Predicate logic as a programming language. 1974. In [86, pp. 569–574].

[70]   A. Krazer, editor. *Verhandlungen des Dritten Internationalen Mathematiker-Kongresses, Heidelberg, Aug. 8–13, 1904*. Verlag von B. G. Teubner, Leipzig, 1905.

[71]   Johann Heinrich Lambert. *Neues Organon oder Gedanken über die Erforschung und Bezeichnung des Wahren und dessen Unterscheidung von Irrthum und Schein*. Johann Wendler, Leipzig, 1764. Vol. I (Dianoiologie oder die Lehre von den Gesetzen des Denkens, Alethiologie oder Lehre

von der Wahrheit) (`http://books.google.de/books/about/Neues_Organon_oder_Gedanken_Uber_die_Erf.html?id=ViS3XCuJEw8C`) & Vol. II (Semiotik oder Lehre von der Bezeichnung der Gedanken und Dinge, Phänomenologie oder Lehre von dem Schein) (`http://books.google.de/books/about/Neues_Organon_oder_Gedanken_%C3%BCber_die_Er.html?id=X8UAAAAcAAj`). Facsimile reprint by Georg Olms Verlag, Hildesheim (Germany), 1965, with a German introduction by Hans Werner Arndt.

[72]   Albert C. Leisenring. *Mathematical Logic and* Hilbert*'s* ε*-Symbol.* Gordon and Breach, New York, 1969.

[73]   Volker Mattick and Claus-Peter Wirth. An algebraic Dexter-based hypertext reference model. Research Report (green/grey series) 719/1999, FB Informatik, Univ. Dortmund, 1999. `http://wirth.bplaced.net/p/gr719`, `http://arxiv.org/abs/0902.3648`.

[74]   Wilfried P. M. Meyer-Viol. *Instantial Logic — An Investigation into Reasoning with Instances.* PhD thesis, Univ. Utrecht, 1995. ILLC dissertation series 1995–11.

[75]   Dale A. Miller. Unification under a mixed prefix. *J. Symbolic Computation*, 14:321–358, 1992.

[76]   Neil V. Murray, editor. *8th Int. Conf. on Tableaux and Related Methods, Saratoga Springs (NY), 1999*, number 1617 in Lecture Notes in Artificial Intelligence. Springer, 1999.

[77]   Guiseppe Peano, editor. *Angelo Genocchi — Calcolo differenziale e principii di calcolo integrale.* Fratelli Bocca, Torino (i.e. Turin, Italy), 1884. German translation: [80].

[78]   Guiseppe Peano. Démonstration de l'intégrabilité des équations différentielles ordinaires. *Mathematische Annalen*, 37:182–228, 1890. Facsimile also in [82, pp. 76–122].

[79]   Guiseppe Peano. Studii di logica matematica. *Atti della Reale Accademia delle Scienze di Torino (i.e. Turin, Italy) — Classe di Scienze Morali, Storiche e Filologiche e Classe di Scienze Fisiche, Matematiche e Naturali*, 32:565–583, 1896f.. Also in Atti della Reale Accademia delle Scienze di Torino (i.e. Turin, Italy) — Classe di Scienze Fisiche, Matematiche e Naturali **32**, pp. 361–397. English translation *Studies in Mathematical Logic* in [67, pp. 190–205]. German translation: [81].

[80]   Guiseppe Peano, editor. *Angelo Genocchi — Differentialrechnung und Grundzüge der Integralrechnung.* B. G. Teubner Verlagsgesellschaft, Leipzig, 1899. German translation of [77].

[81]   Guiseppe Peano. Über mathematische Logik. 1899. German translation of [79]. In [80, Appendix 1]. Facsimile also in [82, pp. 10–26].

[82]   Guiseppe Peano. *Guiseppe Peano — Arbeiten zur Analysis und zur mathematischen Logik.* Number 13 in Teubner-Archiv zur Mathematik. B. G. Teubner Verlagsgesellschaft, 1990. With an essay by Günter Asser (ed.).

[83]   Manfred Pinkal, Jörg Siekmann, and Christoph Benzmüller. Teilprojekt MI 3: DIALOG: Tutorieller Dialog mit einem Mathematik-Assistenten. 2001. In SFB 378 Resource-Adaptive Cognitive Processes, Proposal Jan. 2002 - Dez. 2004, Saarland Univ..

[84] Dag Prawitz. An improved proof procedure. *Theoria: A Swedish Journal of Philosophy*, 26:102–139, 1960. Also in [91, Vol. 1, pp. 159–199].

[85] Willard Van O. Quine. *Mathematical Logic*. Harvard Univ. Press, 1981. 4th rev. edn. (1st edn. 1940).

[86] Jack L. Rosenfeld, editor. *Proc. of the Congress of the Int. Federation for Information Processing (IFIP), Stockholm (Sweden), Aug. 5–10, 1974*. North-Holland (Elsevier), 1974.

[87] Herman Rubin and Jean E. Rubin. *Equivalents of the Axiom of Choice*. North-Holland (Elsevier), 1985. 2nd rev. edn. (1st edn. 1963).

[88] Bertrand Russell. On Denoting. *Mind*, 14:479–493, 1905.

[89] Bertrand Russell. *Introduction to Mathematical Philosophy*. George Allen & Unwin, London, 1919.

[90] Tobias Schmidt-Samoa. Flexible heuristics for simplification with conditional lemmas by marking formulas as forbidden, mandatory, obligatory, and generous. *J. Applied Non-Classical Logics*, 16:209–239, 2006. `http://dx.doi.org/10.3166/jancl.16.208-239`.

[91] Jörg Siekmann and Graham Wrightson, editors. *Automation of Reasoning*. Springer, 1983.

[92] B. Hartley Slater. *Intensional Logic — an essay in analytical metaphysics*. Avebury Series in Philosophy. Ashgate Publ. Ltd (Taylor & Francis), Aldershot (England), 1994.

[93] B. Hartley Slater. *Logic Reformed*. Peter Lang AG, Bern, Switzerland, 2002.

[94] B. Hartley Slater. Completing Russell's logic. *Russell: the Journal of Bertrand Russell Studies, McMaster Univ., Hamilton (Ontario), Canada*, 27:144–158, 2007. Also in [97, Chapter 3 (pp. 15–27)].

[95] B. Hartley Slater. *The De-Mathematisation of Logic*. Polimetrica, Monza, Italy, 2007. Open access publication, `wirth.bplaced.net/op/fullpaper/hilbertbernays/Slater_2007_De-Mathematisation.pdf`.

[96] B. Hartley Slater. Hilbert's epsilon calculus and its successors. 2009. In [28, Vol. 5: Logic from Russell to Church, pp. 365–448].

[97] B. Hartley Slater. *Logic is Not Mathematical*. College Publications, London, 2011.

[98] Raymond M. Smullyan. *First-Order Logic*. Springer, 1968. 1st edn., 2nd extd. edn. is [99].

[99] Raymond M. Smullyan. *First-Order Logic*. Dover Publications, New York, 1995. 2nd extd. edn. of [98], with a new preface and some corrections of the author.

[100] Peter F. Strawson. On Referring. *Mind*, 59:320–344, 1950.

[101] Christian Urban, Murdoch J. Gabbay, and Andrew M. Pitts. Nominal unification. *Theoretical Computer Sci.*, 323:473–497, 2004.

[102] Lincoln A. Wallen. *Automated Proof Search in Non-Classical Logics — efficient matrix proof methods for modal and intuitionistic logics*. MIT Press, 1990. Phd thesis.

[103] Alfred North Whitehead and Bertrand Russell. *Principia Mathematica*. Cambridge Univ. Press, 1910–1913. 1st edn..

[104] Claus-Peter Wirth. Full first-order sequent and tableau calculi with preservation of solutions and the liberalized δ-rule but without Skolemization. Research Report (green/grey series) 698/1998, FB Informatik, Univ. Dortmund, 1998. http://arxiv.org/abs/0902.3730. Short version in Gernot Salzer, Ricardo Caferra (eds.). Proc. 2nd Int. Workshop on First-Order Theorem Proving (FTP'98), pp. 244–255, Tech. Univ. Vienna, 1998. Short version also in [12, pp. 283–298].

[105] Claus-Peter Wirth. A new indefinite semantics for Hilbert's epsilon. 2002. In [17, pp. 298–314]. http://wirth.bplaced.net/p/epsi.

[106] Claus-Peter Wirth. Descente Infinie + Deduction. *Logic J. of the IGPL*, 12:1–96, 2004. http://wirth.bplaced.net/p/d.

[107] Claus-Peter Wirth. lim+, δ^+, and *Non-Permutability of β-Steps*. SEKI-Report SR–2005–01 (ISSN 1437–4447). SEKI Publications, Saarland Univ., 2006. Rev. edn. July 2006 (1st edn. 2005), ii+36 pp., http://arxiv.org/abs/0902.3635. Thoroughly improved version is [110].

[108] Claus-Peter Wirth. Hilbert's epsilon as an operator of indefinite committed choice. *J. Applied Logic*, 6:287–317, 2008. http://dx.doi.org/10.1016/j.jal.2007.07.009.

[109] Claus-Peter Wirth. Herbrand's Fundamental Theorem in the eyes of Jean van Heijenoort. *Logica Universalis*, 6:485–520, 2012. Received Jan. 12, 2012. Published online June 22, 2012, http://dx.doi.org/10.1007/s11787-012-0056-7.

[110] Claus-Peter Wirth. lim +, δ^+, and Non-Permutability of β-Steps. *J. Symbolic Computation*, 47:1109–1135, 2012. Received Jan. 18, 2011. Published online July 15, 2011, http://dx.doi.org/10.1016/j.jsc.2011.12.035.

[111] Claus-Peter Wirth. *Hilbert's epsilon as an Operator of Indefinite Committed Choice*. SEKI-Report SR–2006–02 (ISSN 1437–4447). SEKI Publications, Saarland Univ., 2012. Rev. edn. Jan. 2012, ii+73 pp., http://arxiv.org/abs/0902.3749.

[112] Claus-Peter Wirth. *Herbrand's Fundamental Theorem: The Historical Facts and their Streamlining*. SEKI-Report SR–2014–01 (ISSN 1437–4447). SEKI Publications, 2014. ii+47 pp., http://arxiv.org/abs/1405.6317.

[113] Claus-Peter Wirth. The explicit definition of quantifiers via Hilbert's ε is confluent and terminating. *IFCoLog J. of Logics and Their Applications*, 4:513–533, 2016.

[114] Claus-Peter Wirth, Jörg Siekmann, Christoph Benzmüller, and Serge Autexier. Jacques Herbrand: Life, logic, and automated deduction. 2009. In [28, Vol. 5: Logic from Russell to Church, pp. 195–254].

[115] Claus-Peter Wirth, Jörg Siekmann, Christoph Benzmüller, and Serge Autexier. *Lectures on Jacques Herbrand as a Logician*. SEKI-Report SR–2009–01 (ISSN 1437–4447). SEKI Publications, 2014. Rev. edn. May 2014, ii+82 pp., http://arxiv.org/abs/0902.4682.

Received 23 October 2015

The Explicit Definition of Quantifiers via Hilbert's epsilon is Confluent and Terminating

Claus-Peter Wirth

Dept. of Math., ETH Zurich, Rämistr. 101, 8092 Zürich, Switzerland
wirth@logic.at

Abstract

We investigate the elimination of quantifiers in first-order formulas via Hilbert's epsilon-operator (or -binder), following Bernays' explicit definitions of the existential and the universal quantifier symbol by means of epsilon-terms. This elimination has its first explicit occurrence in the proof of the first epsilon-theorem in Hilbert–Bernays in 1939. We think that there is a lacuna in this proof w.r.t. this elimination, related to the erroneous assumption that explicit definitions always terminate. Surprisingly, to the best of our knowledge, nobody ever published a confluence or termination proof for this elimination procedure; and even myths on non-confluence and the openness of the termination problem are circulating. We show confluence and termination of this elimination procedure by means of a direct, straightforward, and easily verifiable proof, based on a theorem on how to obtain termination from weak normalization.

Keywords: Hilbert–Bernays Proof Theory, History of Proof Theory, Hilbert's epsilon, Quantifier Elimination, (Weak) Normalization, Termination, (Local) Confluence.

1 Introduction

1.1 The Explicit Historical Source of the Problem

With "Hilbert–Bernays" we will designate the "bible of proof theory", i.e. the two-volume monograph *Grundlagen der Mathematik* (*Foundations of Mathematics*) in its two editions [19; 20] and [21; 22].

On p. 19f. of the first edition [20], as well as on p. 20 of the second edition [22], we read:

"Unser zweiter vorbereitender Schritt besteht in der Ausschaltung der All- und Seinszeichen. Wie im vorigen Abschnitt gezeigt wurde, können wir die Anwendung der Grundformeln (a), (b) und der Schemata (α), (β) des Prädikatenkalkuls mit Hilfe der ε-Formel und der expliziten Definitionen (ε_1), (ε_2) entbehrlich machen[1]. Führen wir diese Ausschaltung der Grundformeln und Schemata für die Quantoren an der zu betrachtenden Ableitung der Formel $\mathfrak{E}$ aus und ersetzen wir hernach jeden Ausdruck $(\mathfrak{v})\,\mathfrak{A}(\mathfrak{v})$ durch $\mathfrak{A}\big(\varepsilon_\mathfrak{v}\,\overline{\mathfrak{A}(\mathfrak{v})}\,\big)$, jeden Ausdruck $(E\,\mathfrak{v})\,\mathfrak{A}(\mathfrak{v})$ durch $\mathfrak{A}\big(\varepsilon_\mathfrak{v}\,\mathfrak{A}(\mathfrak{v})\big)$, so gehen die aus (ε_1), (ε_2) durch Einsetzung gewonnenen Formeln in solche über, die durch Einsetzung aus der Formel $A \sim A$ entstehen. Die Quantoren werden durch dieses Verfahren gänzlich ausgeschaltet, so daß *nunmehr gebundene Variablen ausschließlich in Verbindung mit dem ε-Symbol auftreten, und der Beweiszusammenhang nur durch Wiederholungen, Einsetzungen, Umbenennung gebundener Variablen und Schlußschemata stattfindet.*"

"Our second preparatory step consists in the elimination of the universal and existential quantifier symbols. As shown in the previous section, we can dispense with the application of Formulas (a), (b) and Schemata (α), (β) of the predicate calculus if we use the ε-formula and the explicit definitions (ε_1), (ε_2). If we apply this elimination of basic formulas und schemata for the quantifiers to the formula $\mathfrak{E}$ under consideration, and afterwards replace every expression $(\mathfrak{v})\,\mathfrak{A}(\mathfrak{v})$ with $\mathfrak{A}\big(\varepsilon_\mathfrak{v}\,\overline{\mathfrak{A}(\mathfrak{v})}\,\big)$, every expression $(E\,\mathfrak{v})\,\mathfrak{A}(\mathfrak{v})$ with $\mathfrak{A}\big(\varepsilon_\mathfrak{v}\,\mathfrak{A}(\mathfrak{v})\big)$, then the formulas obtained from (ε_1), (ε_2) by substitution are turned into formulas obtained by substitution from the formula $A \sim A$. By this procedure, the quantifiers are completely eliminated, so that *bound variables may occur only in combination with the ε-symbol, and the interconnections of the proof may consist only of repetitions, substitutions, renaming of bound variables, and inference schemata.*"

Note that the "A" is not a meta-variable here (such as "$\mathfrak{A}$" is a meta-variable for a formula, and "$\mathfrak{v}$" for a *bound individual variable*), but a concrete object-level formula variable. In a proof step called *substitution* either such a *formula variable* (which is always free) or a *free individual variable* is replaced everywhere in a formula with an arbitrary formula or term, respectively. Furthermore, note that "Schlußschema" ("inference schema") is nothing but a short name for the inference schema of *modus ponens*.

Moreover, note that Note 1 actually occurs only in the second edition and reads "[1]Vgl. S.15." ("[1]Cf. p.15."). Neither on Page 15 — nor anywhere else in the volumes — can we find any further information, however, regarding the following immediate questions:

- In which order are the final replacements of the two explicitly mentioned forms of expressions to be applied in the elimination of quantifiers?

- Or are such eliminations independent of the order of the replacements in the sense that they always yield unique normal forms?

What we can actually find on Page 15 are the mentioned "explicit definitions (ε_1), (ε_2)", which describe the rewrite relation of these replacements. In the more modern notation we prefer for this paper, these explicit definitions read:

$$\exists x. \, A \quad \Leftrightarrow \quad A\{x \mapsto \varepsilon x. \, A\} \qquad\qquad (\varepsilon_1)$$

$$\forall x. \, A \quad \Leftrightarrow \quad A\{x \mapsto \varepsilon x. \, \neg A\} \qquad\qquad (\varepsilon_2)$$

Note that x is a meta-variable for *individual variables* (in the original: a concrete object-level, bound individual variable), and A is a meta-variable for formulas (in the original: a concrete object-level, singulary formula variable). The original version of (ε_1) literally reads: $(Ex) \, A(x) \sim A\big(\varepsilon_x \, A(x)\big)$.

Note that the formulas considered here and in what follows are always first-order formulas, extended with ε-terms and possibly also with free (second-order) *formula variables*. For our considerations in this paper, it does not matter whether we include such formula variables into our first-order formulas or not.

1.2 Subject Matter

What we will study in this paper is the question how the elimination of first-order quantifiers via their explicit definitions can take place.

Here we should recall that, in *explicit definitions* (contrary to recursive definitions), the symbol to be defined (here: $\exists$ or $\forall$), occurring on the left-hand side of an equation (the *definiendum*), must not re-occur in the term on the right-hand side (*definiens*).

In this standard terminology, (ε_1) and (ε_2) classify as explicit definitions, because $\exists$ and $\forall$ do not occur on the right-hand sides — at least not explicitly.

It is commonplace knowledge that (contrary to recursive or implicit definitions) explicit definitions are analytic (i.e. not synthetic) in the sense that they cannot contribute anything essential to our knowledge base — simply because any notion introduced by an explicit definition can be eliminated from any language (at least in principle) after replacing all *definienda* with their respective *definientia*.

For first-order terms the eliminability is indeed trivial, even for non-right-linear equations such as
$$\mathsf{russell}(x) = \mathsf{mbp}(x, x),$$
where the number of occurrences of defined symbols in x is doubled when rewriting with this equation; i.e., if $n(t)$ denotes the number of explicitly defined symbols in the term t, then $n(\mathsf{russell}(t)) = n(t) + 1$, whereas $n(\mathsf{mbp}(t, t)) \geq 2 * n(t)$.

The termination of a stepwise elimination by applying one equation after the other — until no defined symbols remain — does not crucially depend on whether we rewrite the defined symbols in t before we apply the equation for the defined term $\mathsf{russell}(t)$ or after. Indeed, the difference this alternative can make is only a duplication of the rewrite steps required for the normalization of t.

This argumentation, however, does not straightforwardly apply to our definitions (ε_1), (ε_2). Indeed, the instance of the first occurrence of the meta-variable A on the right-hand side is modified by a substitution that may introduce an arbitrarily large number of copies of the instance of A.

We will show in this paper, however, that rewriting of an arbitrary formula F with (ε_1), (ε_2) is always confluent and terminating. This means that, no matter in which order we eliminate the quantifiers, a resulting quantifier-free formula will always be obtained, and that this formula is a unique normal form for F.

1.3   A Lacuna in Hilbert–Bernays?

The fact that this rewriting is innermost terminating has been well known before, but none of the experts on Hilbert's ε we consulted knew about the strong termination (i.e. termination independent of any rewriting strategy), and one of them even claimed that the rewriting would not be confluent.

As the proofs of the ε-theorems of [20] show, Paul Bernays (1888–1977) was well aware of the influence of strategies on elimination procedures. The mathematical technology of the 1930s, however, makes it most unlikely that he could easily show the strong termination — let alone consider it to be trivial in the context of a textbook (such as Hilbert–Bernays).

Moreover, the actual formula language of Hilbert–Bernays strongly suggests an outermost strategy: A non-outermost rewriting typically requires the instantiation of A to formulas containing variables that are bound by the outer quantifiers and epsilons. Such an instantiation is not permitted in Hilbert–Bernays, however, because these additional variables must come from a set of variables different from the free individual variables, which are called *bound* individual variables and which are not permitted to occur free in a substitution for A. Thus, for an innermost rewriting in the predicate calculus of Hilbert–Bernays, we have to resort to multiple tacit applications of Rule (δ') for a complete reconstruction of the whole outer part of the formula in each innermost rewrite step; for Rule (δ') see e.g. Page 109 in [21; 24].

All in all, the fact that neither the innermost rewriting strategy nor Rule (δ') is mentioned in this context in [20] makes it most likely that Bernays just relied here on his learning that explicit definitions always admit an elimination, which is actually not the case in general for higher-order definitions.

1.4 Alternative Proofs by Applying Theories of First- or Higher-Order Rewriting?

In this paper, we will approach our results directly, without applying the theory of first- or higher-order rewrite systems. Other options for obtaining the crucial termination result are:

Option 1: To map the first-order terms with quantifiers and epsilons to quantifier- and epsilon-free first-order terms, to find a first-order term rewriting system that admits the transitive reduction of the images of any original reduction, and to prove the termination of the first-order term rewriting system, using the powerful theorems and methods to establish termination of first-order term rewriting systems (or even some of the software systems that may show first-order termination automatically, cf. e.g. [40]).

Option 2: To apply some results on termination of higher-order rewriting systems.

Option 3: To map the first-order terms with quantifiers and epsilons to Church's simply-typed λ-calculus (which is known to be terminating), such that the images of each original reduction admit the transitive reduction in simply-typed λ-calculus.

Let us look at second-order formulations of (ε_1), partly because the original formulation of Hilbert's ε as found in [1] and [14; 15] is already a second-order one without binders, and partly to develop options 2 and 3 a bit further.

If we use i to designate the sort (basic type) of individuals and o to designate the sort of formulas (as standard in Church's simply-typed λ-calculus), then the ε gets the typing of $\varepsilon : (i \to o) \to i$, and for a second-order variable $A : i \to o$ and the existential operator $\Sigma : (i \to o) \to o$, we get

$$\Sigma A = A(\varepsilon A),$$

or in η-expanded form

$$\Sigma \lambda x.(Ax) = A(\varepsilon \lambda x.(Ax)).$$

1.4.1 On Option 2

To implement these equations according to option 2, we have to pick one of the three competing higher-order rewriting frameworks, namely *combinatory reduction systems (CRSs)* [27], [28], *higher-order rewrite systems* [35], [36], and *algebraic-functional systems* [25]. We pick the CRS framework because it is the oldest and most popular one (also admitting extension to conditional rewriting straightforwardly, cf. [43, Note 9]).

In CRS syntax (cf. e.g. [28, § 11]), the η-expanded rule reads

$$\Sigma[x](A(x)) = A(\varepsilon[x](A(x))),$$

where x is a variable, A is a singulary *meta-variable* (not only a top-level one, but also w.r.t. the special technical terms used for CRSs, i.e. a meta-variable for a special variable that must not occur in the terms in the range of the rewrite relation), Σ and ε are singulary function symbols (i.e. 1-ary constant symbols), and $[x]$ is an abstraction operator, binding the variable x. In this notation, we indeed have a CRS *rewrite rule* with the intended rewrite relation. We can formulate (ε_2) in a similar way, resulting in a two-rule CRS that is *orthogonal* (called "regular" in [27]), i.e. non-overlapping ("non-ambiguous") and left-linear. Thus, according to [28, Corollary 13.6] ([27, Theorem II.3.11]), the rewrite relation is confluent.

As it is obvious that this rewrite relation is weakly normalizing (as it is innermost terminating), its termination (strong normalization) follows from Theorem II.5.9.3 of [27, p.168], provided that we can show our rewrite relation to be non-erasing. This means that we have to show that the set of free variables is invariant under rewrite steps. Note that the instance of A may contain free variables (such as y), but even if the instance of A is, say, $\underline{\lambda}[x](y = y)$ (i.e. the quantifier is vacuous, binding a variable that does not occur in its scope), it seems that the deletion of the second occurrence of A in the right-hand side does not matter, because all occurrences of free variables are preserved by the first occurrence of A in the right-hand side.

This argumentation, however, forgets that CRSs come without β-reduction. So we may need the rule $(\underline{\lambda}[x](A(x)))B = A(B)$ in addition, which would render the CRS erasing. On the other hand, $\underline{\lambda}$ is different from λ (although some crucial underlining of λ is missing in [28]) and part and parcel of the substitution framework for "meta-variables" in [28]; this means we should get along without the β-rule for λ, provided that we write existential quantification in our formulas as, say, "$\Sigma[x]$" instead of "$\Sigma\lambda x$.".

If the latter is indeed the case, and if our understanding of [27] is the right one, then confluence and termination can be established by applying the theory of CRSs.

1.4.2 On Option 3

To implement option 3, however, we could to take ε as a constant with the above typing and the mapping to Church's simply-typed λ-calculus could replace the previous constant Σ with the λ-term $\lambda A.\ (A(\varepsilon A))$ of the same type as Σ before. Then reduction by the first of the above equations could be done by a first β-reduction, and a second β-reduction on the λ-term A could be used to reduce $A(\varepsilon A)$, such that an original reduction step with (ε_1) results in two β-reduction steps after the mapping to simply-typed λ-calculus. Although this proof plan is most promising, it is not easily accessible in the sense that a working mathematician could verify it without a careful formalization of lots of technical and syntactic details. Moreover, as Bernays in the 1930s could not have known about the termination of simply-typed λ-calculus — first published in 1967 by Tait [39] — this is not a proof plan he could have followed (though he was in correspondence with Church and visiting the Institute for Advanced Study in Princeton during session 1935/36).

1.4.3 Conclusion

Option 1 did not seem to work. Moreover, the contacted experts on higher-order rewriting did not want to help settling the questions arising via option 2 (and no answer was found in [37], [26] either), and the effort to familiarize oneself with the most fascinating and outstanding work documented in Klop's PhD thesis [27] is considerable and disproportionate for our subject matter. Finally, we became aware of option 3 only after our proof was already completed.

Thus, we chose a direct proof, which we present in this paper. Note, that this straightforward and efficiently verifiable proof of termination and confluence of the reduction relation defined directly on first-order terms with quantifiers and epsilons has considerable advantages for the historiographical questions of a possible lacuna in Hilbert–Bernays (where none of the options 1–3 could help, cf. § 1.3), as well as for the accessibility by the community working on Hilbert's ε (where not everybody is familiar with term-rewriting frameworks and higher-order logic).

Indeed, a direct proof is not only more informative on the concrete structure of the particular subject matter than a proof applied after a transformation to a different basic data structure via one of the options 1–3, but it provides also the stronger, more concise, and historiographically more relevant evidence.

2   Background and Tools

2.1   Basic Notions and Notation

We follow standard mathematical writing style, cf. [4].

We try to be self-contained in this this paper. In case we should omit some required information, we refer the reader to the survey [27, § I.5] on abstract rewrite systems.

Let '$\mathbf{N}$' denote the set of natural numbers, and '$<$' the ordering on $\mathbf{N}$. Let $\mathbf{N}_+ := \{\, n \in \mathbf{N} \mid 0 \neq n \,\}$.

For classes R, A, and B we define:

$$\mathrm{dom}(R) := \{\, a \mid \exists b.\ (a,b) \in R \,\} \qquad \textit{domain}$$
$$_A|R \quad := \{\, (a,b) \in R \mid a \in A \,\} \qquad \textit{(domain-) restriction to } A$$
$$\langle A \rangle R \quad := \{\, b \mid \exists a \in A.\ (a,b) \in R \,\} \quad \textit{image of } A,\ \text{i.e. } \langle A \rangle R = \mathrm{ran}(_A|R)$$

And the dual ones:

$$\mathrm{ran}(R) := \{\, b \mid \exists a.\ (a,b) \in R \,\} \qquad \textit{range}$$
$$R|_B \quad := \{\, (a,b) \in R \mid b \in B \,\} \qquad \textit{range-restriction to } B$$
$$R\langle B \rangle \quad := \{\, a \mid \exists b \in B.\ (a,b) \in R \,\} \quad \textit{reverse-image of } B,\ \text{i.e. } R\langle B \rangle = \mathrm{dom}(R|_B)$$

We use 'id' for the identity function, and '$\circ$' for the composition of binary relations. Functions are (right-) unique relations, and so the meaning of "$f \circ g$" is extensionally given by $(f \circ g)(x) = g(f(x))$.

Let $\longrightarrow$ be a binary relation. $\longrightarrow$ is a relation *on* A if

$$\mathrm{dom}(\longrightarrow) \cup \mathrm{ran}(\longrightarrow) \subseteq A.$$

$\longrightarrow$ is *irreflexive* if $\mathrm{id} \cap \longrightarrow = \emptyset$. It is *$A$-reflexive* if $_A|\mathrm{id} \subseteq \longrightarrow$. Speaking of a *reflexive* relation we refer to the largest A that is appropriate in the local context, and referring to this A we write $\xrightarrow{0}$ to ambiguously denote $_A|\mathrm{id}$. With $\xrightarrow{1} := \longrightarrow$, and $\xrightarrow{n+1} := \xrightarrow{n} \circ \longrightarrow$ for $n \in \mathbf{N}_+$, $\xrightarrow{m}$ denotes the m-step relation for $\longrightarrow$. The *transitive closure* of $\longrightarrow$ is $\xrightarrow{+} := \bigcup_{n \in \mathbf{N}_+} \xrightarrow{n}$. The *reflexive closure* of $\longrightarrow$ is $\xrightarrow{=} := \bigcup_{n \in \{0,1\}} \xrightarrow{n}$. The *reflexive transitive closure* of $\longrightarrow$ is $\xrightarrow{*} := \bigcup_{n \in \mathbf{N}} \xrightarrow{n}$. The *reverse* of $\longrightarrow$ is $\longleftarrow := \{\, (b,a) \mid (a,b) \in \longrightarrow \,\}$.

v and w are called *joinable w.r.t.* $\longrightarrow$ if $v \downarrow w$, i.e. if $v \xrightarrow{*} \circ \xleftarrow{*} w$. $\longrightarrow$ is *locally confluent* if $v \downarrow w$ for any v, w with $v \longleftarrow \circ \longrightarrow w$; it is *confluent* if $v \downarrow w$ for any v, w with $v \xleftarrow{*} \circ \xrightarrow{*} w$. a' is an $\longrightarrow$-*normal form* of a if $a \xrightarrow{*} a' \notin \mathrm{dom}(\longrightarrow)$.

A sequence $(s_i)_{i \in \mathbf{N}}$ is *non-terminating in* $\longrightarrow$ if $s_i \longrightarrow s_{i+1}$ for all $i \in \mathbf{N}$. $\longrightarrow$ is *terminating* if there are no non-terminating sequences in $\longrightarrow$. A relation R (on A) is *well-founded* if any non-empty class B ($\subseteq A$) has an R-minimal element, i.e. $\exists a \in B.\ \neg \exists a' \in B.\ a' R a$. Note that well-foundedness of $\longleftarrow$ immediately entails termination of $\longrightarrow$ (via the range of the non-terminating sequence), but the converse requires a weak form of the Axiom of Choice to construct the non-terminating sequence, cf. e.g. [32, § 4.1].

Corollary 2.1 *If a binary relation is well-founded, so is its transitive closure.*

2.2 A Generalized Theorem as the Main Tool

The following Theorem 2.2 is a generalization of Jan Willem Klop's Theorem I.5.18 [27, p. 53], which can be obtained again from Theorem 2.2 by the specialization $\longrightarrow_0 := \emptyset$.

Theorem 2.2
Let $\longrightarrow_0$ and $\longrightarrow_1$ be two binary relations.
Set $\longrightarrow_2 := \overset{*}{\longrightarrow}_0 \circ \longrightarrow_1$.
Set $\longrightarrow_3 := \longrightarrow_0 \cup \longrightarrow_1$.
Let $a \in \mathrm{dom}(\longrightarrow_3)$. Let a' be an $\longrightarrow_3$-normal form of a. Set $A := \langle\!\langle a \rangle\!\rangle \overset{*}{\longrightarrow}_3$.
Set $\longrightarrow_4 := {}_A\!\mid \longrightarrow_3$. If

1. *$\longleftarrow_0 \lfloor_A$ is well-founded;*

2. *there is an upper bound $n \in \mathbf{N}$ on the length of $\longrightarrow_2$-derivations starting from a and reaching a' by $\overset{*}{\longrightarrow}_0$; more formally, this means that we have $m \leq n$ for any $m \in \mathbf{N}$ and any sequence $b_0, \ldots, b_m$ with $a = b_0$, $b_i \longrightarrow_2 b_{i+1}$ for each $i \in \{0, \ldots, m-1\}$, and $b_m \overset{*}{\longrightarrow}_0 a'$;*

3. *for all b_1, b_2 with $b_1 \longleftarrow_4 \circ \longrightarrow_1 b_2$, we have $b_1 \overset{*}{\longrightarrow}_4 \circ \overset{*}{\longleftarrow}_4 b_2$; and*

4. *for all b_1, b_2 with $b_1 \longleftarrow_4 \circ \longrightarrow_0 b_2$, we have $b_1 \overset{*}{\longrightarrow}_4 \circ \overset{=}{\longleftarrow}_4 b_2$;*

then $\longleftarrow_4$ is well-founded.

Proof of Theorem 2.2
<u>Claim 1:</u> For all b_1, b_2 and $n \in \mathbf{N}$ with $b_1 \longleftarrow_4 \circ \overset{n}{\longrightarrow}_0 b_2$, we have $b_1 \overset{*}{\longrightarrow}_4 \circ \overset{=}{\longleftarrow}_4 b_2$.
<u>Proof of Claim 1:</u> By induction on n. In case of $b_1 \longleftarrow_4 \circ \overset{0}{\longrightarrow}_0 b_2$, we have $b_1 \longleftarrow_4 b_2$.
    In case of $b_1 \longleftarrow_4 \circ \overset{n}{\longrightarrow}_0 b_2 \longrightarrow_0 b_3$, by induction hypothesis we have $b_1 \overset{*}{\longrightarrow}_4 b_4 \overset{=}{\longleftarrow}_4 b_2$ for some $b_4 \in A$. In case of $b_4 = b_2$, we have $b_1 \overset{*}{\longrightarrow}_4 b_4 \longrightarrow_0 b_3$, and thus $b_1 \overset{*}{\longrightarrow}_4 b_3$. Otherwise, we have $b_4 \longleftarrow_4 b_2$, and thus $b_4 \overset{*}{\longrightarrow}_4 b_5 \overset{=}{\longleftarrow}_4 b_3$ for some b_5 by item 4, i.e. the desired $b_1 \overset{*}{\longrightarrow}_4 b_5 \overset{=}{\longleftarrow}_4 b_3$.   Q.e.d. (Claim 1)

Set $B := \{ b \in A \mid b \overset{*}{\longrightarrow}_4 a' \}$.
By item 2, we can define a function $l : B \to \{ m \in \mathbf{N} \mid m \leq n \}$ via
$$l(b) := \max \{ m \in \mathbf{N} \mid b \overset{m}{\longrightarrow}_2 \circ \overset{*}{\longrightarrow}_0 a' \}.$$

Claim 2: For all $b \in B$ with $b \overset{*}{\longrightarrow}_4 b'$, we have $b' \in B$.

Proof of Claim 2: By induction on $k := l(b)$ in $<$. The induction hypothesis is that
for all $b'' \in B$ with $b'' \overset{*}{\longrightarrow}_4 b'''$ and $l(b'') < k$, we have $b''' \in B$.
Note that (for $b'' \in B$) $b'' \longrightarrow_4 b'''$ implies $l(b''') \leq l(b'')$. Thus, by another
induction on the length of derivations, the induction conclusion follows from the
induction hypothesis and the proposition that for all $b'' \in B$ with $b'' \longrightarrow_4 b'''$ and
$l(b'') = k$, we have $b''' \in B$.
So let us assume $b \in B$ and $b \longrightarrow_4 b'$. Then, using the induction hypothesis, we
have to show $b' \in B$, for which it suffices to show $b' \overset{*}{\longrightarrow}_4 a'$.
By our assumption, we have $b \overset{*}{\longrightarrow}_4 a'$, which falls into at least one of the following
two cases:

$\underline{b \overset{*}{\longrightarrow}_0 a'}$: By Claim 1: $b' \overset{*}{\longrightarrow}_4 \circ \overset{=}{\longleftarrow}_4 a'$. Because $a' \notin \mathrm{dom}(\longrightarrow_3)$, and a fortiori also
$a' \notin \mathrm{dom}(\longrightarrow_4)$, we actually have $b' \overset{*}{\longrightarrow}_4 a'$.

$\underline{b \overset{*}{\longrightarrow}_0 \hat{b} \longrightarrow_1 b''' \overset{*}{\longrightarrow}_4 a' \text{ for some } \hat{b}, b'''}$: Again by Claim 1, we get $b' \overset{*}{\longrightarrow}_4 b'''' \overset{=}{\longleftarrow}_4 \hat{b}$ for
some $b'''' \in A$.
In case of $b'''' = \hat{b}$, we have $b' \overset{*}{\longrightarrow}_4 b'''' \longrightarrow_1 b''' \overset{*}{\longrightarrow}_4 a'$, i.e. the desired
$b' \overset{*}{\longrightarrow}_4 b''' \longrightarrow_4 b'' \overset{*}{\longrightarrow}_4 a'$.
Otherwise we have $b'''' \longleftarrow_4 \hat{b}$. Thus, by item 3, there is some b'' with
$b'''' \overset{*}{\longrightarrow}_4 b'' \overset{*}{\longleftarrow}_4 b'''$. Because of $b \overset{*}{\longrightarrow}_0 \hat{b} \longrightarrow_1 b''' \overset{*}{\longrightarrow}_4 a'$ we have $b'' \in B$ and $l(b''') <$
$l(b)$. Thus, by the induction hypothesis, we get $b'' \in B$, and then the desired
$b' \overset{*}{\longrightarrow}_4 b'''' \overset{*}{\longrightarrow}_4 b'' \overset{*}{\longrightarrow}_4 a'$.   Q.e.d. (Claim 2)

Claim 3: $A = B$.

Proof of Claim 3: By $a \overset{*}{\longrightarrow}_3 a'$, we also have $a \overset{*}{\longrightarrow}_4 a'$, and so $a \in B$.
Thus, by Claim 2, we get $\langle\!\langle a \rangle\!\rangle \overset{*}{\longrightarrow}_4 \subseteq B$.
All in all, we get: $A = \langle\!\langle a \rangle\!\rangle \overset{*}{\longrightarrow}_3 = \langle\!\langle a \rangle\!\rangle \overset{*}{\longrightarrow}_4 \subseteq B \subseteq A$.   Q.e.d. (Claim 3)

By Claim 3, we get $l : A \to \{ m \in \mathbf{N} \mid m \leq n \}$. Now for every b_1, b_2 with
$b_1 \longleftarrow_4 b_2$, we have $b_1, b_2 \in A$ and, moreover, $(l(b_1), b_1)$ is strictly smaller than
$(l(b_2), b_2)$ in the lexicographic combination of $<$ and $\longleftarrow_0 \lceil_A$, which is well-founded
by item 1. Indeed, in case of $b_1 \longleftarrow_0 b_2$, we have $l(b_1) \leq l(b_2)$ and $b_1 \longleftarrow_0 \lceil_A b_2$,
and in case of $b_1 \longleftarrow_1 b_2$, we have $l(b_1) < l(b_2)$.

                                              Q.e.d. (Theorem 2.2)

536

2.3   Terms, Formulas, Substitutions, Contexts

A straightforward intuitive understanding of terms, formulas, substitutions, and contexts will actually suffice for most working mathematicians to understand the remainder of this paper. For the others, we give an example formalization of these notions here.

Terms and *formulas* are defined inductively as follows:

- An individual variable is a term.

- If A is an n-ary formula variable $(n \in \mathbf{N})$ and $t_1, \ldots, t_n$ are terms, then $A(t_1, \ldots, t_n)$ is a formula.

- If f is an n-ary constant function or predicate symbol $(n \in \mathbf{N})$ and $t_1, \ldots, t_n$ are terms,
  then $\mathrm{f}(t_1, \ldots, t_n)$ is a term or formula, respectively.
  In case of $n = 0$, we simply write "f" instead of "f()".

- If F is a formula, then $\neg F$ is a formula. If F_1 and F_2 are formulas, then $(F_1 \vee F_2)$, $(F_1 \wedge F_2)$, $(F_1 \Rightarrow F_2)$, $\ldots$ are formulas.

- If x is an individual variable and F is a formula,
  then $\varepsilon x.\ F$ is a term and $\exists x.\ F$ and $\forall x.\ F$ are formulas.
  In these terms and formulas, all occurrences of x are *bound*; non-bound occurrences of variables in terms and formulas are called *free*, such as each occurrence of any formula variable, and also of any individual variable y that is not in the scope of a binder on y, such as "$\varepsilon y.$", "$\exists y.$", or "$\forall y.$".

In our definition of terms and formulas we deviate from Hilbert–Bernays in not having an extra set of individual variables for bound occurrences, disjoint from the set to be used for free occurrences. So we have only one set of individual variables, but this does not really make any difference here, in particular because we ignore the variable names in the bound occurrences by the following stipulation:

We equate formulas modulo the renaming of bound variables.

A *substitution* is a mapping of individual variables to terms and of n-ary formula variables to expressions of the form $\underline{\lambda}(x_1, \cdots, x_n).\ F$, respectively, where $x_1, \cdots, x_n$ are mutually distinct individual variables and F is a formula. For $n = 0$, we just write "F" instead of "$\underline{\lambda}().\ F$".

Presupposing the above stipulation of considering formulas only up to renaming of bound variables, we now define the result of an application of a substitution σ to terms and formulas inductively as follows. We use postfix notation with highest operator precedence.

- Let x be an individual variable.
  If $x \notin \mathrm{dom}(\sigma)$, then $x\sigma = x$; otherwise $x\sigma = \sigma(x)$, i.e. the value of x under σ.

- Let A be an n-ary formula variable, and let $t_1, \ldots, t_n$ be terms. If $A \notin \mathrm{dom}(\sigma)$, then $(A(t_1, \ldots, t_n))\sigma = A(t_1\sigma, \ldots, t_n\sigma)$. Otherwise $(A(t_1, \ldots, t_n))\sigma$ is the result of the β-reduction of $\sigma(A)(t_1\sigma, \ldots, t_n\sigma)$, i.e., for $\sigma(A) = \underline{\lambda}(x_1, \cdots, x_n).\, F$, the formula $F\sigma'$, where σ' is the substitution $\{x_1 \mapsto t_1\sigma, \ \ldots, \ x_n \mapsto t_n\sigma\}$.

- If f is an n-ary constant function or predicate symbol and $t_1, \ldots, t_n$ are terms, then $(\mathrm{f}(t_1, \ldots, t_n))\sigma = \mathrm{f}(t_1\sigma, \ldots, t_n\sigma)$.

- If F is a formula, then $(\neg F)\sigma = \neg F\sigma$. If F_1 and F_2 are formulas, then $(F_1 \vee F_2)\sigma = (F_1\sigma \vee F_2\sigma)$, $(F_1 \wedge F_2)\sigma = (F_1\sigma \wedge F_2\sigma)$, $(F_1 \Rightarrow F_2)\sigma = (F_1\sigma \Rightarrow F_2\sigma)$, $\ldots$.

- If x is an individual variable
  — w.l.o.g. neither an element of $\mathrm{dom}(\sigma)$, nor occurring free in $\mathrm{ran}(\sigma)$ —
  and F is a formula,
  then $(\varepsilon x.\, F)\sigma = \varepsilon x.\, F\sigma$, $(\exists x.\, F)\sigma = \exists x.\, F\sigma$, $(\forall x.\, F)\sigma = \forall x.\, F\sigma$.

Corollary 2.3 *If σ is a substitution, then, for any formula or term G whose free variables are in the set A, the following syntactic equality holds:*
$$G\sigma \;=\; G(_A|\sigma).$$

By induction on the construction of G_1 we easily get:

Corollary 2.4 *Let G_1 be a formula or a term. Let X and G_2 be either an individual variable and a term, or a nullary formula variable and a formula. Then, for any substitution σ where $X \notin \mathrm{dom}(\sigma)$ and where X does not occur free in $\mathrm{ran}(\sigma)$, the following syntactic equality holds:*
$$(G_1\{X \mapsto G_2\})\sigma \;=\; (G_1\sigma)\{X \mapsto G_2\sigma\}.$$

Finally, let $H_0, \ldots, H_n$ $(n \in \mathbf{N})$ be mutually distinct, nullary formula variables, reserved for the following definition: A *context* written "$G[\cdots]$" (a formula or term with holes) is actually a formula or term G with one single (free) occurrence of each of the formula variables $H_1, \ldots, H_n$. Moreover, "$G[F_1, \ldots, F_n]$" denotes $G\{H_1 \mapsto F_1, \ \ldots, \ H_n \mapsto F_n\}$, for formulas $F_1, \ldots, F_n$.

Corollary 2.5 *For any context $G[\cdots]$, any formula F, and any substitution σ, the following syntactic equality holds:*
$$(G[F])\sigma \;=\; G\sigma[F\sigma].$$

3 The Concrete Rewrite Relation

By writing "$\neg^\forall$" for "$\neg$" and "$\neg^\exists$" for the empty string " ", we can unify the two formulas (ε_1) and (ε_2) to the single formula

$$Qx.\ A \quad \Leftrightarrow \quad A\{x \mapsto \varepsilon x.\ \neg^Q A\} \qquad\qquad (\varepsilon_Q)$$

for $Q \in \{\exists, \forall\}$, and x a meta-variable for an individual variable, and A a meta-variable for a formula.

Let $\longrightarrow$ be the rewrite relation resulting from rewriting with the equivalence (ε_Q) as a rewrite rule from left to right. Explicitly, this means that $F_1 \longrightarrow F_2$ if there are a context $G[\cdots]$, a quantifier symbol Q, an individual variable x, and a formula A, such that $F_1 = G[Qx.\ A]$ and $F_2 = G[A\{x \mapsto \varepsilon x.\ \neg^Q A\}]$.

Let $\longrightarrow_0$ and $\longrightarrow_1$ be the partition of $\longrightarrow$ for the case of a *vacuous* quantifier (i.e. for the case that x does not occur in the formula A in (ε_Q)), and for the case that the quantifier is not vacuous.

Let $\longrightarrow_{\mathcal{I}}$ be the innermost rewrite relation given by innermost rewriting with the equivalence (ε_Q).

Let $\twoheadrightarrow$ be the version of $\longrightarrow$ for the rewriting of parallel redexes. Explicitly, this means that $F_1 \twoheadrightarrow F_2$ if there are a context $G[\cdots]$ with $n \in \mathbf{N}$ holes, quantifier symbols $Q_1, \ldots, Q_n$, individual variables $x_1, \ldots, x_n$, and formulas $A_1, \ldots, A_n$, such that

$$\begin{aligned} F_1 &= G[Q_1 x_1.\ A_1,\ \ldots,\ Q_n x_n.\ A_n], \\ F_2 &= G[A_1\{x_1 \mapsto \varepsilon x_1.\ \neg^{Q_1} A_1\},\ \ldots,\ A_n\{x_n \mapsto \varepsilon x_n.\ \neg^{Q_n} A_n\}]. \end{aligned}$$

From these definitions, we immediately get the following corollaries.

Corollary 3.1 $\quad \longrightarrow_{\mathcal{I}} \subseteq \longrightarrow$ *and* $\operatorname{dom}(\longrightarrow_{\mathcal{I}}) = \operatorname{dom}(\longrightarrow)$.

Corollary 3.2 $\quad \twoheadrightarrow \subseteq \overset{*}{\longrightarrow}$.

3.1 Local Confluence

Note that the technical terms of the following lemma are clarified and formalized in its proof.

Lemma 3.3 *If we have a peak $F_1 \longleftarrow F_0 \longrightarrow F_2$ of local divergence and the redex of the rewrite step to F_1 is properly inside the one of the rewrite step to F_2, which is on top of F_0, then there are formulas F_3, F_4 satisfying all the following items:*

1. $F_1 \longrightarrow F_4 \longleftarrow F_3 \longleftarrow\!\!\!\!\!+\!\!\!\!\!- F_2$.

2. *If the initial step to the left is actually applied to a non-vacuous quantifier (i.e. if $F_1 \longleftarrow_1 F_0$), then we have $F_4 \longleftarrow_1 F_3 \longleftarrow\!\!\!\!\!+\!\!\!\!\!-_1 F_2$.*

3. *If the initial step to the right is actually applied to a non-vacuous quantifier (i.e. if $F_0 \longrightarrow_1 F_2$), then we have $F_1 \longrightarrow_1 F_4$.*

4. *If the initial step to the right is actually applied to a vacuous quantifier (i.e. if $F_0 \longrightarrow_0 F_2$), then we have $F_3 = F_2$.*

Proof of Lemma 3.3

Suppose we have a peak $F_1 \longleftarrow F_0 \longrightarrow F_2$ of local divergence and the redex of the rewrite step to F_1 is properly inside the one of the rewrite step to F_2, which is on top of F_0. Then F_0 has the form

$$Q_1 x_1.\ G_1[Q_2 x_2.\ G_2]. \tag{F_0}$$

We may in particular assume here that x_2 is different from x_1 and does not occur free in the context $G_1[\cdots]$ (if we consider the dots "$\cdots$" to be empty). Moreover we may assume that the formulas F_1 and F_2 are the following:

$$Q_1 x_1.\ G_1[G_2\{x_2 \mapsto \varepsilon x_2.\ \neg^{Q_2} G_2\}]. \tag{F_1}$$
$$\left(\ G_1[Q_2 x_2.\ G_2]\ \right)\!\{\ x_1 \mapsto \varepsilon x_1.\ \neg^{Q_1} G_1[Q_2 x_2.\ G_2]\ \}. \tag{F_2}$$

If we rewrite the outermost redex in F_1, we obtain the formula

$$\left(\ G_1[G_2\{x_2 \mapsto \varepsilon x_2.\ \neg^{Q_2} G_2\}]\ \right)\sigma$$

written with the help of the substitution σ given as

$$\{\ x_1 \mapsto \varepsilon x_1.\ \neg^{Q_1} G_1[G_2\{x_2 \mapsto \varepsilon x_2.\ \neg^{Q_2} G_2\}]\ \}. \tag{σ}$$

If we propagate this substitution, by Corollary 2.5 we obtain a formula given by the context

$$G_1 \sigma[\cdots] \tag{C}$$

where we read the dots "$\cdots$" as

$$(G_2\{x_2 \mapsto \varepsilon x_2.\ \neg^{Q_2} G_2\})\sigma.$$

Because x_2 occurs free in none of $\mathrm{dom}(\sigma)$, $G_1[\cdots]$, $G_1[G_2\{x_2 \mapsto \varepsilon x_2.\ \neg^{Q_2} G_2\}]$, $\mathrm{ran}(\sigma)$, by Corollary 2.4 we can propagate σ further to write the inner formula as

$$G_2 \sigma\{x_2 \mapsto \varepsilon x_2.\ \neg^{Q_2} G_2 \sigma\}. \tag{I}$$

Putting (C) and (I) together again, we can choose formula F_4 with the property $F_1 \longrightarrow F_4$ as follows:

$$G_1 \sigma\big[\ G_2 \sigma\{x_2 \mapsto \varepsilon x_2.\ \neg^{Q_2} G_2 \sigma\}\ \big]. \tag{F_4}$$

If we now rewrite all occurrences of the redex mentioned at the end of the notation of the formula F_2 in parallel, then we obtain the formula
$$(\ G_1[Q_2x_2. \ G_2] \ )\sigma.$$
Before we can rewrite the remaining redex, we have to propagate σ to obtain a clear description of it. By Corollary 2.5, this results again in a context as given in (C) above, where, however, we now read the "$\cdots$" as
$$Q_2x_2. \ G_2\sigma.$$
Note that, in this formula, the substitution σ has passed the quantifier "Q_2x_2." soundly. Indeed, as mentioned above, x_1 is different from x_2, and x_2 cannot occur free in $\operatorname{ran}(\sigma)$. Putting this formula and its context together again, we can choose as F_3 with the property $F_3 \longleftrightarrow\hspace{-0.7em}\shortmid\hspace{0.3em} F_2$ as follows:
$$G_1\sigma\big[\ Q_2x_2. \ G_2\sigma \ \big]. \qquad\qquad (F_3)$$
If we now rewrite the remaining redex, we again obtain the formula F_4, as was to be shown for item 1.

For item 2, it suffices to note that, if x_2 occurs free in G_2, then x_2 also occurs free in $G_2\sigma$ because x_1 and x_2 are different.

For item 3, it suffices to note that, if x_1 occurs free in $G_1[Q_2x_2. \ G_2]$, then x_1 also occurs free in $G_1[G_2\{x_2 \mapsto \varepsilon x_2. \ \neg^{Q_2}G_2\}]$.

For item 4, it suffices to note that, if x_1 does not occur free in $G_1[Q_2x_2. \ G_2]$, then both F_2 and F_3 are actually $G_1[Q_2x_2. \ G_2]$. **Q.e.d. (Lemma 3.3)**

As overlaps are trivial and as peaks of local divergence with parallel redexes are joinable in one step at each side trivially, we get as corollaries of Lemma 3.3(1,4) and Corollary 3.2:

Corollary 3.4 $\longrightarrow$ *is locally confluent.*

Corollary 3.5 *For all F_1, F_2 with $F_1 \longleftarrow \circ \longrightarrow_0 F_2$, we have $F_1 \stackrel{*}{\longrightarrow} \circ \stackrel{=}{\longleftarrow} F_2$.*

3.2 Well-Foundedness

As every $\longrightarrow_0$-step (vacuous quantifiers) and every $\longrightarrow_{\mathcal{I}}$-step (innermost quantifiers) reduces the number occurrences of quantifiers by 1, we have:

Corollary 3.6 $\longleftarrow_0 \cup \longleftarrow_{\mathcal{I}}$ *is well-founded.*

Theorem 3.7 $\longleftarrow$ *is well-founded.*

Proof of Theorem 3.7

Assume that B is a non-empty class. Then there is some $a \in B$. We just have to find an $\longleftarrow$-minimal element in B.

If a is $\longleftarrow$-minimal in B, then we have succeeded. Thus suppose that a is not $\longleftarrow$-minimal in B. Then $a \in \mathrm{dom}(\longrightarrow)$.

Set $A := \langle\!\langle\{a\}\rangle\!\rangle \stackrel{*}{\longrightarrow}$. Set $\longrightarrow_4 := {}_A|\!\longrightarrow$. It now suffices to show that $\longleftarrow_4$ is well-founded (because an $\longleftarrow_4$-minimal element of $A \cap B$ is also an $\longleftarrow$-minimal element of B).

By Corollary 3.6, A has an $\longleftarrow_{\mathcal{I}}$-minimal element a'. As $a' \notin \mathrm{dom}(\longrightarrow)$ by Corollary 3.1, a' is an $\longrightarrow$-normal form of a. To obtain the well-foundedness of $\longleftarrow_4$, we are now going to apply Theorem 2.2.

Set $\longrightarrow_2 := \stackrel{*}{\longrightarrow}_0 \circ \longrightarrow_1$. Set $\longrightarrow_3 := \longrightarrow_0 \cup \longrightarrow_1$. Then $\longrightarrow = \longrightarrow_3$.

It now suffices to show items 1 to 4 of Theorem 2.2. Item 1 holds by Corollary 3.6. Item 3 holds by Corollary 3.4. Item 4 holds by Corollary 3.5. As the number of occurrences of the ε is invariant under $\longrightarrow_0$ and is increased at least by 1 by every $\longrightarrow_1$-step, it increases at least by 1 by every $\longrightarrow_2$-step. Thus, to satisfy item 2, we can choose the upper bound n to be the number of occurrences of ε in a' (minus the number in a).   **Q.e.d. (Theorem 3.7)**

3.3   Confluence

By the Newman Lemma (cf. [34] or, for a formal proof, [41, § 3.4]), we obtain from Corollary 3.4 and Theorem 3.7:

Theorem 3.8 $\longrightarrow$ *is confluent.*

3.4   On the Length of Derivations

By Theorems 3.7 and 3.8, we now know for certain that the rewrite relation is confluent and terminating (as its reverse is even well-founded), which means that we can eliminate the quantifiers in any order — but this does not mean that this is efficient.

Here is a serious warning to the contrary: The nesting depth of the occurrences of the ε-symbols introduced by the normalization can be exponential in the number of quantifiers in the input formula, and the number of steps of an outermost normalization is even higher and seems to be non-elementary, cf. [44, Example 4.7], [42, Example 8].

As any innermost rewrite step reduces the number of quantifiers exactly by 1, and as no rewrite step can reduce the number of quantifiers by more than 1, we immediately get:

Theorem 3.9
Let F be a formula with n quantifiers. Innermost rewriting of F by $\longrightarrow_{\mathcal{I}}$ obtains the (unique) $\longrightarrow$-normal form F' of F in exactly n steps, which is the minimal number of steps to reach F' by $\longrightarrow$ from F.

4 Conclusion

With Theorems 3.7 and 3.8, we have shown confluence and termination of the elimination of quantifiers via their explicit definition by Hilbert's ε. This means in particular that any first-order term with quantifiers and epsilons (and formula variables), has a unique normal form w.r.t. this elimination of quantifiers, which has its first explicit occurrence in Hilbert–Bernays [20] in 1939, namely in the proof of the 1st ε-theorem on Page 19f.

Moreover, the directness, self-containedness, and easy verifiability of the proofs should settle the questions on confluence and termination here once and for all — at least for working mathematicians. Formalists and rewriters, however, may see the need to develop a more formal verification of our proof and write a short paper that our results are all trivial in some higher-order rewriting theory. Writing or helping to find a good textbook on higher-order rewriting, however, seems to be in more urgent demand.

Furthermore, we hope that some philosophers will be stimulated by this paper to pick up the subject of the non-triviality of higher-order *explicit definition* and write or help to find a book on that subject.

Finally, the starting point of our interest in the subject, namely the question whether there is a lacuna in Hilbert–Bernays as discussed in § 1.3, needs further discussion by the experts on Hilbert's ε and the history of mathematical logic in the 20th century. On basis of our current knowledge, we would clearly answer this question positively.

References

[1] Wilhelm Ackermann. Begründung des „tertium non datur" mittels der Hilbertschen Theorie der Widerspruchsfreiheit. *Mathematische Annalen*, 93:1–36, 1925. Received March 30, 1924. Inauguraldissertation, 1924, Göttingen.

[2] Anon, editor. *Festschrift zur Feier der Enthüllung des* Gauß-Weber-*Denkmals in Göttingen, herausgegeben von dem Fest-Comitee*. Verlag von B. G. Teubner, Leipzig, 1899.

[3] Michael Codish and Aart Middeldorp. 7th Int. Workshop on Termination (WST), 2004. Technical Report AIB–2004–07, RWTH Aachen, Dept. of Computer Sci., 2004. ISSN 0935–3232. `http://sunsite.informatik.rwth-aachen.de/Publications/AIB/2004/2004-07.ps.gz`.

[4] Leonard Gillman. *Writing Mathematics Well*. The Mathematical Association of America, 1987.

[5] Jean van Heijenoort. *From* Frege *to* Gödel: *A Source Book in Mathematical Logic, 1879–1931*. Harvard Univ. Press, 1971. 2nd rev. edn. (1st edn. 1967).

[6] David Hilbert. Grundlagen der Geometrie. 1899. In [2, pp. 1–92]. 1st edn. without appendixes. Reprinted in [18, pp. 436–525].

[7] David Hilbert. Über den Zahlbegriff. *Jahresbericht der Deutschen Mathematiker-Vereinigung*, 8:180–184, 1900. Received Dec. 1899. Reprinted as Appendix VI of [10; 11; 12; 13; 17].

[8] David Hilbert. *Grundlagen der Geometrie. — Zweite, durch Zusätze vermehrte und mit fünf Anhängen versehene Auflage. Mit zahlreichen in den Text gedruckten Figuren.* Druck und Verlag von B. G. Teubner, Leipzig, 1903. 2nd rev. extd. edn. of [6], rev. and extd. with five appendixes, newly added figures, and an index of notion names.

[9] David Hilbert. Über die Grundlagen der Logik und der Arithmetik. 1905. In [29, pp. 174–185]. Reprinted as Appendix VII of [10; 11; 12; 13; 17]. English translation *On the foundations of logic and arithmetic* by Beverly Woodward with an introduction by Jean van Heijenoort in [5, pp. 129–138].

[10] David Hilbert. *Grundlagen der Geometrie. — Dritte, durch Zusätze und Literaturhinweise von neuem vermehrte und mit sieben Anhängen versehene Auflage. Mit zahlreichen in den Text gedruckten Figuren.* Number VII in Wissenschaft und Hypothese. Druck und Verlag von B. G. Teubner, Leipzig, Berlin, 1909. 3rd rev. extd. edn. of [6], rev. edn. of [8], extd. with a bibliography and two additional appendixes (now seven in total) (Appendix VI: [7]) (Appendix VII: [9]).

[11] David Hilbert. *Grundlagen der Geometrie. — Vierte, durch Zusätze und Literaturhinweise von neuem vermehrte und mit sieben Anhängen versehene Auflage. Mit zahlreichen in den Text gedruckten Figuren.* Druck und Verlag von B. G. Teubner, Leipzig, Berlin, 1913. 4th rev. extd. edn. of [6], rev. edn. of [10].

[12] David Hilbert. *Grundlagen der Geometrie. — Fünfte, durch Zusätze und Literaturhinweise von neuem vermehrte und mit sieben Anhängen versehene Auflage. Mit zahlreichen in den Text gedruckten Figuren.* Verlag und Druck von B. G. Teubner,

Leipzig, Berlin, 1922. 5th extd. edn. of [6]. Contrary to what the sub-title may suggest, this is an anastatic reprint of [11], extended by a very short preface on the changes w.r.t. [11], and with augmentations to Appendix II, Appendix III, and Chapter IV, § 21.

[13] David Hilbert. *Grundlagen der Geometrie. — Sechste unveränderte Auflage. Anastatischer Nachdruck. Mit zahlreichen in den Text gedruckten Figuren.* Verlag und Druck von B. G. Teubner, Leipzig, Berlin, 1923. 6th rev. extd. edn. of [6], anastatic reprint of [12].

[14] David Hilbert. Über das Unendliche — Vortrag, gehalten am 4. Juni 1925 gelegentlich einer zur Ehrung des Andenkens an Weierstraß von der Westfälischen Math. Ges. veranstalteten Mathematiker-Zusammenkunft in Münster i. W. *Mathematische Annalen*, 95:161–190, 1926. Received June 24, 1925. Reprinted as Appendix VIII of [17]. English translation *On the infinite* by Stefan Bauer-Mengelberg with an introduction by Jean van Heijenoort in [5, pp. 367–392].

[15] David Hilbert. Die Grundlagen der Mathematik — Vortrag, gehalten auf Einladung des Mathematischen Seminars im Juli 1927 in Hamburg. *Abhandlungen aus dem mathematischen Seminar der Univ. Hamburg*, 6:65–85, 1928. Reprinted as Appendix IX of [17]. English translation *The foundations of mathematics* by Stefan Bauer-Mengelberg and Dagfinn Føllesdal with a short introduction by Jean van Heijenoort in [5, pp. 464–479].

[16] David Hilbert. Probleme der Grundlegung der Mathematik. *Mathematische Annalen*, 102:1–9, 1930. Vortrag gehalten auf dem Internationalen Mathematiker-Kongreß in Bologna, Sept. 3, 1928. Received March 25, 1929. Reprinted as Appendix X of [17]. Short version in *Atti del congresso internationale dei matematici, Bologna, 3–10 settembre 1928*, Vol. 1, pp. 135–141, Bologna, 1929.

[17] David Hilbert. *Grundlagen der Geometrie. — Siebente umgearbeitete und vermehrte Auflage. Mit 100 in den Text gedruckten Figuren.* Verlag und Druck von B. G. Teubner, Leipzig, Berlin, 1930. 7th rev. extd. edn. of [6], thoroughly revised edition of [13], extd. with three new appendixes (now ten in total) (Appendix VIII: [14]) (Appendix IX: [15]) (Appendix X: [16]).

[18] David Hilbert. *David Hilbert's Lectures on the Foundations of Geometry, 1891–1902.* Springer, 2004. Ed. by Michael Hallett and Ulrich Majer.

[19] David Hilbert and Paul Bernays. *Grundlagen der Mathematik — Erster Band.* Number XL in Grundlehren der mathematischen Wissenschaften. Springer, 1934. 1st edn. (2nd edn. is [21]). Reprinted by J. W. Edwards Publ., Ann Arbor (MI), 1944. English translation is [23; 24].

[20] David Hilbert and Paul Bernays. *Grundlagen der Mathematik — Zweiter Band.* Number L in Grundlehren der mathematischen Wissenschaften. Springer, 1939. 1st edn. (2nd edn. is [22]). Reprinted by J. W. Edwards Publ., Ann Arbor (MI), 1944.

[21] David Hilbert and Paul Bernays. *Grundlagen der Mathematik I.* Number 40 in Grundlehren der mathematischen Wissenschaften. Springer, 1968. 2nd rev. edn. of [19]. English translation is [23; 24].

[22] David Hilbert and Paul Bernays. *Grundlagen der Mathematik II*. Number 50 in Grundlehren der mathematischen Wissenschaften. Springer, 1970. 2nd rev. extd. edn. of [20].

[23] David Hilbert and Paul Bernays. *Grundlagen der Mathematik I — Foundations of Mathematics I, Part A: Title Pages, Prefaces, and §§ 1–2*. Springer, 2017. First English translation and bilingual facsimile edn. of the 2nd German edn. [21], incl. the annotation and translation of all differences of the 1st German edn. [19]. Ed. by Claus-Peter Wirth, Jörg Siekmann, Volker Peckhaus, Michael Gabbay, Dov Gabbay. Translated and commented by Claus-Peter Wirth &al. Thoroughly rev. 3rd edn. (1st edn. College Publications, London, 2011; 2nd edn. `http://wirth.bplaced.net/p/hilbertbernays`, 2013).

[24] David Hilbert and Paul Bernays. *Grundlagen der Mathematik I — Foundations of Mathematics I, Part B: §§ 3–5 and Deleted Part I of the 1st Edn.*. Springer, 2017. First English translation and bilingual facsimile edn. of the 2nd German edn. [21], incl. the annotation and translation of all deleted texts of the 1st German edn. [19]. Ed. by Claus-Peter Wirth, Jörg Siekmann, Volker Peckhaus, Michael Gabbay, Dov Gabbay. Translated and commented by Claus-Peter Wirth &al. Thoroughly rev. 3rd edn. (1st edn. College Publications, London, 2012; 2nd edn. `http://wirth.bplaced.net/p/hilbertbernays`, 2013).

[25] Jean-Pierre Jouannaud and Mitsuhiro Okada. A computation-model for executable higher-order algebraic specification languages. 1991. In [30, pp. 350–361].

[26] Jeroen Ketema and Femke van Raamsdonk. Erasure and termination in higher-order rewriting. 2004. In [3, pp. 30–33]. Also available at `www.prove-and-die.org/publ/wst04.pdf`.

[27] Jan Willem Klop. Combinatory reduction systems. Mathematical centre tracts 127, Mathematisch Centrum (since 1983: Centrum Wiskunde & Informatica), Amsterdam, 1980. PhD thesis, Utrecht Univ.., `http://libgen.io/get.php?md5=a80df1dda74dffd97700bda2277e8bc4`.

[28] Jan Willem Klop, Vincent van Oostrom, and Femke van Raamsdonk. Combinatory reduction systems: introduction and survey. *Theoretical Computer Sci.*, 121:279–308, 1993.

[29] A. Krazer, editor. *Verhandlungen des Dritten Internationalen Mathematiker-Kongresses, Heidelberg, Aug. 8–13, 1904*. Verlag von B. G. Teubner, Leipzig, 1905.

[30] *Proc. 6th Annual IEEE Symposium on Logic In Computer Sci. (LICS), Amsterdam, 1991*. IEEE Press, 1991. `http://lii.rwth-aachen.de/lics/archive/1991`.

[31] Aart Middeldorp, editor. *12th Int. Conf. on Rewriting Techniques and Applications (RTA), Utrecht (The Netherlands), 2001*, number 2051 in Lecture Notes in Computer Science. Springer, 2001.

[32] J Strother Moore and Claus-Peter Wirth. *Automation of Mathematical Induction as part of the History of Logic*. SEKI-Report SR–2013–02 (ISSN 1437–4447). SEKI Publications, 2014. Rev. edn. July 2014 (1st edn. 2013), ii+107 pp., `http://arxiv.org/abs/1309.6226`.

[33] Paliath Narendran and Michaël Rusinowitch, editors. *10th Int. Conf. on Rewriting Techniques and Applications (RTA), Trento (Italy), 1999*, number 1631 in Lecture Notes in Computer Science. Springer, 1999.

[34] Max H. A. Newman. On theories with a combinatorial definition of equivalence. *Annals of Mathematics*, 43:223–242, 1942.

[35] Tobias Nipkow. Higher-order critical pairs. 1991. In [30, pp. 342–349].

[36] Femke van Raamsdonk. Higher-order rewriting. 1999. In [33, pp. 221–239].

[37] Femke van Raamsdonk. On termination of higher-order rewriting. 2001. In [31, pp. 261–275].

[38] Femke van Raamsdonk, editor. *24th Int. Conf. on Rewriting Techniques and Applications (RTA), Eindhoven (The Netherlands), 2013*, number 24 in Leibniz Int. Proc. in Informatics. Dagstuhl (Germany), 2013.

[39] William W. Tait. Intensional interpretations of functionals of finite type I. *J. Symbolic Logic*, 32:198–212, 1967.

[40] Sarah Winkler, Harald Zankl, and Aart Middeldorp. Beyond Peano Arithmetic — automatically proving termination of the Goodstein sequence. 2013. In [38, pp. 335–351].

[41] Claus-Peter Wirth. Descente Infinie + Deduction. *Logic J. of the IGPL*, 12:1–96, 2004. http://wirth.bplaced.net/p/d.

[42] Claus-Peter Wirth. Hilbert's epsilon as an operator of indefinite committed choice. *J. Applied Logic*, 6:287–317, 2008. http://dx.doi.org/10.1016/j.jal.2007.07.009.

[43] Claus-Peter Wirth. Shallow confluence of conditional term rewriting systems. *J. Symbolic Computation*, 44:69–98, 2009. http://dx.doi.org/10.1016/j.jsc.2008.05.005.

[44] Claus-Peter Wirth. A Simplified and Improved Free-Variable Framework for Hilbert's epsilon as an Operator of Indefinite Committed Choice. *IFCoLog J. of Logics and Their Applications*, 4:421–512, 2017.

Received 23 October 2015

www.ingramcontent.com/pod-product-compliance
Lightning Source LLC
Chambersburg PA
CBHW080509090426
42734CB00015B/3012